Thermal Applications of Solar Energy

Thermal Applications of Solar Energy

A Professional Guide

Klaus W. Heinemann

Garland STPM Press
New York & London

Library of Congress Cataloging in Publication Data

Heinemann, Klaus W. 1941–
Thermal applications of solar energy.

Includes index.
1. Solar heating—Handbooks, manuals, etc.
2. Solar air conditioning—Handbooks, manuals, etc.
I. Title.
TH7413.H44 697′.78 80-29577
ISBN 0-8240-7278-2 AACR1

Published by Garland STPM Press
136 Madison Avenue, New York, New York 10016

Printed in the United States of America

15 14 13 12 11 10 9 8 7 6 5 4 3 2 1

Contents

Appendixes

Acknowledgments

This handbook could not have been written without the continuing support of my former colleagues at the Alten Corporation, in particular Mr. Terryl Buffum and Mr. Winston Boone, who provided encouragement during the experimental work that became the basis for this book. Mr. Robert Fogelsong contributed most of the section on control systems. Mr. Robert Pelikan of Pelikan & Lindsay, Inc., Palo Alto, California, reviewed the manuscript and made helpful suggestions.

I am indebted to my wife Adelgund and our sons Jörg and Mike for their patience during the countless evening and weekend hours spent in writing this manuscript, writing the computer programs, and testing the calculations.

Preface

In the middle 1970s, following significant increases in the price of Middle East oil, it was predicted that the West would turn to solar energy as the solution to its thermal needs for space heating, air conditioning, and hot water. However, these predictions of rapid growth in the infant solar industry have not been realized. In the initial period of enthusiasm for more widespread applications of solar energy there were few industry standards; any solar collector, no matter how poorly constructed or inefficient, could be presented as acceptable, and unfortunately many such collectors were sold and installed. Such practices have understandably led to considerable caution and skepticism on the part of the general public toward solar energy applications. Nevertheless, solar heating can provide a viable alternative as part of a new solution to the worldwide energy shortage. It is practicable for most regions of the world and almost anywhere in the United States. Thus it is in global and in our national interest to explore the possible applications of solar energy and take this opportunity to save our nonrenewable energy sources by using solar energy to the fullest extent. The efforts in this direction will be most effective if high system efficiency, high quality, and reasonably low cost are the leading goals of the solar industry. This handbook is intended as a contribution toward these goals. It was written with the specific aim of providing the reader enough detailed state-of-the-art information to design effective solar energy systems without having to consult much other literature and without having had prior experience in solar design. To accomplish this objective, two aspects are emphasized: thorough description of the design details for the selected system and thorough explanation of the reasons behind the suggested approaches.

A solar energy system is not a standardized product. It must be tailored to individual requirements and to the circumstances of the dwelling it is supposed to service. The solar designer must view all these requirements and circumstances holistically before starting with the details of a solar system design. This manual is intended to provide the necessary background and tools for tailoring solar systems. In accordance with this objective, no attempt has been made to describe all possible approaches to the design of solar systems; there is an excellent body of literature available for more comprehensive questions. Herein, a few approaches have been selected, and those that are described have been demonstrated to work satisfactorily in practice.

For the few approaches and methods selected, a ranking has been attempted, with descriptions of their advantages and disadvantages. In doing so, we have not hesitated to label certain approaches and products clearly inferior to other approaches and products for given applications.

We have chosen to confine the detailed descriptions of solar energy systems to those that use flat-plate collectors and a liquid heat-transfer medium. However, a special type of stationary concentrating collectors (insolation augmentation by reflective surfaces) is also discussed, and a brief overview of air-type collectors and tracking concentrators is given in the first chapter.

Much emphasis is placed on determining the performance of any solar energy system one designs. A detailed calculation procedure has been developed; with it, the reader can make performance calculations that will lead to exceptionally accurate performance and payback predictions. These calculation procedures are presented in the appendixes, programmed for a Hewlett-Packard 67/97 hand-held programmable calculator. These programs, which can be easily translated for other calculators such as the Texas Instruments TI-59, cover solar space heating, domestic water heating, process water heating, absorption-chilling air conditioning, and pool heating systems. The appendixes also include programs to calculate the building heat-loss coefficient and the direct solar heat gain through the windows. Finally, the appendixes include two programs for solar system financial analysis; one is based on the payback period, the other on cash flow.

A rapidly increasing number of architects and engineers who are getting involved with solar installations have access to minicomputers that are much more powerful and versatile than "conventional" programmable calculators. Programs for TRS-80 computers, written in Basic Level II (16k memory requirement), are now available from the author. These programs follow the general, highly versatile, and accurate computation method outlined in Chapter 5 and Appendix I; yet they are simple to use, and they produce results after only a few minutes of data entry and actual computing time.

The British system of units was used in all sections dealing with solar energy applications. This was done so as to build on units that are familiar to most readers and to make for clarity and direct applicability of the material presented, saving the reader the trouble of adjusting to new concepts in content and form simultaneously. Conversion to the international system of units (SI) is, in most instances, numerically quite convenient. For example, 1 kBtu (the energy unit used in many graphic representations) is approximately equivalent to 1 MJ (1×10^3 Btu $= 1.055 \times 10^6$ J $\approx$ 1 MJ). Another rule of thumb approximates 1 Btu-hr^{-1}-ft^{-2} with πWm^{-2} (1 Btu-hr^{-1}-ft^{-2} = 3.152 W-m^{-2} $\approx$ πW-m^{-2}), and 1 kBtu-ft^{-2} is approximately 11 MJ-m^{-2} (1×10^3 Btu-ft^{-2} $= 1.136 \times 10^7$ J-m^{-2} $\approx$ 11 MJ-m^{-2}). A more complete conversion table follows. We have attempted to be consistent in the use of symbols throughout the book. A partial list of these symbols follows.

List of Symbols

A heat-exchanger area, ft^2

A_0 solar-collector area, ft^2

A_p pool surface area, ft^2

L_B building heat-loss/gain coefficient (conductive and convective), $Btu\text{-}hr^{-1}\text{-}°F^{-1}$

C_0 constant, 1×10^{-3} Btu-gal-hr-°F-min^{-1}

C_w specific heat of heat-transfer fluid, $Btu\text{-}lb^{-1}\text{-}°F^{-1}$

E_c efficiency of glazed collectors

E_u efficiency of unglazed collectors

E_x efficiency of heat exchanger (capacity), $Btu\text{-}hr^{-1}\text{-}°F^{-1}\text{-}ft^{-2}$

E_{ch} efficiency of chiller

F fin efficiency factor

I total insolation into collector, $Btu\text{-}hr^{-1}\text{-}ft^{-2}$

I_p total insolation into pool, $Btu\text{-}hr^{-1}\text{-}ft^{-2}$

I_0 direct component of insolation at normal incidence, $Btu\text{-}hr^{-1}\text{-}ft^{-2}$

I_1 diffuse component of insolation, $Btu\text{-}hr^{-1}\text{-}ft^{-2}$

I_2 total reflector augmentation of insolation, $Btu\text{-}hr^{-1}$ per ft^2 of collector area

I_3 (diffuse) radiation during overcast periods, $Btu\text{-}hr^{-1}\text{-}ft^{-2}$

M_1, M_2 incident-angle modifier constants

P_a ambient water-vapor pressure, $lb\text{-}in.^{-2}$

P_w water-vapor pressure immediately above pool surface, $lb\text{-}in.^{-2}$

Q total clear-day insolation, $Btu\text{-}ft^{-2}\text{-}day^{-1}$

Q_p pool heating demand, $Btu\text{-}hr^{-1}$

R insulation value, $Btu^{-1}\text{-}hr\text{-}ft^2\text{-}°F$

R_B radiation loss, $Btu\text{-}hr^{-1}\text{-}ft^{-2}$

$R^{(0)}$ ratio of gross collector to net absorber area

ΔT difference between collector inlet temperature and ambient temperature, °F

ΔT_0 difference between pipe temperature and ambient temperature, °F

δT mean collector temperature increase due to fin efficiency loss, °F

T_a ambient temperature, °F

$T(h{:}00)$ temperature at time $h{:}00$, °F

T_c fluid temperature entering the collector, °F

T_{out}	fluid temperature exiting the collector, °F
T_0	mean storage-tank temperature, °F
T_{sky}	effective sky temperature (for black-body radiation), °F
T_2	temperature inside of a building
V_0	volume of storage tank, gal
V_p	volume of pool, gal
W_1	direct (window) solar heat gain of a building, Btu-hr^{-1}
W_2	system heat losses and domestic water-heating consumption, Btu-hr^{-1}
W_0	heat taken from storage tank for space heating and/or air conditioning, Btu-hr^{-1}
a	linear temperature-loss coefficient for glazed collectors, $\text{Btu-hr}^{-1}\text{-ft}^{-2}\text{-°F}^{-1}$
a_w	linear temperature-loss coefficient for unglazed collectors, $\text{Btu-hr}^{-1}\text{-ft}^{-2}\text{-°F}^{-1}$
b	second-order (quadratic) temperature-loss coefficient, $\text{Btu}^2\text{-hr}^{-2}\text{-ft}^{-4}\text{-°F}^{-2}$
c	difference between ambient temperature near collectors and ambient temperature distant from collectors, °F
d_0	vertical collector extension (Fig. 5.2)
f	flow rate of heat-transport fluid, gpm
g	reflector augmentation coefficient
h	time between sunrise and sunset, hr
i	flow rate of fluid in collector loop, gpm
k	insulation constant, $\text{Btu-hr}^{-1}\text{-ft}^{-2}\text{-°F}^{-1}\text{-in.}$
k_1	I_1/I_0
k_2	I_2/I_0
m	time integer
n	number of calculations (iterations) per hour during collection computations
p	sunshine probability, $0 < p < 1$
q	collected solar energy, $\text{Btu-hr}^{-1}\text{-ft}^{-2}$
q_0	heat loss in pipes, Btu-hr^{-1}
r_1, r_2	outside and inside radii of pipe insulation, respectively
r_0	reflectivity of reflector, $0 < r_0 < 1$
t_g	glazing transmission, $0 < t_g < 1$
Δt	difference between collector inlet and outlet temperatures, °F
v	wind speed, mph
α	solar altitude, deg
β	declination, deg
γ	geographic latitude, deg

ϵ	transmission-absorption coefficient $0 < \epsilon < 1$
ϵ_0	emissivity of absorber, $0 < \epsilon_0 < 1$
ϵ_p	emissivity of pool, $0 < \epsilon_p < 1$
ρ	absorptivity of absorber, $0 < \rho < 1$
ρ_p	absorptivity of pool water, $0 < \rho_p < 1$
σ	black-body radiation constant, $0 < \sigma < 1$
ϑ	sun-panel angle, deg
ϕ	solar azimuth, deg
ϕ_0	collector deviation from due south, deg
ψ	collector tilt angle, deg
ω	tilt angle of reflector, deg

Useful Data and Conversions

Length

1 ft = 0.3048 m	1 m = 3.281 ft
1 in. = 25.4 mm	1 mm = 0.0394 in.

Area

1 ft^2 = 0.0929 m^2	1 m^2 = 10.764 ft^2
1 in.^{2} = 645 mm^2	1 mm = 0.00155 in.^{2}

Volume

1 ft^3 = 0.02832 m^3	1 m^3 = 35.32 ft^3
= 28.32 liter	= 264.2 U.S. gal
1 qt = 0.9464 liter	

Mass

1 lb = 453.59 g	1000 g = 2.2046 lb

Force

1 lb = 4.4482 N	1 N = 0.22481 lb

Energy

1 Btu = 1055 J	1 J = 9.478×10^{-4} Btu
1 Btu = 1.055 kJ	1 kJ = 0.9478 Btu
1 Btu = 2.931×10^{-4} kW-hr	1 kW-hr = 3412 Btu
1 Btu = 242 cal	1 cal = 3.968×10^{-3} Btu

Power

1 hp = 745.7 W	1 W = 1.341×10^{-3} hp
1 Btu/hr = 2.931×10^{-4} kW	1 kW = 3412 Btu/hr

Power Density

1 Btu/hr/ft^2 = 3.152 W/m^2	1 W/m^2 = 0.317 Btu/hr/ft^2

Energy Density

1 Btu/ft^2 = 11.36 kJ/m^2	1 kJ/m^2 = 0.088 Btu/ft^2

Pressure

$1 \text{ psi} = 6894.76 \text{ Pa}$ $\qquad$ $1 \text{ Pa} = 1.450 \times 10^{-4} \text{ psi}$

$1 \text{ Pa} = 1 \text{ N/m}^2$

Temperature

1°R difference = 1°F difference = (5/9)°C difference = (5/9)°K difference

$T(°\text{F}) = \frac{9}{5}T(°\text{C}) + 32$ $\qquad$ $T(°\text{C}) = \frac{5}{9}[T(°\text{F}) - 32]$

$T(°\text{R}) = 459 + T(°\text{F})$ $\qquad$ $T(°\text{K}) = 273 + T(°\text{C})$

Velocity

$1 \text{ ft/sec} = 0.3048 \text{ m/sec}$ $\qquad$ $1 \text{ m/sec} = 3.281 \text{ ft/sec}$

$1 \text{ mph} = 1.6093 \text{ km/hr}$ $\qquad$ $1 \text{ km/hr} = 0.6214 \text{ mph}$

Mass Density

$1 \text{ lb/ft}^3 = 16.018 \text{ kg/m}^3$ $\qquad$ $1 \text{ kg/m}^3 = 0.06243 \text{ lb/ft}^3$

Volume Flow Rate

$1 \text{ gpm} = 6.31 \times 10^{-5} \text{ m}^3/\text{sec}$ $\qquad$ $1 \text{ m}^3/\text{sec} = 1.585 \times 10^4 \text{ gpm}$

Thermal Conductivity

$1 \frac{\text{Btu-in.}}{\text{hr-ft}^2\text{-°F}} = 0.1442 \frac{\text{W}}{\text{m-°K}}$ $\qquad$ $1 \frac{\text{W}}{\text{m-°K}} = 6.934 \frac{\text{Btu-in.}}{\text{hr-ft}^2\text{-°F}}$

$1 \frac{\text{Btu}}{\text{hr-ft-°F}} = 1.7307 \frac{\text{W}}{\text{m-°K}}$ $\qquad$ $1 \frac{\text{W}}{\text{m-°K}} = 0.5778 \frac{\text{Btu}}{\text{hr-ft-°F}}$

Heat Transfer

$1 \frac{\text{Btu}}{\text{hr-ft}^2\text{-°F}} = 5.678 \frac{\text{W}}{\text{m}^2\text{-°K}}$ $\qquad$ $1 \frac{\text{W}}{\text{m}^2\text{-°K}} = 0.1761 \frac{\text{Btu}}{\text{hr-ft}^2\text{-°F}}$

Specific Heat

$1 \frac{\text{Btu}}{\text{lb-°F}} = 4186.8 \frac{\text{J}}{\text{kg-°K}}$ $\qquad$ $1 \frac{\text{J}}{\text{kg-°K}} = 2.389 \times 10^{-4} \frac{\text{Btu}}{\text{lb-°F}}$

$1 \frac{\text{cal}}{\text{g-°C}} = 1 \frac{\text{Btu}}{\text{lb-°F}}$

Black-Body Radiation Constant

$$\begin{aligned}\sigma &= 1.73 \times 10^{-9} \text{ Btu-ft}^{-2}\text{-hr}^{-1}\text{-°R}^{-4} \\ &= 4.92 \times 10^{-8} \text{ kcal-m}^{-2}\text{-hr}^{-1}\text{-°K}^{-4} \\ &= 5.71 \times 10^{-8} \text{ N-m}^{-1}\text{-sec}^{-1}\text{-°K}^{-4}\end{aligned}$$

1

Solar-powered Heating and Cooling Equipment

1.1 Solar Collectors

1.1.1 Overview of Collector Types

The equipment that is being marketed to utilize thermal energy from the sun can be subdivided into four categories: flat-plate collectors that use air as the heat-transport medium (commonly known as "air collectors"); flat-plate collectors that use a liquid as the heat-transport medium (known as "liquid collectors"); stationary concentrating collectors; tracking concentrating collectors. The basic characteristics of these four collector categories will be briefly summarized in this section. Almost the entire subsequent discussion of systems to utilize solar thermal energy will then focus on the one category that has found the most widespread application: flat-plate collectors that use a liquid as the heat-transfer medium.

Flat-Plate Collectors, Air Type

A typical flat-plate collector based on air as the heat-transport medium is shown in Fig. 1.1. The incoming solar heat passes through one or two layers of a glazing material[1] and is absorbed at the top surface of the absorber unit. The collected solar heat passes (virtually instantaneously) by conduction to the inside surface of the absorber unit, where it is picked up by air, which is used as the heat-carrier medium to the heat-storage unit. The purpose of the air deflectors shown in Fig. 1.1 is to cause sufficient air turbulence within the absorber unit for efficient heat transfer.

Many different designs have been proposed for absorber units for air-type flat-plate collectors. It is beyond the scope of this handbook (which will concentrate mainly on liquid-type solar collectors) to discuss these design differences in detail. An excellent survey of the existing literature on this subject is provided by de Winter [15].

The integration of an air-type solar-collector system into a building service system is shown schematically in Fig. 1.2. The major components of the service

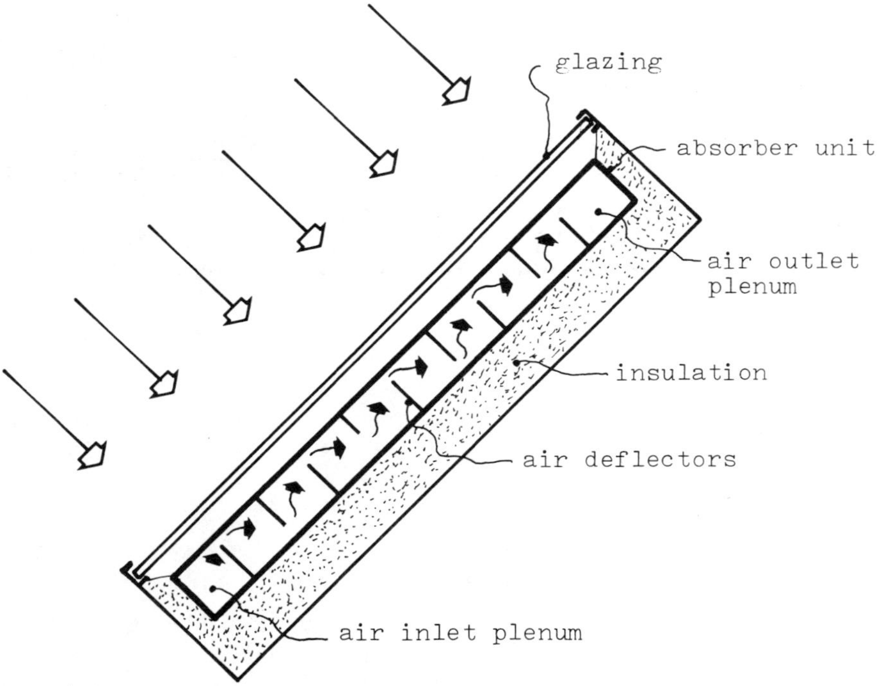

Fig. 1.1. Generalized cross section of a flat-plate collector using air as the heat-transport medium.

system are (a) the air-handling unit, containing a fan and two motor-driven dampers, (b) the heat-storage unit (rock bed[2]), the temperature-control system (Section 1.2), and (c) the solar collectors. Depending on the position of the dampers A and B, three modes of system operation can be achieved:

1. Dampers A and B open (i.e., in the horizontal position in Fig. 1.2). This is the normal daytime direct solar heating mode. The storage unit is bypassed. If the temperature sensor in the top of the collector array is below a necessary limit required for effective space heating, the auxiliary furnace is automatically turned on (Section 2.4.1).
2. Damper A open and damper B closed (this is the position shown in Fig. 1.2). This mode is used whenever solar heat is collected but no space heating is required at the same time. The fan blows the solar-heated air through the rock bed for thermal storage.
3. Damper A closed and damper B open. This mode is used whenever no solar heat is being collected (i.e., during cloudy periods or during the night hours). The return air from the building is now pulled through the rock bed, where it picks up stored solar heat. The auxiliary furnace is activated automatically if the temperature pickup in the rock bed is not sufficient to meet the demand.

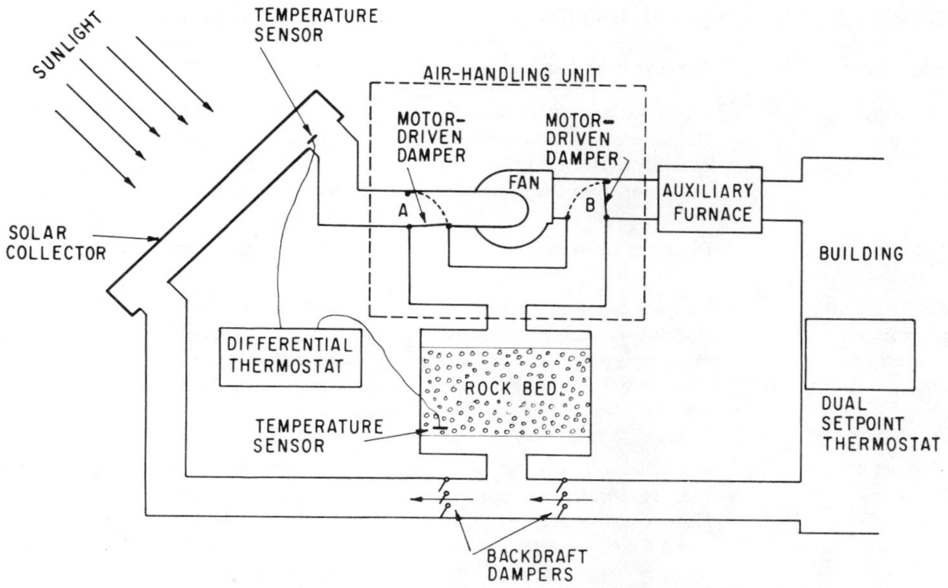

Fig. 1.2. Schematic representation of a typical solar space-heating system with air-type collectors [14]. Dampers are indicated for the solar rock-bed charging mode (mode 2). Solar domestic water-heating subsystem not included.

The question whether a liquid-based collector system or an air-based collector system is preferable is still a matter of controversy. There is no general answer to this question; the decision depends on a number of parameters that may be unique to a given installation situation. However, the fact that the vast majority of all solar heating systems installed to date use liquid-type collectors suggests that the advantages of these collectors generally outweigh the advantages of air-type collectors. In Table 1.1 an attempt is made to compare air-based and liquid-based collector systems in regard to some key aspects.

A solar heating system using air-type collectors is usually designed in such a way that the temperature of the air entering the collectors is near room temperature. A considerable temperature gain occurs during a one-time pass of the air through the collectors (20°F–40°F). This is possible and feasible primarily because of the good temperature-stratification properties of rock-bed storage systems (Section 1.3.2) and because of the circumstance that the density and heat capacity of air are low (otherwise, uneconomically large fans would be required). Air-type systems differ in this regard from liquid-type systems, which typically operate at high rates of mass flow and concomitantly low temperature increases during a single pass through the collectors.

A considerable amount of the information given in this book, although tailored for liquid-type collectors, can also be applied to air-type collectors if this basic difference in operation of the two collector types is understood and appropriately considered. For example, the collector efficiency equations 1.1 and 1.4

TABLE 1.1 Comparison of Air-Type and Liquid-Type Flat-Plate Collectors

Parameter	Air-Type Collectors	Liquid-Type Collectors
Leakage	Air leaks do not cause damage. However, it is difficult to detect air leaks, and for efficient system operation the system must be leakproof.	The system must be leakproof at all times. Leakage of the heat-transport fluid is easily detectable. Leakage may in certain cases cause damage to the building and the (then stagnating) collectors
Freezing protection	Does not apply	Special care must be taken (see section 2.2)
Ducts for heat-transport medium	Large cross section, conventional-type air ducts. Ducts must be insulated very well (R-11 to R-19), which is expensive and increases the bulkiness of the ducts.	Conventional water pipes. Normally, 0.75-in. fiberglass or foam insulation is sufficient. Pipes with insulation are much less costly than insulated ducts for air systems.
Collector efficiency	Fin efficiency is often poorer than for liquid-type collectors. However, the average operating ΔT (see equation 1.1) is usually lower than for liquid-type collectors because of the nature of rock bed storage systems, which have a high degree of temperature stratification.	High fin efficiency is warranted in most designs (see section 1.1.2 on fin efficiency and Appendix G). However, in most cases the average collector operating temperature is higher than in air-type collectors; this approximately cancels the advantage due to higher fin efficiency.
System electric power consumption	Friction losses in collector and duct system necessitate substantial increase in power rating for fan (by 0.5–1 hp in residential systems)	Typical residential systems require two pumps (1/12 hp each), with an electricity consumption slightly less than the additional power rating for the fan in air-type systems
Adaptability to retrofit installation	Poor, because of bulky duct systems and large storage volume	Good
Adaptability to new construction projects	Good, if it is an integral part of building design phase	Good; normally does not have to be integrated in the entire building design phase
Storage of solar heat	Rock beds (see Section 1.3.2); requires three times the volume of a water-storage system	Water tanks (see Section 1.3.1)
Solar domestic water-heating subsystem	Because of rock-bed storage principle, not as congenial as in the case of liquid-based systems; requires air-to-water heat exchanger	Requires simple water-to-water heat exchanger (see Section 1.4.3)
Solar space-heating subsystem	Straightforward for forced-air-type heating systems	Straightforward for radiating floor heating systems; forced-air systems require (simple) water-to-air heat exchanger (see Section 2.4)
Nocturnal cooling	Rock-bed storage is well suited for direct-air nocturnal cooling, bypassing collectors	Not congenial to glazed collector systems; however, unglazed collector systems do offer the possibility of effective radiative night-sky cooling (see Section 3.6)

(Section 1.1.2) are directly applicable to air-type collectors if the parameter ΔT is appropriately redefined. For liquid-type collectors, ΔT is the difference between the collector inlet temperature and the ambient temperature; the temperature rise between inlet and outlet can be considered small compared with ΔT. In air-type collectors, this temperature rise is substantial, and an appropriate redefinition of ΔT will have to take this into account. Frequently, the difference between the mean collector temperature $[T_{in} - T_{out})/2]$ and the ambient temperature is defined as ΔT. The collector testing equation 1.5 remains valid if the appropriate values for the (air) flow rate f and the (air) heat capacity C_w are used. The entire subsection Determination of Efficiency Parameters by Design (see Section 1.1.2) remains applicable, with the exception of the material covered in the topics (6) Absorber Geometry and (7) Fin Efficiency, which deal specifically with the geometry of liquid-type collectors. Other subsections that also are basically valid for air-type collectors are Materials Aspects (with the exception of the topic Absorber) and Example of a Commercial Collector (see Section 1.1.2) and Efficiency Characteristic (see Section 1.1.3).

However, computation of solar system heat collection, as presented in Chapter 5, is based specifically on low or negligible temperature stratification in the heat-storage tank, and therefore, in principle, it is valid only for liquid-based systems. It should also be noted that air-type collectors cannot be used in connection with absorption-chilling air-conditioning systems because they generally do not deliver high enough temperatures.

Flat-Plate Collectors, Liquid Type

A typical flat-plate collector using a liquid as the heat-transport medium is shown in Fig. 1.3. Many variations are possible for the absorber plate. Some frequently used designs are discussed under the topic Materials Aspects in Section 1.1.2. A comparison with air-type collectors is given in Table 1.1

Liquid-type flat-plate collectors are discussed in detail in Sections 1.1.2 and 1.1.3, and their integration into systems is discussed in Chapter 2 (space heating), Chapter 3 (space cooling), and Chapter 4 (water heating). The sections on control systems (Section 1.2), heat-storage equipment (Section 1.3), and related equipment (Section 1.4) relate mainly to systems that are based on liquid-type collectors, and the extensive chapter on solar system heat-collection computation, as well as Appendixes I–K, which give specific calculator/computer programs for such computations, were written for liquid-type collectors.

Stationary Concentrating Collectors

A collector is defined as concentrating if its fin (absorber) width d is smaller than the aperture width d' and if reflective surfaces are used to reflect a portion of the incident sunlight into the absorber. The ratio of d'/d is the concentration ratio.

A concentrating collector with a low concentration ratio ($2 \lesssim d'/d \lesssim 5$) can be designed in such a way that its absorbers intercept a major portion of the incident sunlight not only at one fixed solar altitude but also within a certain range of sun angles, as indicated in Fig. 1.4. This "acceptance range" may be as wide as 40 to 60 deg. Such collectors can then be operated in a stationary mode (i.e., "tracking" the sun is not required). In some instances, not even seasonal changes of the

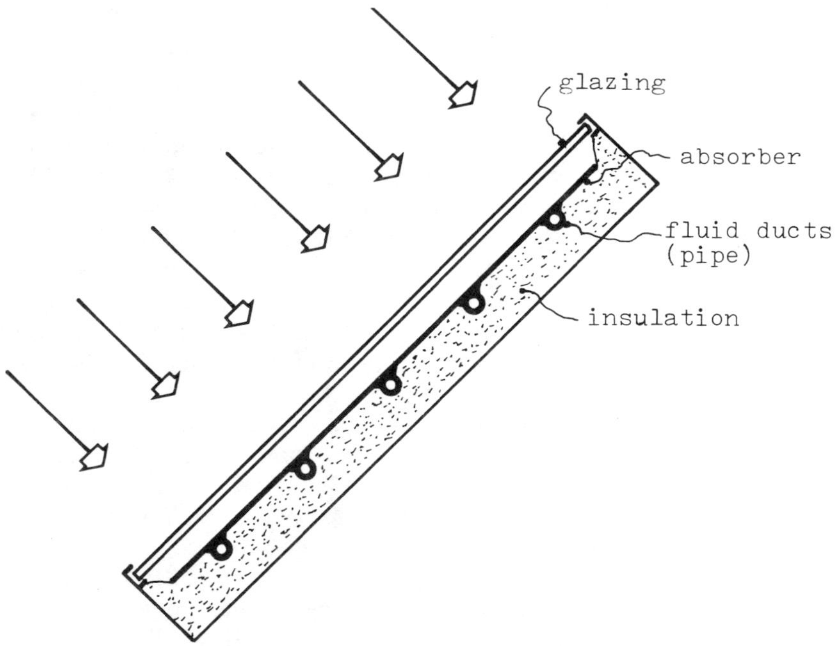

Fig. 1.3. Generalized cross section of a flat-plate collector using a liquid as the heat-transport medium.

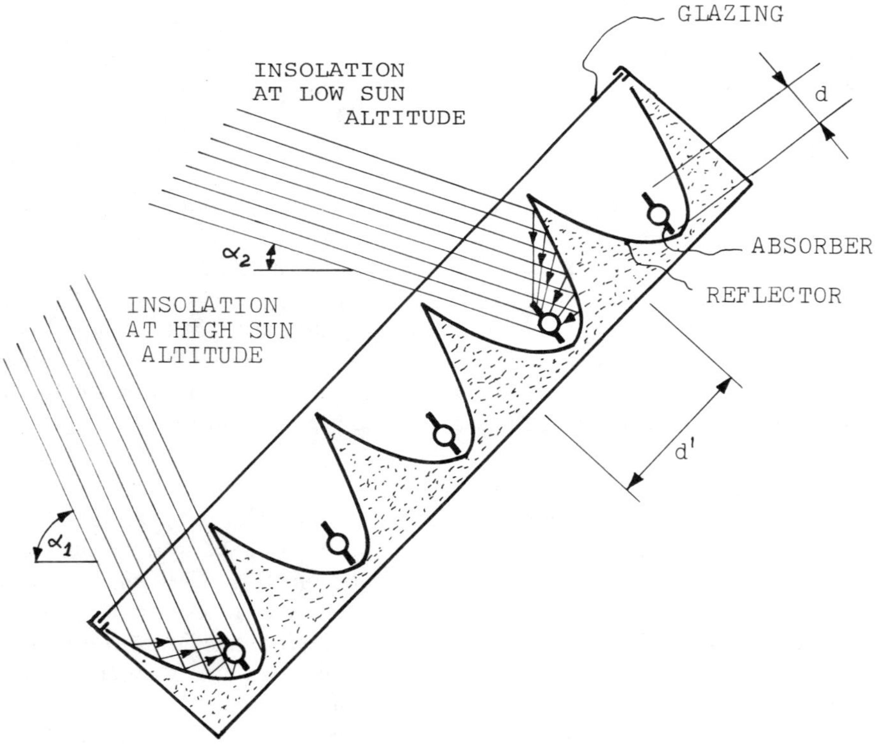

Fig. 1.4. Schematic horizontal cross section through a stationary concentrating solar collector.

collector tilt angle are essential. This is in contrast to the situation with collectors that have high concentration ratios (10 or more); these require precise tracking.

Stationary concentrating collectors have a number of apparent advantages over flat-plate collectors because their thermal-loss coefficients, a and b (see equation 1.1 and following), are considerably lower.[3] This, in turn, lowers the required standards for back and side insulation, a factor that may reduce overall collector cost. However, because a concentrating collector relies on reflection processes (usually at least one) that occur with appreciable losses,[4] and because only the direct component of the incoming solar radiation can be fully utilized (the concentration effect does not apply to the randomly directed diffuse radiation), the collection efficiency at low values of ΔT (see equation 1.1) is low ($\epsilon \approx 0.6$), as compared with flat-plate collectors ($\epsilon \approx 0.8$). Thus a concentrating collector is generally less efficient than a flat-plate collector in the low-ΔT range (see curves A and C in Fig. 1.7). The crossover point at which the two collector types perform equally well usually lies well within (or sometimes even beyond) the temperature range for domestic water heating and space heating. Solar-powered air conditioning usually calls for concentrating-type collectors, and high-quality collectors of the tracking concentrating type are necessary for applications in which hot steam must be produced (e.g., in solar-powered generation of electricity).

One drawback of stationary concentrating collectors is their unaesthetic appearance. They are bulkier than flat-plate collectors, having a thickness-to-width ratio greater than that for flat-plate collectors, and therefore they cannot be integrated into roof systems as unobtrusively as flat-plate collectors.

A regular flat-plate collector can become a stationary concentrating collector if it works in conjunction with a reflector that augments the insolation actually incident on the absorber plate. Such collector/reflector arrangements are feasible in large flat-plate collector installations on flat surfaces, such as flat rooftops of commercial buildings, where the collector banks are arranged in sawtooth fashion. Figure 3.1 shows an example of such an arrangement, a solar-driven absorption-chilling air-conditioning system. Such systems require hot-water temperatures between 180°F and 200°F; thus, reflector augmentation, which increases the efficiency of the collectors at such high temperatures, is highly desirable. A classic solar installation with reflector augmentation was described by Duncan and Doering [18] for a school building in Atlanta, Georgia.

The actual increase in solar heat collection due to reflector augmentation is discussed in Section 5.2.7 and in Appendix F, as well as in other works [22, 24]. A general result of these discussions is that the amount of augmentation achieved with reflectors in front of collectors peaks during the summer months if the reflectors are in the described sawtooth arrangement and are facing north, whereas a collection peak is realized in the winter months if the reflectors are horizontal. In both cases, a relatively steep collector tilt angle (45–60 deg) yields optimum results.

Figure 1.5 illustrates the similarities and differences of flat-plate collector systems and reflector-augmented collector systems. In each case, as is indicated, a shaft of sunlight of relatively constant width is intercepted by the overall solar system. However, in the case of horizontally arranged flat-plate collectors, twice as much collector (absorber) area is needed. Because the temperature-dependent collection losses increase with increasing absorber area, the reflector/collector arrangement is much more efficient (if the operating ΔT is above a moderately low limit of 20°F–50°F). Note that the reflectors are hardly used at all in the winter because of the shadowing effect of the tilted collectors at the low sun altitudes prevailing in the winter. Also note that the number of the sun's rays indicated in Fig. 1.5 is proportional to the actually incident solar radiation. It is evident that the roof receives considerably more insolation during the summer months.

Tracking Concentrating Collectors

If the concentration factor exceeds approximately 5, the acceptance angle for varying solar altitudes becomes so narrow that tracking becomes necessary. For this to be possible, the collectors must be mounted free-standing, with an appreciable space between them, so that they will not shade each other when tilted for early morning and late evening collection conditions. This has several conse-

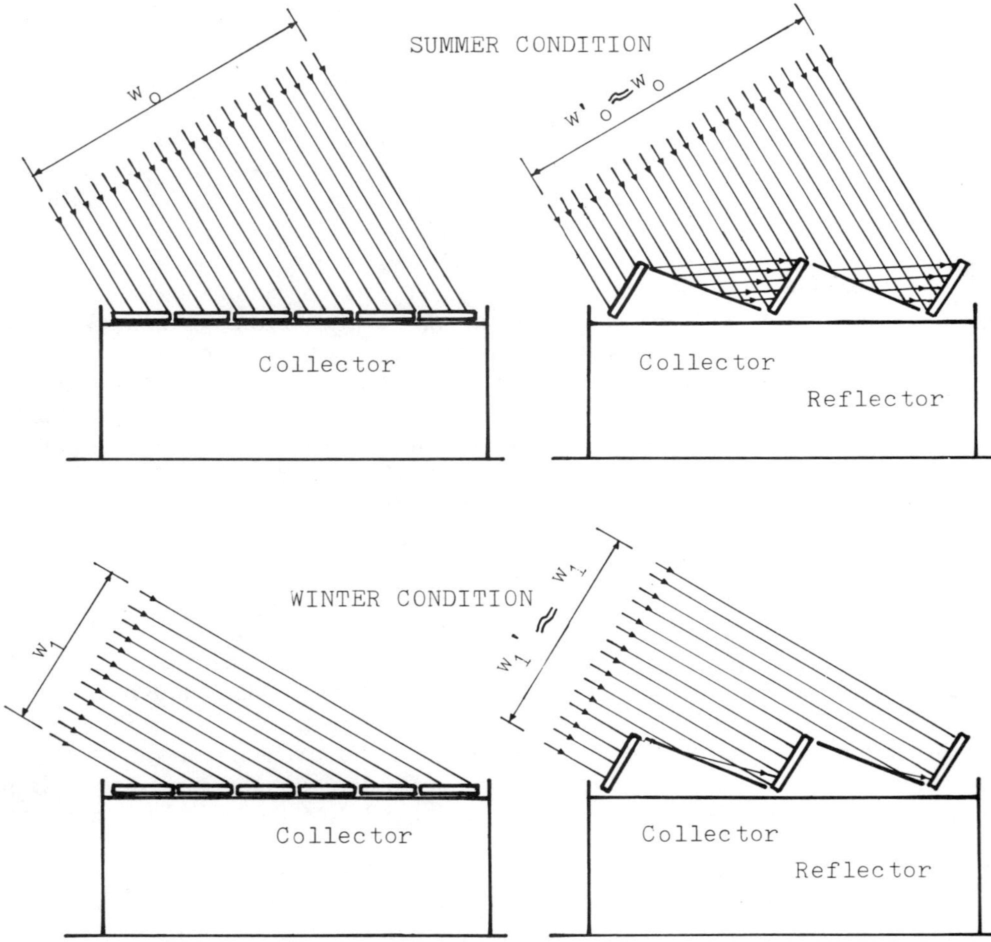

Fig. 1.5. Schematic representation of the amount of sunlight intercepted by horizontally arrayed flat-plate collectors (left) and by tilted flat-plate collectors augmented by reflectors (right) on a flat rooftop. The alternative on the right can be considered a stationary concentrating collector array.

quences that should be noted. First, although it is correct that each tracking concentrating collector is exposed to considerably more incoming radiation,[5] the required spacing between the individual collectors (Fig. 1.6) offsets this increase when we consider the total solar radiation incident on a given area available for the installation for the solar collectors. Second, it is very difficult to design tracking concentrating collector arrays that are architecturally pleasing (Fig. 1.6). It is virtually impossible to design an aesthetically pleasing tracking collector system for use on the rooftop of a conventional single-family residence. Third, tracking collectors have a number of moving parts that are exposed to the elements and require periodic service and/or replacement. In most cases the moving parts involve the pipe connections to the passages for the heat-transport fluid, and leakage problems are likely to occur during the service life of the solar installation.

For these reasons, tracking concentrating collector systems have been proposed primarily for applications in which high output water temperatures are essential, where flat-plate or stationary concentrating collectors would be appreciably inferior.

(a) (b)

Fig. 1.6. Tracking concentrating solar collector system on a federal government building in California shown in the morning (a) and at noon (b); arrow indicates two rows of collectors with malfunctioning tracking.

1.1.2 Glazed Flat-Plate Collectors, Liquid Type

Efficiency Characteristic

Solar heat collection q with flat-plate collectors can be approximated with the following equation:[6]

$$q = I\left[\epsilon - a\,\frac{\Delta T - c}{I} - b\left(\frac{\Delta T - c}{I}\right)^2\right] \tag{1.1}$$

with

$$I = I_0(\cos\vartheta + k_1) \tag{1.2}$$

and

$$k_1 = I_1/I_0 \tag{1.3}$$

In these equations, ϵ is the instantaneous efficiency of the collector for the case in which the fluid inlet temperature is the same as the ambient temperature near the collector (i.e., for $\Delta T - c = 0$). The factor ϵ contains the glazing transmission losses and the absorption losses of the absorber coating; it is the product of the glazing transmision t_g and the absorber absorptivity ρ. The temperature difference $\Delta T = T_c - T_a$ between the collector fluid inlet temperature T_c and the ambient temperature T_a influences greatly the amount of solar heat collected. The effect is attenuated by c, which accounts for a somewhat higher ambient temperature near the actual solar-collector array (found in some installations, e.g., in sawtooth collector arrangements on flat roofs) and denotes the difference between that higher ambient temperature and the average, tabulated ambient temperature.[7] The factors a and b are the first- and second-order temperature-loss coefficients. I_0 is the direct component of insolation at normal incidence (see Section 5.2.1), and ϑ is the angle between the sun's rays and the collector normal. It depends on the time of day, the collector tilt angle, the geographic latitude, and the date (see Section 5.2.2). I_1 is the diffuse component of insolation (see Section 5.2.2).

The instantaneous efficiency, $E_c = q/I$, can be expressed with equation 1.1 as

$$E_c = \epsilon - a\,\frac{\Delta T - c}{I} - b\left(\frac{\Delta T - c}{I}\right)^2 \tag{1.4}$$

As can be seen directly from this equation, E_c is least affected by the first- and second-order temperature-loss terms if I is large. This has several consequences that will be used throughout the following sections, which deal with solar collection equipment and principles.

The collector tilt angle is very important. The sun's altitude varies significantly with the seasons,[8] and $\cos\vartheta$ is quite low, for example, during the winter months when the collector tilt is low. Thus it is not advisable to use low collector tilt angles for solar installations that are designed primarily for winter space heating. On the other hand, systems designed for summer air conditioning require low collector tilt angles for optimum performance.[9]

The concentration factor is important. Concentrating solar collectors work with effective I values that are considerably greater than I_0, depending on the degree of concentration.[10] It should be noted that the considerable advantage of concentrating collectors over flat-plate collectors (concentration $\equiv$ 1), when operated under high-ΔT conditions, is counteracted by ϵ values that are usually substantially inferior to those of flat-plate collectors. This is because concentrators work only with the direct component of the incident solar radiation and because concentrating collectors are in principle dependent on reflection of the solar radiation into the absorbing element, each reflection process being subject to a "reflection loss" that typically can amount to 5%–10% per reflection. Thus concentrating collectors are advantageous only if operated under high-ΔT conditions (typically $\Delta T > 100°F$) (see Fig. 1.7, curve C compared to curves A and B).

Collection during overcast periods is usually negligible. In the presence of overcast conditions, when I is very low,[11] the temperature-loss terms become so high that in most practical applications they exceed the gain term ϵ; that is, the effective solar collection is zero[12] (or negative, if an automatic control system is not available to shut off the circulation of the heat-transport fluid if $E_c < 0$) (see Sections 5.2.2 and 5.2.8 and Appendix C).

Figure 1.7 shows typical efficiency curves for three types of solar collectors: single-glazed collectors (curve A), double-glazed collectors (curve B), and concentrators (curve C). The trend is that single-glazed collectors are preferable for low-grade heat applications (range R_2) and double-glazed collectors should be used for higher-ΔT operations, such as space heating and domestic water heating in moderate climates. The temperature range required for absorption-chilling air conditioning usually justifies some type of concentrating collector arrangement, or at least highly efficient double-glazed collectors.

In the low-ΔT range, double-glazed collectors are typically less efficient than single-glazed collectors because of the increased "front-end" losses during trans-

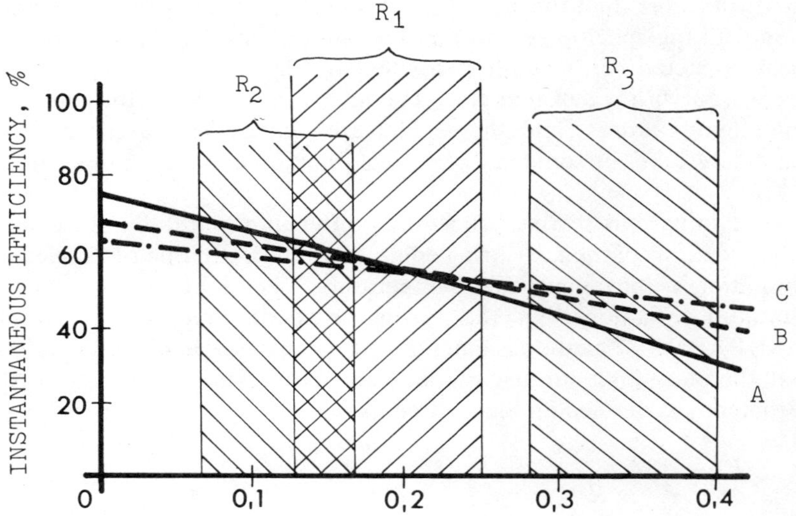

Fig. 1.7. Typical efficiency characteristics for a single-glazed collector (curve A), a double-glazed collector (curve B), and a concentrating collector (curve C). R_1 indicates the typical temperature operating range for domestic water heating and space heating; R_2 is the range for low-grade heating systems for process water; R_3 indicates the range for absorption air conditioning.

mission through the glazings.[13] The reason that the concentrating collector is typically even less efficient during low-ΔT operation was mentioned earlier in this section.

The instantaneous efficiency ϵ in equation 1.1 and equation 1.4 varies with the incident angle of the direct component of the insolation I_0 onto the collector. This behavior can be expressed as

$$\epsilon = \epsilon_n \left[M_1 - M_2(1/\cos \vartheta - 1)\right]$$

In this equation, ϵ_n is the instantaneous efficiency at normal incidence ($\vartheta = 0$), and M_1 and M_2 are the incident angle modifier parameters. Virtually all flat-plate solar collectors have $M_1 = 1$. In rare cases, particularly when the absorber geometry differs from a flat surface, M_1 may increase to 1.01 or even slightly above that value. A typical value for M_2 is 0.1; most collectors fall between $M_2 = 0.05$ and $M_2 = 0.2$. Using $M_1 = 1$ and $M_2 = 0.1$, one obtains $\epsilon = 0.985\ \epsilon_n$ for $\vartheta = 30$ deg, $\epsilon = 0.959\ \epsilon_n$ for $\vartheta = 45$ deg, and $\epsilon = 0.90\ \epsilon_n$ for $\vartheta = 60$ deg. This relative minor influence seems to justify the practice of often neglecting the incident angle modifier because it is considered as of a second-order influence. It has been our experience from comparison with high-accuracy computer calculations [25] that the solar heat collection for typical solar domestic water heating systems is lowered by approximately 3% due to the incident-angle modifier, all other system parameters being the same.

Determination of Efficiency Parameters by Testing

Various testing procedures have been devised and are being used by collector manufacturers and by impartial accredited testing laboratories. Common to all valid methods is the following basic testing procedure: The temperature rise of the heat-transport fluid during a pass through the collector is measured. Multiplication with the specific heat of the fluid and the flow rate gives the amount of solar heat collected. This result is divided by the incident solar radiation, which is measured with a pyronometer. The measurement is performed for various fluid inlet temperatures, and the results are plotted versus the collector operating parameter, ΔT, or some mathematical function containing this parameter (usually $\Delta T/I$).

The various testing procedures differ in a multitude of details, such as the specifications of the measuring equipment, the definition of "collector size,"[14] the definition of "mean absorber temperature," and several other, mostly less important, specifications. Of the testing methods that have been proposed, the ASHRAE 93–77 method is emerging as the procedure specified most often [6]. We will therefore limit our discussion of collector testing methods to ASHRAE 93–77. Efficiency is defined and measured as

$$E_c = a_0 f C_w \frac{\Delta t}{I A_0} \tag{1.5}$$

where f is the flow rate of the heat-transport fluid, C_w is the heat capacity of that fluid, I is the insolation (see equation 1.2), a_0 is a constant factor,[15] Δt is the temperature rise during one pass of the fluid through the collector (i.e., $\Delta t = T_{out} - T_c$), and A_0 is the collector area. Most collector manufacturers specify flow rates for their products that maintain $\Delta t \lesssim 15°F$ under maximum collection conditions. This is reasonable, because an excessive temperature rise would unduly reduce the average ΔT operating point of the collector and hence reduce its efficiency.[16] For a typical solar collector of 30 ft^2 gross area operating under 200-Btu/hr-ft^2 collection conditions, a water flow rate of 0.8 gpm would be required to maintain ΔT at 15°F. Unless the collector manufacturer specifies a quite different flow rate, a rate close to this value (0.03 gpm/ft^2 collector area) is used during the tests.

The inlet fluid (water) temperature T_c is varied to provide, in sequential tests, results for $\Delta T - c \cong 10°F$, 25°F, 50°F, 75°F, and 100°F.[17] For some collectors that are designed to be used at high-ΔT conditions, which are necessary for absorption air conditioning, the tests are extended to higher ΔT values.

The collector is mounted, at 30 to 45 deg of tilt, in a structure in such a way that manufacturer-recommended mounting conditions are optimally simulated. Each data point is taken no earlier than 15 min after outlet-temperature equilibrium is reached (after a temperature or flow-rate change, or after the sun emerges from behind a cloud), and all parameters are averaged over at least 10 min for each data point.

Other specifications of the ASHRAE 93–77 test procedure are as follows: (a) air-velocity measurements at a specified distance from the collector; (b) direct and diffuse insolation measurements performed with an Eppley model NIP pyrheliometer; (c) a minimum insolation requirement of $I_0 + I_1 > 200$ Btu/hr-ft^2 on the collector surface; (d) measurement of the incident-angle modifier, which accounts for changes in the glazing transmission and absorber absorptivity ϵ due to shallow angles of incidence on the collector. Details of these specifications can be found in the appropriate references [6, 17].

The test results are plotted as a graph of E_c versus $(\Delta T - c)/I$, from which the parameters ϵ, a, and b of equation 1.4 are determined analytically. Quite frequently the collector efficiency equation 1.4 is approximated by the simpler equation

$$E_c = \epsilon - a \frac{\Delta T - c}{I} \tag{1.6}$$

which is different from equation 1.4 in that the second-order loss term is omitted. Equation 1.6 fits the general linear expression $y = mx + b$, where y, the ordinate, is the efficiency E_c, m is the slope of the line and is equivalent to $-a$ (the overall thermal-loss coefficient), and x is the abscissa, $(\Delta T - c)/I$. The ordinate intercept of the line for $(\Delta T - c)/I = 0$ is equivalent to ϵ, and the abscissa intercept $E_c = 0$ is equivalent to $(\Delta T - c)/I = \epsilon/a$.

The test points for flat-plate collectors usually fit a straight line so well that often this linear fit is quite an acceptable representation of the collector performance. However, care must be taken to apply the linear fit especially well in that range of $(\Delta T - c)/I$ in which the collector is usually operated. An example is given in Fig. 1.8, where the official test curve for a single-glazed flat-plate collector[18] is plotted (solid line). For this (second-order) fitted curve, the coefficients ϵ, a, and b were determined to be[19]

$$\begin{aligned} \epsilon^{(2)} &= 0.736 \\ a^{(2)} &= 0.820 \text{ Btu-hr}^{-1}\text{-ft}^{-2}\text{-°F}^{-1} \\ b &= 0.718 \text{ Btu}^{2}\text{-hr}^{-2}\text{-ft}^{-4}\text{-°F}^{-2} \end{aligned} \tag{1.7}$$

The dot-dash line in Fig. 1.8 gives the linear fit applicable to the main collector use range indicated in the graph. This straight line corresponds to

$$\epsilon^{(1)} = 0.762 \tag{1.8}$$

$$a^{(1)} = 1.068 \text{ Btu-hr}^{-1}\text{-ft}^{-}\text{-}^{\circ}\text{F}^{-1}$$

Either of the two sets of values may be used for solar system collection computations (as long as the collector use range is maintained, if the linear fit is used), but it is evident that it would be wrong to use the second-order-fitted coefficients and neglect b, while claiming that $\epsilon^{(2)}$ and $a^{(2)}$ alone represent the collector performance.

If the collector is to be used under high-ΔT conditions, the corresponding linear fit will be the dash line in Fig. 1.8. This line corresponds to

$$\epsilon^{(1)} = 0.818$$

$$a^{(1)} = 1.30 \text{ Btu-hr}^{-1}\text{-ft}^{-2}\text{-}^{\circ}\text{F}^{-1} \tag{1.9}$$

The linear fit is sometimes performed by using a straight line through $\epsilon^{(1)} = \epsilon^{(2)}$ and that point on the second-order loss curve which is most representative for the specific use range. We prefer, however, the somewhat more cumbersome method described above.

Although efficiency is an important criterion, certainly it should not be the only criterion for selection of a solar collector for a specific application. Other important considerations are (a) the corrosion resistance of the absorber plate and collector housing, (b) the leakage probability after many years of operation, (c) the longevity of the absorptive coating (chalking, peeling, change in absorptivity);[20]

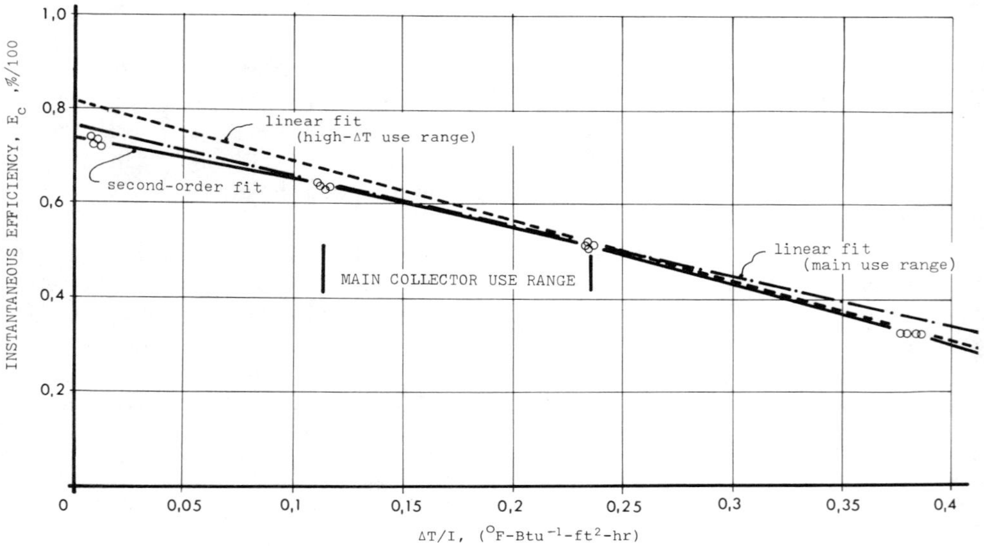

Fig. 1.8. Typical flat-plate collector test curve with quadratic fit (solid line) and linear fit (high-ΔT use range, dash line; main use range, dot-dash line).

(d) the ease of mounting the collector and plumbing, and (e) the general outer appearance of the collector. Any one of these considerations may well be worth an efficiency loss of a few percent. The solar collection system should supply solar heat for 20 years or more of trouble-free operation. In many cases this justifies the choice of a well-built high-quality product, even if its efficiency parameters are slightly inferior to those of a less expensive product that may be of inferior quality and workmanship.

Determination of Efficiency Parameters by Design

Many solar collectors that are being marketed have never been subjected to critical performance tests by an impartial testing laboratory. Although the trend is toward test certification being a requirement for any sale of solar collectors, at this time such certification is mandatory only for some government-funded projects. Thus it is essential that designers of solar systems be able to evaluate competing products on the basis of their design.[21] Such evaluation is possible with fair accuracy if absolute values for $\epsilon^{(1)}$ and $a^{(1)}$ are derived. It is usually possible with high accuracy if relative differences in the parameters $\epsilon^{(1)}$ and $a^{(1)}$ with respect to the $\epsilon^{(1)}$ and $a^{(1)}$ values of a standard tested collector are derived, which can be done by careful comparison of those collector design characteristics that determine collection efficiency.

The following comparison is made with respect to a well-designed single-glazed collector in the medium-to-moderate price range. This reference collector is of the aluminum-fin/copper-tube design, which is quite common for solar collectors and also is well suited for discussion of fin efficiency losses. Figure 1.9 shows a simplified cross-sectional drawing of the collector. The efficiency parameters $\epsilon^{(2)}$, $a^{(2)}$, and b for this reference collector, as tested at DSET Laboratories, Phoenix, Arizona, are as follows:

$$\begin{aligned} \epsilon^{(2)} &= 0.736 \\ a^{(2)} &= 0.820 \text{ Btu-hr}^{-1}\text{-ft}^{-2}\text{-}^\circ\text{F}^{-1} \\ b &= 0.718 \text{ Btu}^2\text{-hr}^{-2}\text{-ft}^{-4}\text{-}^\circ\text{F}^{-2} \end{aligned} \tag{1.10}$$

These efficiency parameters correspond to the following linear-fit efficiency parameters applicable for the average collector ΔT use range:

$$\begin{aligned} \epsilon^{(1)} &= 0.762 \\ a^{(1)} &= 1.068 \text{ Btu-hr}^{-1}\text{-ft}^{-2}\text{-}^\circ\text{F}^{-1} \end{aligned}$$

In the following comparisons, only the linear-fit efficiency parameters are considered. Data and results for the reference collector are designated by the subscript r; data for the new collector to be evaluated bear the subscript n. The steps necessary to evaluate the new collector, i.e., to generate the parameters ϵ_n and a_n from $\epsilon^{(1)} = \epsilon_r$ and $a^{(1)} = a_r$, are the following.[22]

(1) Ratio of Gross Collector Area to Net Absorber Area The linear-fit efficiency curve changes its slope and its ordinate intercept directly proportionally to $R_r^{(0)}/R_n^{(0)}$, where $R^{(0)}_{r,n}$ = (gross collector area)/(net absorber area):

$$\epsilon_n = \epsilon_r \frac{R_r^{(0)}}{R_n^{(0)}}, \qquad a_n = a_r \frac{R_r^{(0)}}{R_n^{(0)}}$$

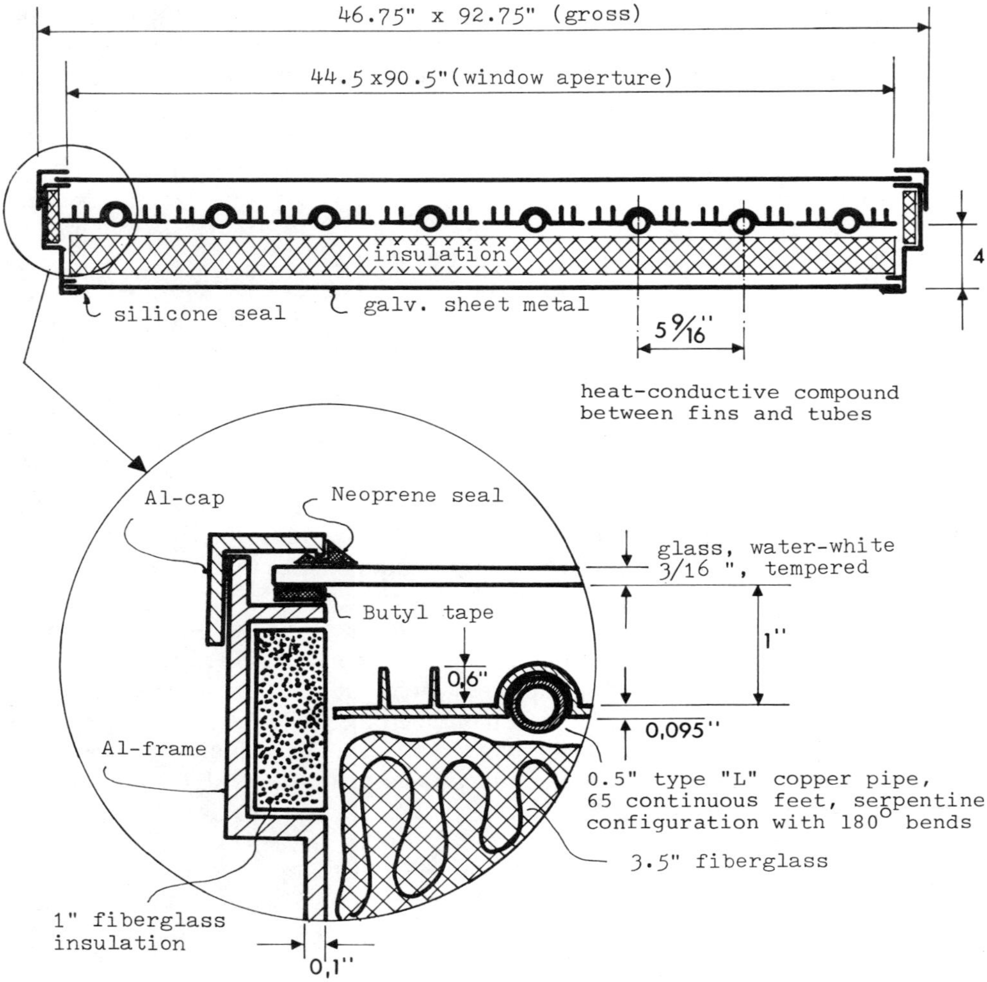

Fig. 1.9. Cross-sectional drawing of reference collector with data and explanations relevant to efficiency considerations.

Example. In accordance with Fig. 1.9, $R_r^{(0)} =$ (46.75 × 92.75 in.)/(44.5 × 90.5 in.) = 1.077. A test collector having a gross area of 17 ft^2 and a net absorber area of 14.5 ft^2, has $R_n^{(0)}$ 17/14.5 = 1.172. Because $R_r^{(0)}/R_n^{(0)} = 0.918$, the new efficiency parameters (all other aspects being identical with those of the reference collector) will be

$$\epsilon_n = 0.762 \times 0.918 = 0.700$$

$$a_n = 1.068 \times 0.918 = 0.980 \text{ Btu-hr}^{-1}\text{-ft}^{-2}\text{-°F}^{-1}$$

(2) Glazing Transmission The solar-collector efficiency curve changes its ordinate intercept (while maintaining its slope) directly proportionally to t_{g_n}/t_{g_r}, where t_g is the glazing transmission:[23]

$$\epsilon_n = \frac{t_{g_n}}{t_{g_r}}, \qquad a_n = a_r$$

Example. The solar transmission of the glazing of the reference collector is 91% (very-low-iron glass). A collector with a glazing of only 81% transmissivity will have the following efficiency parameters under otherwise identical conditions:

$$\epsilon_n = \frac{0.81}{0.91} 0.762 = 0.89 \times 0.762 = 0.678$$

$$a_n = 1.068 \text{ Btu-hr}^{-1}\text{-ft}^{-2}\text{-}^\circ\text{F}^{-1} \quad \text{(unchanged)}$$

(3) Number of Glazings For best accuracy it is recommended that double-glazed collectors with unknown efficiency parameters be compared with a double-glazed reference collector, not with a single-glazed reference collector. As a less accurate alternative, a comparison with the single-glazed collector of Fig. 1.9 is offered here. The improvement in collector performance in going from single glass glazing to double glass glazing is quite comparable to the difference in heat loss between single- and double-glazed windows. This is typically 0.4 to 0.5 Btu-ft.$^{-2}$-hr^{-1}-°F^{-1}. Because the ratio $R^{(0)}$ between gross collector area and net absorber area must also be accounted for, the slope of the efficiency characteristic changes to approximately

$$a_n = a_r - 0.45/R_n^{(0)} = a_r \left(1 - \frac{0.45}{a_r R_n^{(0)}}\right)$$

when a second glass glazing is added. The improvement will be less impressive if the second glazing uses a material with low absorptivity (high transmittance) in the near infrared (IR) wavelength region, because such a glazing, while reducing the convective heat losses, will not significantly reduce the radiative heat losses.[24]

Example. Because the second glazing always results in reduced glazing transmission, our example must allow for this circumstance. A second glazing is assumed to be added to the reference collector. The glazing is Sunadex (very-low-iron glass), with 91% transmission. Because $R_r^{(0)}$ $(= R_n^{(0)}) = 1.077$ (see preceding example),

$$a_n = 1.068(1 - 0.45/1.150) = 0.609 \times 1.068 = 0.650$$

$$\epsilon_n = 0.91\epsilon_r = 0.693$$

(4) Absorber Absorptivity The ordinate intercept changes directly proportionally to ρ_n/ρ_r, where ρ is the absorptivity for solar radiation. The slope remains unchanged:

$$\epsilon_n = \epsilon_r \frac{\rho_n}{\rho_r}, \qquad a_n = a_r$$

Example. The effective absorptivity[25] of the reference collector is 98.5%. A collector with only 95% absorptivity will, under otherwise identical conditions, have the following efficiency parameters:

$$\epsilon_n = \frac{95}{98.5} \times 0.762 = 0.964 \times 0.762 = 0.735$$

$$a_n = 1.068 \text{ Btu-hr}^{-1}\text{-ft.}^{-2}\text{-}^\circ\text{F}^{-1} \quad \text{(unchanged)}$$

TABLE 1.2 Radiation Loss R_B of an Absorber/Emitter[a] at Temperature T_c

T_c (°F)	R_B (Btu-hr^{-1}-ft^{-2})
80	62
100	82
120	105
140	130
160	158
180	189
200	223

[a] The absorber is radiating into a sky temperature of 0°F and is assumed to have no glazing. A glass glazing would reduce R_B considerably (to some 10% to 20%).

(5) Absorber Emissivity The emissivity of the reference collector is 89%, i.e., the coating is "flat black." It is difficult to assess quantitatively what the effect of turning to a low emissivity (i.e., 15%–30%) will be. Qualitatively, the re-radiation losses will be decreased. Because these losses increase with the fourth power of the absolute absorber temperature,[26] the improvement achieved with "selective" (low infrared or low IR emissivity) coatings over flat-black coatings is most noticeable at high collector temperatures; it is much less dramatic at low collector operating temperatures.

The effects of selective coatings also depend on the IR transmittance of the glazing. Glass has a low IR transmittance (i.e., most emitted radiation is intercepted by the glass glazing). During the interception, the IR radiation is converted into thermal energy and manifests itself in an increased glazing temperature (i.e., glass glazings of flat-black-coated collectors "run hotter" than the glass glazing of selectively coated collectors). The higher glazing temperature, in turn, means faster convective heat transfer at the outside surface of the glazing. This transfer is strongly dependent on the ambient wind conditions and cannot be generalized. It is for this reason that the advantages of selective coatings over flat-black coatings cannot be unequivocally quantified.[27]

A greater advantage of selective coatings is seen in collectors that are glazed with materials that have high transmissivity for IR radiation. Such glazing materials include Teflon (FEP, about 90% IR transmission), Tedlar (PVF, about 50% IR transmission), and most other polymers used as collector glazing materials.

The least advantage of selective coatings over nonselective coatings is seen with double-glass-glazed collectors, except if the collector is used under high-absolute-temperature operating conditions.

(6) Absorber Geometry The reference collector has upright fins, approximately 5/8 in. high and spaced at 1.5 in., on the absorber plate. If the fins are in the horizontal position, which corresponds to the collector mounting specification of the manufacturer [2], they interrupt the internal-convection loops and effectively decrease the temperature-loss coefficient. This increase in efficiency is claimed to be approximately 3%, based on tests performed by the manufacturer using collectors with and without fins under otherwise comparable conditions. According to these results, a 3% increase in the a_n value should be assumed for collectors that do not have upright fins.

Other geometric factors to be considered include shading of the absorber by the collector rim (if the absorber is excessively recessed in the housing) and absorber arrangements in which, for some reason, the sun does not intercept the absorber at certain incidence angles.

(7) Fin Efficiency A solar absorber plate that is based on the tube-on-fin construction principle may be subject to additional collection losses. All heat that is collected at the outside edge of an absorber fin must be transported through the fin material to the water passage. The thinner the fin material, the higher the temperature that the outer fin portions will reach in order to fulfill the heat-transfer condition. Consequently, the mean collector operating temperature will be increased by an amount that can be computed as

$$\delta T \approx 0.5 \frac{qd_1^2}{Cd_2} \tag{1.11}$$

As explained in detail in Appendix G, q in equation 1.11 is the collected solar heat (Btu-hr^{-1}-ft^{-2}), d_1 is the fin width parameter (fin width minus outer tube diameter, ft), d_2 is the fin thickness (ft), and C is the thermal conductivity of the fin material (Btu-hr^{-1}-ft^{-2}-°F^{-1}-in.).

The significance of equation 1.11 is that in the efficiency equation, $E_c = \epsilon - a(\Delta T - c)/I$ (equation 1.6), the temperature difference ΔT is increased by δT. This will transform equation 1.6 into

$$E_c = \epsilon - a\left[\frac{\Delta T - c}{I} + \frac{\delta T}{I}\right] \tag{1.6a}$$

If we substitute equation 1.11 for δT in equation 1.6a and use $q = I\epsilon - a(\Delta T - c)$, which is derived from equation 1.6, we obtain

$$\begin{aligned} E_c &= \epsilon - a\left[\frac{\Delta T - c}{I} + 0.5\frac{d_1^2}{Cd_2}\left[\epsilon - \frac{a}{I}(\Delta T - c)\right]\right] \\ &= \left(1 - 0.5a\frac{d_1^2}{Cd_2}\right)\epsilon - \left(1 - 0.5a\frac{d_1^2}{Cd_2}\right)a\,\frac{\Delta T - c}{I} \\ &= F\epsilon - Fa\,\frac{\Delta T - c}{I} \end{aligned} \tag{1.12}$$

This means that we have obtained new values for ϵ and a, changed by the factor $F = 1 - 0.5ad_1^2/Cd_2$. If we assume that a reference collector, having ϵ_r and a_r, has ideal fin efficiency, then the parameters ϵ_n and a_n of another collector with less perfect fin efficiency but otherwise identical conditions will have the efficiency characteristic shown by the dash in Figure 1.10. This indicates that the strongest effect of fin efficiency is seen at low-ΔT operating conditions and that the influence of fin efficiency decreases proportionally with increasing ΔT.

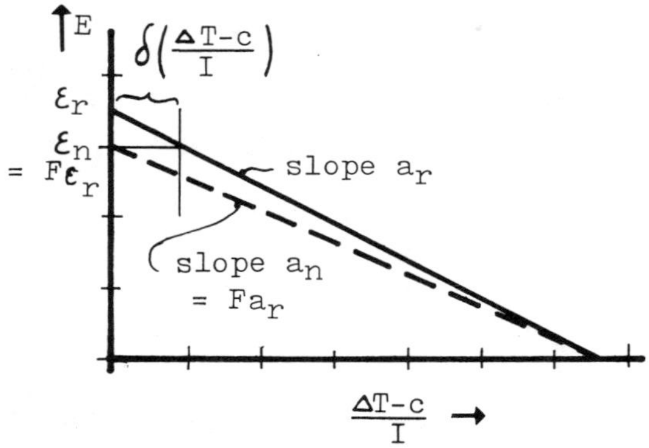

Fig. 1.10. Comparison of efficiency characteristics for a collector with high fin efficiency (solid line) and a collector with low fin efficiency (dash line).

If, as is the case in the present examples, the reference collector does not have perfect fin efficiency, we must compute the new parameters, ϵ_n and a_n, as follows:

$$\epsilon_n = \epsilon_r \left[1 - 0.5a_r \left(\frac{d_1^{(n)2}}{C_{(n)}d_2^{(n)}} - \frac{d_1^{(r)2}}{C_{(r)}d_2^{(r)}} \right) \right] \tag{1.13}$$

$$a_n = a_r \left[1 - 0.5a_r \left(\frac{d_1^{(n)2}}{C_{(n)}d_2^{(n)}} - \frac{d_1^{(r)2}}{C_{(r)}d_2^{(r)}} \right) \right] \tag{1.14}$$

Example. For the reference collector, the numerical values of d_1, d_2, and C are the following: $d_1^{(r)} = 0.40$ ft; $d_2^{(r)} = 0.008$ ft; $C_{(r)} = 1500$ Btu-hr^{-1}-ft^{-2}-°F^{-1}-in. If a collector to be tested has $d_1^{(n)} = 0.46$ ft, $d_2^{(n)} = 0.00042$ ft, and $C_{(n)} = 2700$ Btu-hr^{-1}-ft^{-2}-°F^{-1}-in. (e.g., a copper-tube/copper-sheet absorber with ⅜-in.-ID tubes soldered to 0.005-in. copper sheet, 6-in. spaced), the quantity

$$1 - 0.5a_r \left(\frac{d_1^{(n)2}}{C_{(n)}d_2^{(n)}} - \frac{d_1^{(r)2}}{C_{(r)}d_2^{(r)}} \right)$$

computes to

$$1 - 0.5 \times 1.068 \left(\frac{0.46^2}{2700 \times 0.00042} - \frac{0.4^2}{1500 \times 0.008} \right) = 0.907$$

The new efficiency parameters for the case that all other aspects of the two collectors are unchanged are

$$\epsilon_n = 0.762 \times 0.907 = 0.691$$

$$a_n = 1.068 \times 0.907 = 0.969$$

(8) Insulation The sides of the reference collector are insulated with 1-in. fiberglass insulation material. It is assumed that the collector to be examined also has reasonable side insulation, in which case the heat losses through the collector sides are quite comparable and need not be further examined with respect to ϵ and a.

The heat loss through the back of the collector is determined by the R value between the absorber and the collector housing. The slope of the efficiency characteristic a depends on this insulation effectiveness in the following manner:

$$a_n = a_r \left[1 + \frac{1}{a_r R_n} - \frac{1}{a_r R_r}\right] \tag{1.15}$$

$$\epsilon_n = \epsilon_r \quad \text{(unchanged)}$$

Example. The reference collector has $R_r = 12$. If the collector to be examined has $R_n = 8$, its efficiency parameters are

$$\epsilon_n = 0.762 \quad \text{(unchanged)}$$

$$a_n = 1.068(1 + 1/8.54 - 1/12.82) = 1.068 \times 1.039 = 1.11$$

(9) Combined Changes The evaluation of a solar collector with unknown efficiency characteristics, on the basis of the preceding considerations (1) to (8), must be performed in such a way that the comparison is made with the reference collector for each step (i). Calculation of the correction factors $\epsilon_n^{(i)}/\epsilon_r^{(i)}$ and $a_n^{(i)}/a_r^{(i)}$ is the final result for each step, and multiplication of all correction factors gives the efficiency parameters for the collector being evaluated.

Example. A collector having the characteristics of (1), (2), (4), (7), and (8) combined will have the following efficiency parameters:

$$\begin{aligned}\epsilon_n = 0.762 &\times 0.918 \text{ [from (1)]}\\ &\times 0.890 \text{ [from (2)]}\\ &\times 0.964 \text{ [from (4)]}\\ &\times 0.907 \text{ [from (7)]}\\ &\times 1.0 \quad \text{[from (8)]} = 0.544\end{aligned}$$

$$\begin{aligned}a_n = 1.068 &\times 0.918 \text{ [from (1)]}\\ &\times 1.0 \quad \text{[from (2)]}\\ &\times 1.0 \quad \text{[from (4)]}\\ &\times 0.907 \text{ [from (7)]}\\ &\times 1.039 \text{ [from (8)]} = 0.924 \text{ Btu-hr}^{-1}\text{-ft}^{-2}\text{-°F}^{-1}\end{aligned}$$

Materials Aspects

Absorber The following types of solar flat-plate absorbers are most frequently used: (a) all-copper plates with integrated water passages (Rollbond type) (the same type can be made of aluminum); (b) all copper (copper tube on copper sheet); (c) copper tube/aluminum fin; (d) iron or steel; (e) plastic (polymers) (Fig. 1.11). Some of the characteristics of these collector types are summarized in Table 1.3.

Absorptive Coatings Many varieties of absorber coatings are being used, ranging from simple flat-black paint to baked enamel. Two categories of coatings are distinguished: flat-black (i.e., nonselective) coatings and selective coatings. Flat-black absorber coatings have high absorptivity in the entire visible range and the medium IR range (wavelengths from 0.3 μm to 20 μm). Whereas high absorptivity is desirable in the visible range and the near IR range (wavelength

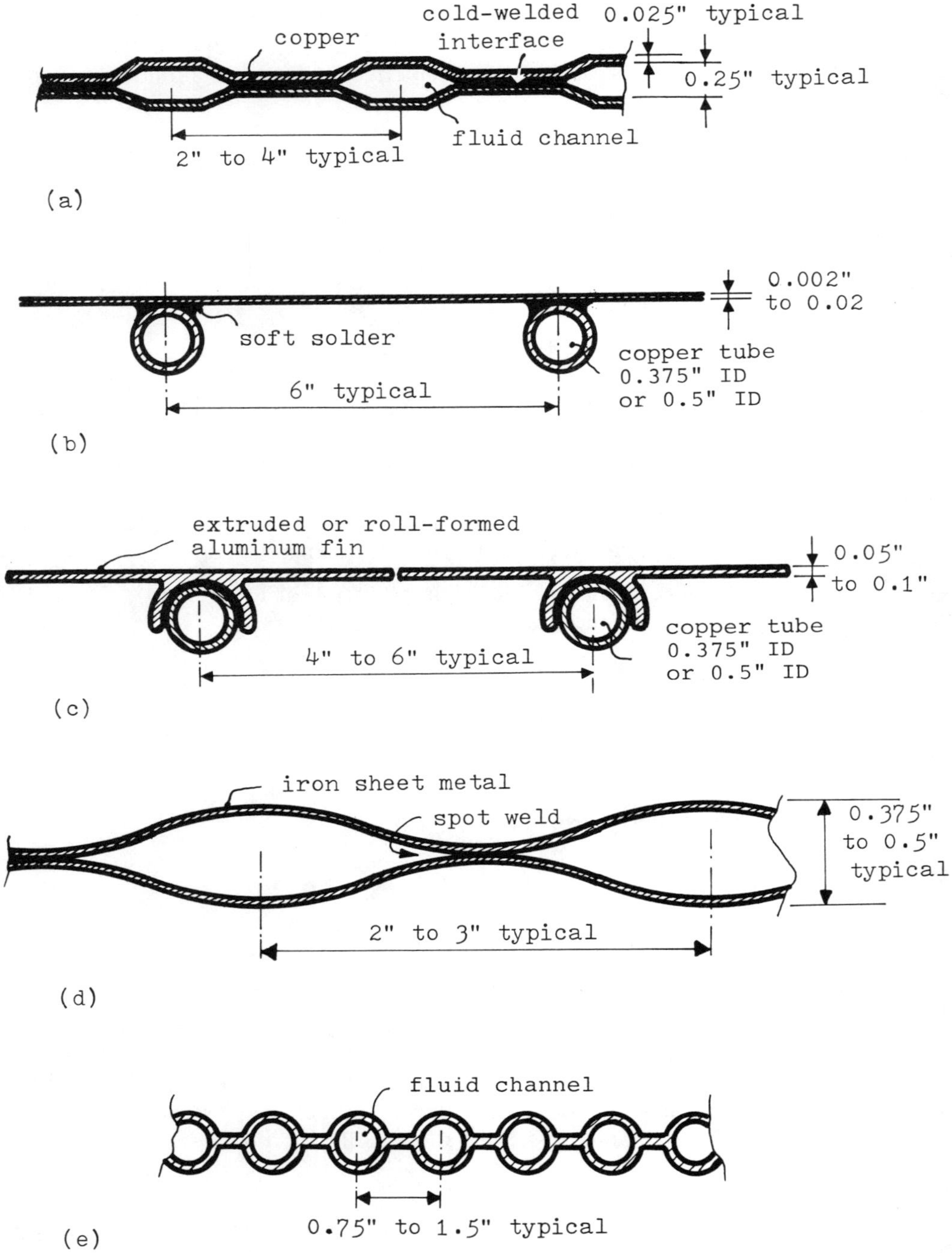

Fig. 1.11. Cross sections through typical absorbers: (a) Rollbond type; (b) copper tube on copper sheet; (c) copper tube on aluminum fin; (d) iron sheet sandwich; (e) polymer tubes.

TABLE 1.3 Characteristics of Various Types of Absorber Plates

	Collector Type					
Characteristic	Rollbond Copper	Rollbond Aluminum	Copper Tube on Copper Sheet	Copper Tube on Aluminum Sheet	Steel	Plastic
Corrosion of water passages	Low risk, often even open systems acceptable; most designs require care in flow velocity [37].	Risk of pitting corrosion and erosion corrosion (closed-loop design only, with corrosion-inhibiting fluid) [37]	Even better than Rollbond copper (thicker walls); also open systems are acceptable	Even better than Rollbond copper (thicker walls); also open systems are acceptable	Can pit and scale corrode and erode (must be designed as closed-loop system with corrosion-inhibiting fluid) [37]	None
Factors questioning 20-year service life	None (except corrosion if open system)	Corrosion (serious, even in closed-loop systems; use extreme care)	None	None[a]	Corrosion (in thin-wall designs)	Stagnation heat damage; accidental overpressure; fatigue fracturing
Is city water pressure permissible?	Not advised (use 30-psi relief valve)	Not advised (use 30-psi relief valve)	Yes	Yes	No (some designs allow 30-psi relief, others less)	No (only open systems with low static head)
Factors questioning good fin efficiency	Excellent fin efficiency	Excellent fin efficiency	Pipe spacing, sheet thickness [see section 1.1.2(7)]	Pipe spacing, fin thickness [see section 1.1.2(7)]	Excellent fin efficiency	Good fin efficiency only for close tube spacings
Heat resistivity	Excellent	Excellent	Excellent	Excellent	Excellent	Maximum 180°F to 250°F (use unglazed only)
Cost	High	Moderate/low	Moderate	Moderate	Moderate	Low

[a] It should be noted that there is no corrosion problem at the copper/aluminum interface. Even if the gap between these two dissimilar metals is not protected by a water-repellent agent (such an agent might also be recommended for increased heat conductivity), corrosion will attack only the aluminum and will not endanger the copper water passages. It should also be noted that construction of copper-tube/aluminum-fin heat exchangers has been common practice for many years in the heating and ventilating industry.

spectrum 0.4–1.5 μm), high absorptivity for longer wavelengths is not desirable because the thermal-radiation losses[28] that occur in this wavelength range are high and are proportional to the absorptivity[29] in this range. Selective coatings are selective in their absorptivity ratings. In the visible range, their absorptivities are high (>90%), whereas their absorptivities decrease for $\lambda > 1.5$ μm (λ is the wavelength of the thermal radiation) [28, 33, 36].

Many optical coatings that have selective characteristics have been developed in recent years. Areas of concern with regard to most types of selective coatings are high cost and longevity (ultraviolet resistivity, thermal stability, corrosion resistance). However, at least one type of selective coating has been found to be reasonable in cost and longevity, and it is used in several high-quality collectors that are available commercially. The coating is known as "black chrome," and it is applied in a sequence of chemical/electrolytic baths. It can be used on copper, aluminum, and steel absorbers.

The absorptivity of black chrome is 92%–95% in the visible spectrum and 10%–20% in the IR range. The decision to use a selective coating should be carefully considered. Generally, the benefit from a selective coating is highest if the collector is used at high temperatures.[30] At low temperatures the disadvantage of a somewhat lower absorptivity in the visible range (some flat-black coatings have 97%–98% absorptivity) may outweigh the advantage stemming from the selectivity.

The following specifications should be met by an absorber coating for a flat-plate solar collector: (a) It must not degrade under ultraviolet (UV) exposure. (b) It must withstand (stagnation) temperatures up to 400°F. (c) It must withstand many temperature cycles over ±100°F. (d) It must withstand many cycles of low-to-high relative humidity.[31] (e) It must not chalk, peel, or chip.[32] (f) It must not be so thick that heat conduction through the paint to the metal absorber is impeded.[33] These specifications should be met for the entire service life of the collector.

Glazings The glazing on a solar collector has three important functions. First, it must reduce convective heat losses from the absorber plate (see Section 1.1.3). Second, it must suppress radiative heat losses from the absorber plate. Third, it must protect the absorber from the elements and from excessive UV exposure. Glass meets all three of these requirements; furthermore, it is compatible with the general longevity requirement that applies to all solar collectors. Various other types of glazing materials are also being used in commercially available solar collectors. Some of these materials do not meet the second and/or third of the preceding requirements. Most non-glass glazings are also questionable with respect to longevity. Table 1.4 shows specific details for various glazing materials.

The following requirements must be met by glazing materials, in addition to the optical qualities mentioned earlier:

1. A glazing material must be reasonably impact-resistant. Thin or nontempered glass panes are questionable because of the risk of damage from hail, birds, and vandalism. Plastic materials of low tensile strength (e.g., Teflon) are not advisable as outer glazings.
2. A glazing material must be resistant to significant temperature shocks. A leaf on a stagnant collector can cause high localized thermal stresses on the glazing; sudden rain will cause rapid overall temperature changes. Thus heat-tempered

TABLE 1.4 Characteristics of Various Glazing Materials

Characteristic	Very-Low-Iron Glass	Low-Iron Glass	Regular Glass	FEP (Teflon)	PVF (Tedlar)	Tedlar-coated Fiberglass
Iron oxide content	$<0.01\%$	0.04%–0.06%	$\approx 0.1\%$	n.a.	n.a.	n.a.
Solar transmittance	91%–92%	88%–89%	80%–84%	96%–98%	85%–92%	75%–82%
Impact resistance	Good	Good	Good	Low	Good	Good
UV resistivity	Excellent	Excellent	Excellent	Excellent	Good	Good
Mechanical stability	Excellent	Excellent	Excellent	Excellent as inner glazing; poor as outer glazing	Fatigue fractures if unsupported	Good
Long-term optical stability	Excellent	Excellent	Excellent	Very good	Good	Good after initial 3%–6% decrease in transmittance
IR transmissivity	Low	Low	Low	High	Absorbs about 50%	Low
Temperature stability	Excellent	Excellent	Excellent	Excellent	Heat-shrinks above 250°F	Poor (do not use as inner glazing)
Suitability as outer glazing	Excellent	Good	Medium (low transmissivity)	Poor (low impact resistance)	Poor (fatique fractures)	Medium (low transmissivity)
Suitability as inner glazing	Excellent (fair transmissivity but high IR absorptivity)	Medium	Poor (too low transmissivity)	Excellent (fair IR absorptivity but high transmissivity)	Not suitable (temperature limitation)	Not suitable (severe temperature limitation)

glass is an absolute necessity for the outer collector glazing. Plastic glazings generally can easily withstand the temperature shocks that occur during solar-collector operation.

3. A glazing material must be resistant to UV radiation. Non-UV-inhibited plastic materials are not acceptable, and even many inhibited materials are questionable. An excellent polymer with respect to UV damage is Teflon (FEP), which has high transmissivity for UV radiation. Polyvinylfluoride (PVF, Tedlar) is known to withstand UV radiation exceptionally well while absorbing it. Therefore Tedlar is often used to protect other materials underneath from UV radiation. When bonded to fiberglass, for example, it makes an excellent UV-stable glazing material.
4. A glazing material must be able to withstand wind conditions. Whereas glass and fiberglass meet this requirement, Tedlar is prone to fatigue fracture if stretched over larger open frames such as are used in solar collectors. This rules out Tedlar as a long-lasting glazing material if used by itself; but as mentioned earlier, Tedlar is a good material if it is bonded to a fiberglass support.
5. A glazing material's optical rating must not change during its service life. This is a requirement that can hardly be met by any plastic glazing material. However, some fiberglasses show a reduction in transmissivity by a few percent during the first few months or years of service and then remain constant in their optical ratings for the rest of their service lives. Such behavior may be acceptable, but it must be known and accounted for during the design of a solar installation.

Many collectors have two glazings. This enhances the three functions that glazings serve (reductions in convective heat losses and radiative heat losses and UV protection for the absorber), but concomitantly it decreases the instantaneous efficiency of the collector in the low-ΔT operating range.[34]

The inner glazing of a double-glazed collector must meet an additional and highly important requirement: It must be able to withstand stagnation temperatures well above 300°F. Most plastic materials fail to meet this specification. Polyvinylfluoride, for example, although more heat-resistant than most polymers, shrinks at temperatures above 250°F. An exception is Teflon, which remains inert at temperatures up to 450°F. This makes Teflon an excellent candidate as an inner glazing material, because its low tensile strength (important for outer glazings) is no drawback for an inner glazing. Teflon also has the advantage of an exceptionally high solar transmittance (97%),[35] although its high IR transmittance is a distinct disadvantage.[36]

Insulation, Shell A solar flat-plate collector must be insulated against excessive heat losses on its back side and on its edges. Because the heat-loss values for a collector through its front side (i.e., through the glazing) are of the order of R-1 for a single-glazed type and R-2 for a double-glazed type, it is evident that it is easy to construct a solar flat-plate collector that has relatively negligible heat losses through its back side and through its edges. In practice, an R-value of 10–12 is quite adequate for the back side (i.e., 3.5 in. of fiberglass insulation or 2 in. of foam insulation). In fact, often it may be more cost-effective to consider an R-10 to an R-12 as sufficient, rather than designing a thicker collector that can house a substantially heavier insulation.[37]

The side insulation for a collector is usually quite adequate if 1 in. of fiberglass or 0.5 to 0.75 in. of foam insulation is used. It should be noted that larger collectors are more desirable with respect to edge heat losses because of their higher ratios of collector area to edge length. The insulation material must meet the following requirements:

1. It must withstand the maximum collector stagnation temperature rating (400°F) without damage. Most foam materials do not meet this requirement. Excessive heat will cause these materials to shrink (excessive heat is unavoidable during stagnation conditions, which usually occur during the collector installation period but also may occur afterward during occasional electricity outages). This reduces the effective insulation value. Usually such failure is not obvious; it may be difficult to detect without actually taking the collector apart.
2. The maximum stagnation temperature must not cause evaporation or sublimation of substances in the insulation material, such as the binder of fiberglass insulation. Such vapors can condense on a relatively cool surface, particularly at the glazing. Thus they can coat the interior surface of the glazing with a thin film that may look unattractive and may actually impede solar transmission through the glazing.[38] Special fiberglass materials are available that have quite satisfactory outgassing rates.

The collector housing requires attention to the following construction details:

1. No materials should be used that have the potential to develop a flash temperature below the rated stagnation temperature. In particular, no wood should be used in the interior of the collector. Repeated stagnation causes disruption of the hydrocarbon bonds in wood fibers, and this reduces the flash point to a dangerously low level.
2. The design of the collector must ensure that expansion and contraction of the absorber plate, as well as of the glazing, do not cause any problems during the service life of the collector. This requirement eliminates most standard window-glazing techniques for the mounting of the glass coverplate of a collector. Expansion and contraction of as much as $\pm 1/8$ in. can occur, and this must be accounted for in such a way that no leakage and no glazing breakage can occur even after thousands of temperature cycles.

Specifications of a Commercial Collector

Figures 1.12 to 1.15 provide an example of the amount of data and the kind of information that should be given in a solar-collector brochure. Rather than giving the explicit efficiency graphs, the more analytic method of giving the values for ϵ, a, and b (see Section 1.1.2 for definitions) is used by some manufacturers. It is also desirable to have information about the type and IR emissivity of the absorber coating and information on the iron content of the glass glazing.

1.1.3 Unglazed Flat-Plate Collectors, Liquid Type

Efficiency Characteristic

Because unglazed collectors are used only at low-temperature (or, better, low-ΔT) conditions, it is not necessary and is not customary to express their efficiencies in the form of second-order equations (such as equation 1.4). A simpler linear expression of the type discussed in Section 1.1.2 (equation 1.6) is used instead.

Fig. 1.12. Cross section of absorber fin (aluminum) with copper tube (right) and cross section of side extrusion of collector housing. Typical material thickness 0.09 in.

Model 200G/100G

Size	92-3/4" x 46-3/4" x 4-3/4"
Weight – 200G	220 lbs (dry); 225 lbs (water filled)
– 100G	180 lbs (dry); 185 lbs (water filled)
Absorber	Extruded aluminum fins pressfitted on copper tubing, silicone-base heat transfer compound for increased heat transfer. Net absorber area: 28 square feet.
Absorptivity	99%
Insulation	Back: 2-1/2" fiberglass Sides: 1" fiberglass
Glazing	3/16" tempered, double strength glass. Model 200G has double glazing. Model 100G has a single glazing.
Plumbing	1/2" Type "L" copper pipe, protruding 2" outside panel frame. Optional inlet/outlet locations: both on one short side or one on each long side.
Flow Requirements	1 gpm minimum, 2.5 gpm maximum. Pressure drop at 1 gpm is 0.7 psi (1.6 ft. of head).
Stagnant Temperature Rating	350° F continuous
Flammability	>600° F for all panel components
Warranty	2 year full warranty on parts and labor; 10 year guarantee against internal leaks. See complete warranty statement for terms and conditions.

Fig. 1.13. Specifications of a solar collector.

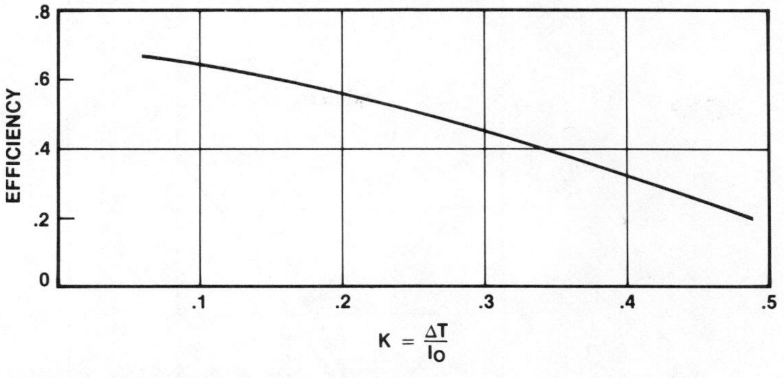

Fig. 1.14. Model 200G efficiency characteristic.

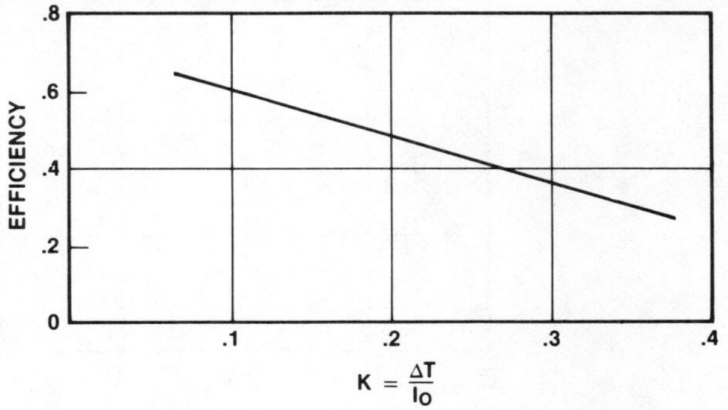

Fig. 1.15. Model 100G efficiency characteristic.

However, several additions and changes must be made to reflect deletion of the glazing and the back-side insulation:

1. The transmission-absorption coefficient ϵ becomes equal to the absorption coefficient ρ.[39]
2. The temperature-loss coefficient a is no longer a constant factor; it now depends strongly on the wind velocity prevailing over the collectors (this dependence is expressed in equation 1.16, and it makes any further use of the factor c in equation 1.6 unnecessary).
3. Black-body radiation into the sky becomes a significant additional loss factor, and it must be included in the equation (see footnote 26).
4. Radiation losses to the surroundings (other than the sky) also become significant. This effect includes, in particular, the radiation from the back side of the (uninsulated) collector to the mounting surface (roof or ground). This radiation loss can be expressed simply as a linear function of $\Delta T = T_c - T_a$, with the numerical value of 1.0 as proportionality constant,[40] and added to the wind-velocity-dependent temperature-loss coefficient a_w.

The efficiency equation for unglazed collectors now becomes

$$E_u = \epsilon - (a_w + 1)\frac{\Delta T}{I} - \frac{\epsilon_0 \sigma}{I}(T_c^4 - T_{\text{sky}}^4) \tag{1.16}$$

with $I = I_0(\cos\vartheta + k_1)$, as defined in equation 1.2. The dependence of a_w on the wind velocity [16] is shown in Fig. 1.16 (solid curve).

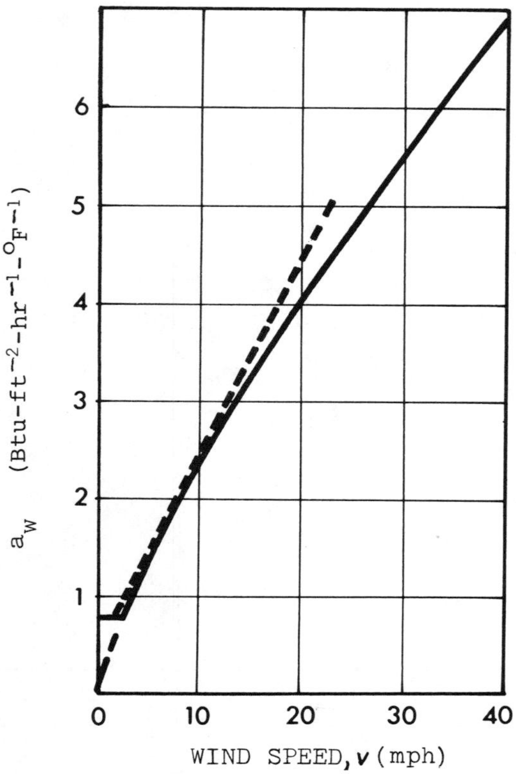

Fig. 1.16. Dependence of the temperature-loss coefficient a_w on wind speed v. Solid curve: data of de Winter [16]. Dash curve: linear approximation of $a_w = 0.2v + 0.5$ to be used for system efficiency calculations with small computers (see Appendix K).

The value of the black-body-radiation term $\epsilon_0\sigma(T_c^4 - T_{sky}^4)$ depends on the effective sky temperature T_{sky} and can vary considerably with the relative humidity [10] and with the concentrations of CO_2 and particulates in the airspace above the collector. It can reach average values as high as 60 Btu-ft^{-2}-hr^{-1} in desert climates; other sources list 25 Btu-ft^{-2}-hr^{-1} as an average value [9, 16].[41] See Table H.5 in Appendix H for daily collection rates with unglazed collectors at various temperatures, tilt angles, geographic latitudes, and wind velocities.

Materials Aspects

All the types of absorber plates listed earlier for glazed collectors can be used as unglazed collectors if the proper measures are taken to avoid corrosion (internal and external). Table 1.3 lists some of the properties of the various collector types.

The specifications for absorber coatings for unglazed collectors are different from those suggested earlier for glazed collectors in terms of absolute temperature requirements (200°F maximum temperature stability is satisfactory) and in that they must meet more severe weathering resistance requirements.

1.2 Solar Control Systems

1.2.1 General Description and Hardware

In general, control systems for solar heating installations are simple and comparatively inexpensive. The cost of the entire control system rarely exceeds 5% of the cost of the installation. More likely, in particular if the solar installation is large, the cost of the solar control system will range between 1% and 3% of the cost of the installation.

A typical residential solar heating system contains three controlled loops: (a) the solar-collector loop, (b) the furnace or space-heating loop, and (c) the domestic hot-water (DHW) loop. These loops are independently controlled, and the more general aspects of their control are discussed in Section 2.4. In this chapter, some practical details will be described and some examples given. In some instances these examples concern less sophisticated control functions than those described in Section 2.4.

The key element for each controlling function is a differential thermostat that senses the temperatures at two locations, such as the absorber plate and the heat-storage tank, or the DHW vessel and the storage tank, and switches pumps and valves accordingly. The following are some of the manufacturers of differential controllers: Heliotrope General [27], with a broad line of compact controllers having many features and ranging in price between $40 and $80; Rho Sigma (Van Nuys, California), offering several sturdily built models in the $75 to 110 price range; Hawthorn, Inc. (West Palm Beach, Florida), with several models in the $30 to $70 price range. Robertshaw, Barber-Coleman, and Honeywell have also introduced solar-system controllers. An example of the variety of control systems offered by one manufacturer is presented in Table 1.5.

1.2.2 Solar-Collector Loop

The objective here is to turn on the collector-loop pump when the sun is providing enough energy for the collector panels to deliver heat to the storage tank. Because the storage-tank temperature is variable, the best method of doing this is to use a differential thermostat that compares the absorber temperature with the tank temperature. When the sensed temperature difference is right, the collector-loop pump is turned on. Figure 1.17 shows how this is done. When the controller senses that the collector panels are hotter by a preset amount (4°F–20°F, depending on the model) than the solar storage tank, the relay in the controller closes, delivering electrical power to the pump. When the collectors cool off and the temperature difference decreases to 1°F–5°F (depending on the model), the relay opens and the pump shuts off.

Some control systems have a third temperature sensor that is calibrated to sense either freezing conditions or very high temperature conditions; i.e., it will also activate the pump for "hot-shot" freezing protection (see Section 2.2.3) or for overheating protection by night-sky radiation (see Section 2.2.5). This option is indicated in Fig. 1.17 by the dash lines.

The additional overheating protection can also be obtained by using an additional thermostat connected in parallel with the collector temperature sensor of a Heliotrope General controller. This is indicated by dotted lines in Fig. 1.17. The

TABLE 1.5 Differential Thermostats and Sensors Offered by Heliotrope General [27][a]

MODEL APPLICATION

With forty UL listed models, there is a unit to fit any application. Each model series is available in **four different ranges of differentials** "ON" and "OFF", as well as adjustable. The four ranges are then further broken down into several "MODE" catagories:

NONE: The 80 and 90 series have only the basic differential temperature function.

MODE 1: The 100 series is "OFF" when the collector is below 80°F. This feature prevents the pump turning on at night when a large amount of heat has been extracted from storage.

MODE 2: Known as the "Freeze-Recycle" method of collector freeze protection, the 200 and 290 series turn the pump on at 36°F and off at 37.5°F. This pushes a small amount of water through the collector to prevent freezing. This model utilizes a third thermistor sensor for the freeze detection and it should be placed at the bottom inlet to the collector where the freezing temperatures will be first detected. A very small amount of stored heat is lost with this method due to the precise control with the low thermal mass sensors.

MODE 4: The 400 series is a High Limit tank protection and does not allow the pump to activate once the storage is above 160°F. A third thermistor for this sensing allows this temperature detection to be made at the top of the storage tank while the differential sensor is at the bottom.

This function can also be added to the MODE 1 and MODE 2 models by ordering a separate sensor, HLS-2, which adds this 160° "OFF" feature.

MODE 4 is also available as adjustable from 60 to 200°F in the 3410 series.

MODE 6: The 690 series has two separate outputs. One output controls the pump and incorporates the "OFF" above 160°F of the MODE 4, and the second output controls valves for collector freeze protection. This method known as "Freeze-Dump" protection, electrically holds valves in a circulation postion. Upon reaching 42°F or a power failure, the valves are de-energized and return to a position which shuts off circulation and drains the collector. They are re-energized at 52°F.

The sensor included for this MODE 6, the FS-1, is a low voltage, hermetically sealed bi-metallic element designed for military/space applications under U.S. Govt. Spec. MIL-S-24236.

SPECIFICATIONS, PRICING

	DTT-Model Numbers according to "ON/OFF" differentials				Electrical Hook-Up			
	9/3 F	4.5/1.5 F	15/5 F	15/1.5 F	line cord/ cover receptacle	hard wire ½" box knockouts	MODES INCORPORATED OTHER THAN BASIC DIFFERENTIAL FUNCTION	PRICE Including Standard Sensors
NON ADJUSTABLE MODELS	DTT-80	DTT-81	DTT-82	DTT-83		X	NONE	$38.00
	DTT-90	DTT-91	DTT-92	DTT-93	X			44.50
	DTT-100	DTT-101	DTT-102	DTT-103		X	MODE 1. "OFF" below 80°F	43.00
	DTT-200	DTT-201	DTT-202	DTT-203		X	MODE 2. "ON" below 36°F	43.00
	DTT-290	DTT-291	DTT-292	DTT-293	X			49.50
	DTT-400	DTT-401	DTT-402	DTT-403		X	MODE 4. "OFF" above 160°F	43.00
	DTT-690	DTT-691	DTT-692	DTT-693	X		MODE 6. Valves "OFF" below 42°F	68.00
ADJUSTABLE MODELS	DTT-3410	DTT-3411	DTT-3412	DTT-3413		X	MODE 4. Hi-limit "OFF" ADJUSTABLE 60-200 F	$53.00
	DTT-70	Differential "ON" ADJUSTABLE from 3 to 24 F. with "OFF" fixed at 1.5 F				X	NONE	48.00
	DTT-71					X	MODE 1	53.00
	DTT-72					X	MODE 2	53.00
	DTT-74					X	MODE 4	53.00

All models shown above have "Normally Open" relay contacts. For "Normally Closed" relay contact function specify "NC" after part number and add $5.00 to the price. All models shown above are 120VAC, 10 Amp for use with up to a ⅓HP motor. For 240VAC up to a ½HP motor specify "240V" after part number and add $7.50 to the price. Up to 3HP 240V controllers are available, consult factory for details.

SENSORS, Thermistor type 3,000 ohm @ 25°C

TES-1, STANDARD SENSOR

$4.00
10" long, hi-temp wire, operational to 400°F, extremely low thermal mass. Includes wire nuts for electrical hook-up and silicone fiberglass tape for surface mounting.

BP-25, BRASS PLUG

¼" MPT $6.75

BP-50, BRASS PLUG

½" MPT $7.00

DBP-25

¼" MPT, with two sensors. $10.50

DBP-50

½" MPT, with two sensors. $11.00

Note: Substitution of thermistor sensors can be made for the standard TES-1 sensor when ordering any HELIOTROPE GENERAL products. Net difference in price will be charged.

RBP-50, REVERSE BRASS PLUG

½" MPT, $7.00
The RBP series is utilized for direct top immersion into a non-pressurized tank. Utilize a pipe coupling and the necessary length of pipe to fabricate a drop tube with the open end above the tank top. The wire connections should be made before the RBP is connected to the pipe coupling.

SAS-3, SCREW ATTACHED SENSOR

$5.00
Screw-attached sensor in copper housing providing minimum thermal mass.

IS-6.5, STAINLESS STEEL IMMERSION SENSOR

½" MPT, 6 inches deep immersion, $11.00

IS-6.75, STAINLESS STEEL IMMERSION SENSOR

¾" MPT, 6 inches deep immersion, $11.00

SENSORS, Bi-Metallic type

FS-1 $20.00

Includes stainles steel mounting hardware for pipe installation. Closed with lowering temperatures until 42°F when sensor opens and re-closes at 52°F. Included with DTT-690 models.

HLS-2 $3.50

Closed with rising temperatures until 160°F is reached when sensor opens and recloses at 130°F. Adds MODE 4 function when used with DTT-100, 200 and 290 Series.

[a] Prices may vary.

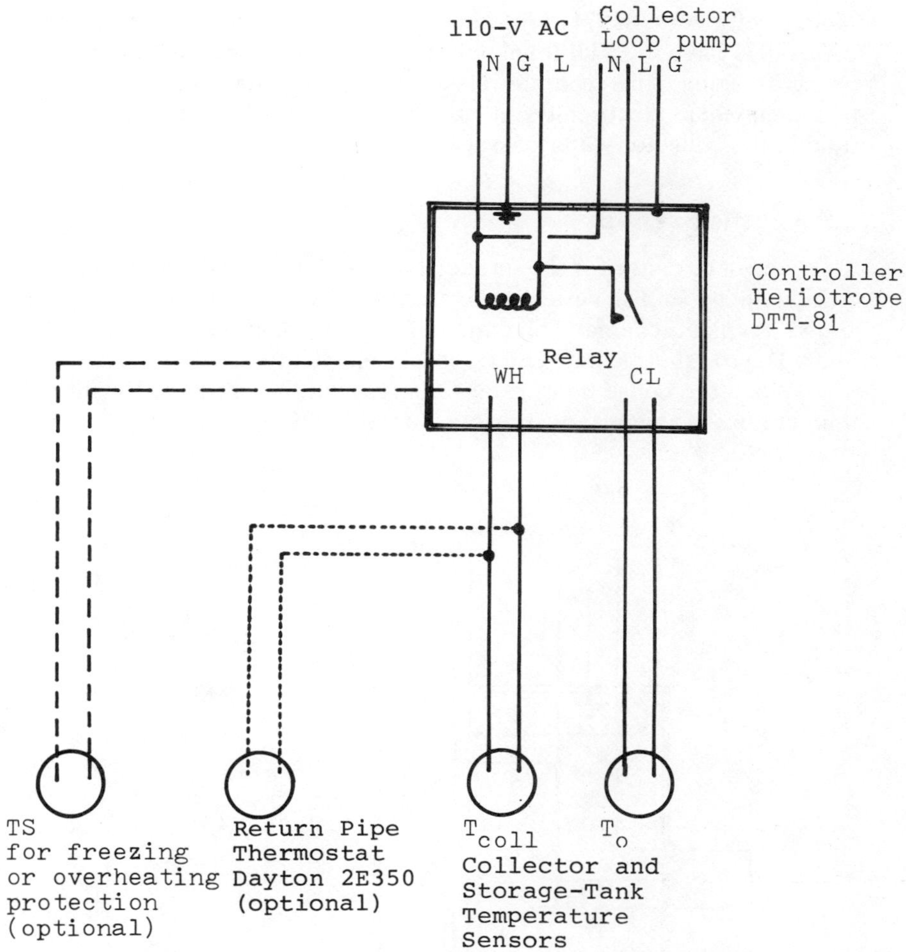

Fig. 1.17. Typical control schematic for a solar-collector loop. If $T_{coll} > T_0 + \delta t$, where δt is a positive temperature difference of a few degrees, the relay is closed, and it activates the collector-loop pump. Also indicated are two possible methods for freezing and/or overheating protection.

thermostat should be mounted on the return pipe from the collectors. If the water temperature exceeds a certain adjustable value, the contacts will close, shorting the collector temperature sensor, which simulates a very high temperature and therefore causes the controller to deliver power to the pump even when there is no sunlight. This will cause the collectors to lose heat to the sky, lowering the water temperature in the storage tank. When the temperature in the tank has fallen several (preset) degrees below the turn-on value, the thermostat will open again, turning off the pump. The thermostat set point should be the maximum temperature for the tank[42] less the possible temperature increase during one day. Note that at high tank temperatures (high-ΔT values) the collector efficiency is usually so low that there will not be much temperature rise. Thermostat models acceptable for this purpose include the Dayton 2E350 (cost about $30) and the

Honeywell LA409B1071 Aquastat (cost about $40). Some commercially available controllers have the additional optional feature of separate relay closure (or opening) at freezing. The separate relay action can be used to drain down the collectors, heat them electrically, or otherwise prevent freezing of the heat-transport fluid in the collector water passages.

1.2.3 Furnace Loop

The control circuitry for the furnace loop is more complicated than the other control circuitry, and it varies more from one application to another. Figure 1.18 shows a typical schematic diagram based on a Honeywell fan center (R8239B1043) and a Dayton thermostat (2E051) as key elements.[43]

When the house thermostat calls for heat, it connects W with R. If the tank temperature is above some preset value, say 120°F (indicating that enough heat is

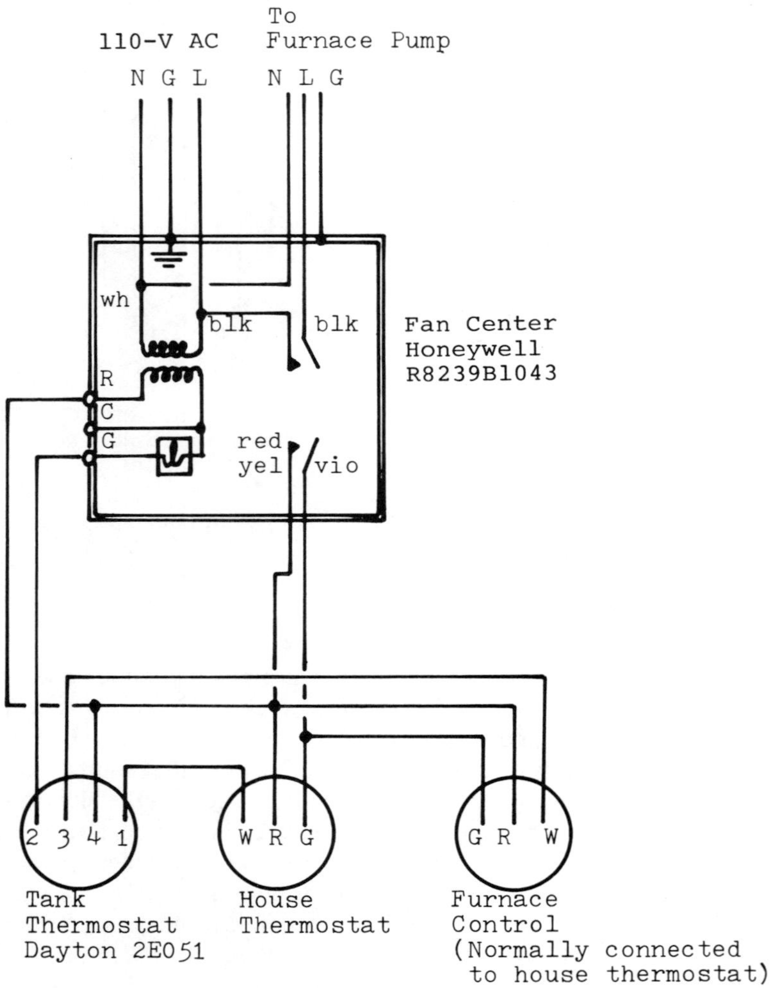

Fig. 1.18. Typical control schematic for the space-heating loop of a forced-air-heated building. (Note: not applicable if the house thermostat controls an air conditioner.)

stored for solar rather than auxiliary space heating), the contacts 1 and 2 of the tank thermostat are connected. This causes the fan center to close both its relay contacts. One relay contact turns on the furnace pump, which pumps hot water from the tank to a heat exchanger (or a fan coil) located inside a conventional furnace (see Section 2.4.1). The other relay contact in the fan center connects R with G in the furnace control unit (to which the house thermostat would be directly connected if this were not a solar heating system). This causes the furnace fan to run, pulling heat off the fan coil and delivering it to the house for space heating. However, the furnace remains off.

If the tank is not hot enough, then the contacts 1 and 2 in the tank thermostat will remain open, and the fan center will be inoperative. Instead, the tank thermostat contacts 3 and 4 will be connected, which will connect R and W in the house thermostat with R and W in the furnace control unit, turning on conventional auxiliary heat.[44]

An elegant variation of the control system just discussed is shown in Fig. 1.19. This scheme allows two separate temperature settings in the house: one at a comfort level of 70°F to 75°F, strictly operating the "free" solar-heat portion of the system, the other one set at an energy-conserving level of 60°F to 65°F, operating the auxiliary furnace.

Figure 1.20 shows an example of an automatic control system for a radiant floor heating subsystem. The furnace pump is directly controlled by the house thermostat, and it runs whenever space heat is demanded,[45] irrespective of the availability of heat in the storage tank. The thermostats regulate when to use auxiliary heat.[46] $T_{a(\min)}$ will typically be preset at $T_a \approx 30$°F.

1.2.4 Domestic Hot-Water Loop

The domestic hot-water (DHW) loop is simple to control, as is shown schematically in Fig. 1.21. The function of the controller is to cause recirculation from the DHW tank to the heat-storage tank and back (see Fig. 2.23) whenever the DHW temperature is lower than the temperature of the heat-storage medium. Thus the controller tries to keep the DHW vessel as hot as the storage-tank water.[47]

Because the storage-tank water can become too hot to be touched (more than 160°F), it is essential that some overtemperature protection be provided. One way is to use a controller with an optional antiscalding feature that keeps the DHW pump off when the DHW temperature exceeds 160°F. Another method to prevent scalding is to incorporate a tempering valve in the plumbing to mix cold water with the hot water as it leaves the DHW tank.

1.2.5 Subtleties

This section is concerned with some problems that are not immediately obvious, such as those related to sensor location and electrical codes. The solar-collector sensor (to detect collector heating by the sun) is best located inside the collector panel, between the glazing and the absorber plate. It should be mounted under a black-painted surface, thermally isolated from the collector housing and from the absorber plate. All wire inside the collector housing must withstand temperatures as high as 400°F (Teflon insulation is acceptable). An alternate location for the sensor is on the collector outlet pipe, outside the panel. This involves two prob-

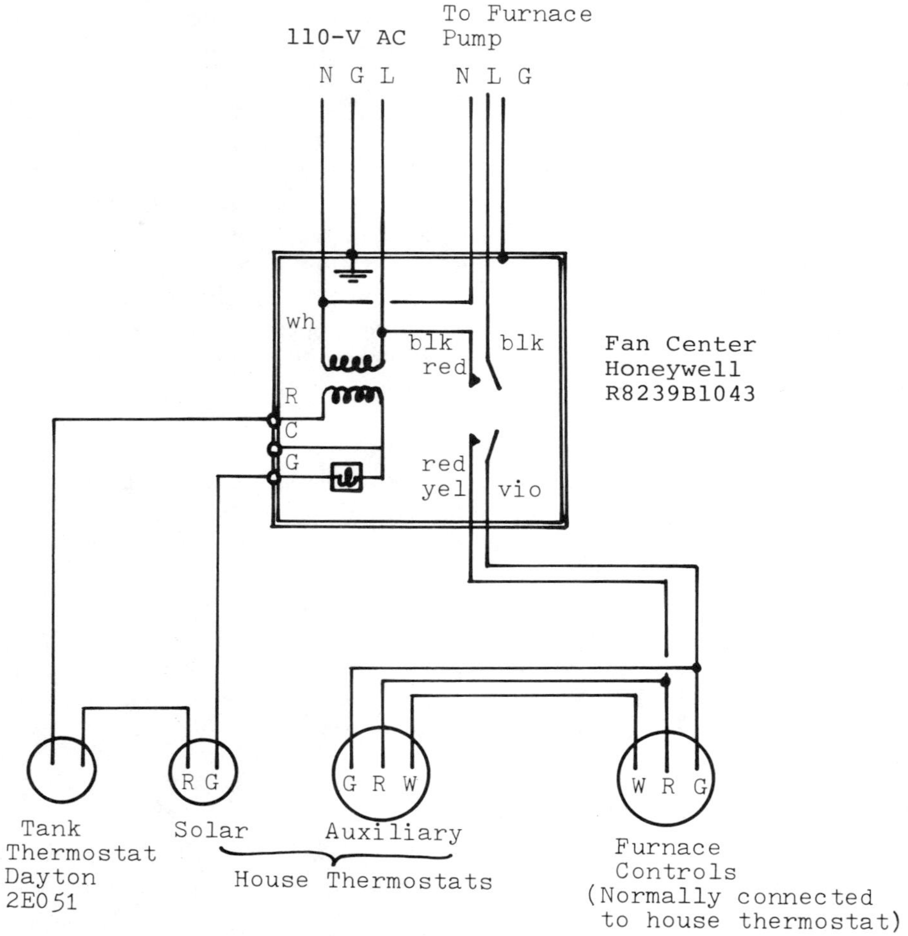

Fig. 1.19. Variation of the control scheme for a solar-driven forced-air heating subsystem allowing two independent house-temperature settings and making use of a "free" comfort level when possible. (Note: not applicable if the house thermostat controls an air conditioner.)

lems: (a) It is somewhat cooler than the absorber plate; thus it senses less accurately the condition in which heat is added to the heat-storage tank. (b) During start-up conditions (e.g., in the morning), when the initial surge of cold water in the pipes flows through the collector, the absorber plate is immediately strongly cooled again, which causes the pump to cycle and, in turn, shortens the life of the pump. To avoid cycling, a controller will be needed to delay start-up until a large temperature difference has been built up, but this may decrease overall system efficiency.

Tank-mounted sensors must be heavily covered with insulation to minimize the temperature gradient near the tank surface. This includes the DHW sensor, which must be mounted inside the DHW tank enclosure under the fiberglass insulation.

The controller should be located as close to the pumps as possible. If the distance is less than 6 ft, most codes permit the use of flexible metal conduit,

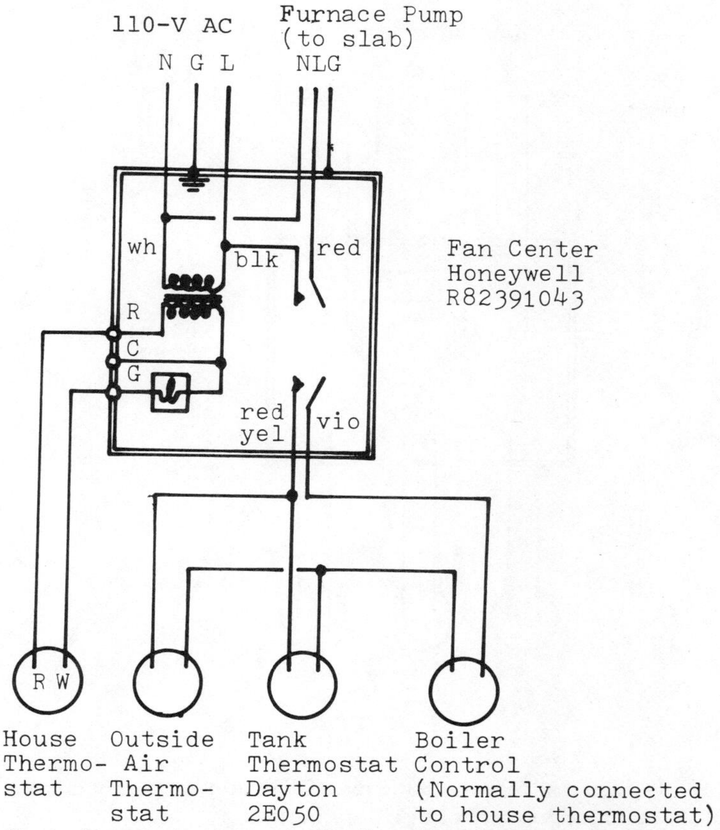

Fig. 1.20. Typical control schematic for the space-heating loop of a radiant slab-heated building.

which is easier to work with than regular rigid-metal conduit. All controllers and pumps must be grounded. The pumps must have a separate ground wire. Low-voltage and high-voltage wires must not be in the same boxes or conduits unless the low-voltage wires use high-voltage insulation (600 V).

The temperature sensors can be damaged by overheating during the soldering of connections. Damage to the sensors may be prevented if the sensor head is immersed in cold water during careful, short-duration soldering efforts. A low-wattage circuit-card soldering iron should be used if possible.

1.3 Heat-Storage Equipment

There are three basic categories of heat storage for solar applications: (a) liquid heat-storage medium (water), (b) solid heat-storage medium (rock bed), and (c) heat storage by phase change. Because the majority of the sections in this book deal with solar energy applications involving solar collectors with a liquid heat-transfer medium, the first category is most appropriately discussed here. However, we shall also be discussing some of the design principles of rock beds.[48]

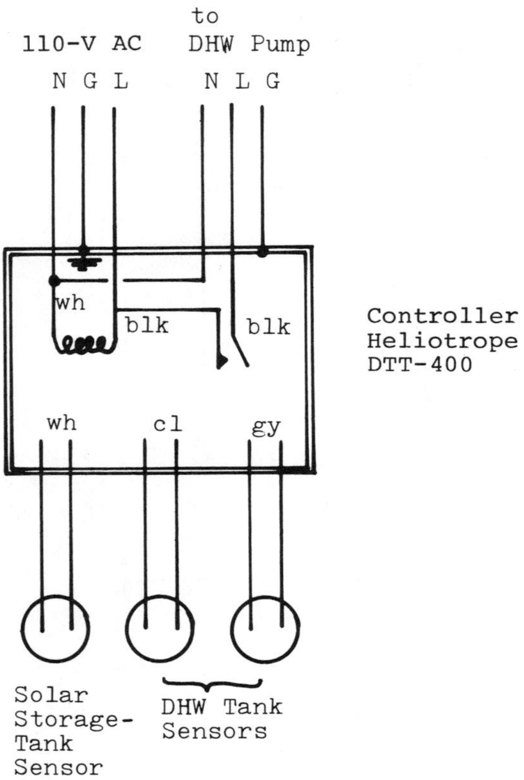

Fig. 1.21. Typical control schematic for a DHW loop as part of a solar space-heating system (not included: control of auxiliary heat).

Heat storage by phase change (latent thermal storage) is still in the research-and-development stage and therefore is not discussed in this book. Good reviews of the state of the art of latent heat-storage facilities have been published by Lorsch and Kauffman [31] and Telkes [40], and the interested reader is advised to consult that literature for further details.

1.3.1 Liquid-based Systems

Tank Types

Steel Tanks The main advantages of steel tanks are relatively low cost, ease of fabrication to custom dimensions and to the specifications of the ASME pressure-vessel code, and the considerable experience already available with steel tanks. The main disadvantage of steel tanks is that they are subject to corrosion. The following have been used to protect steel tanks from failure due to corrosion:

1. A protective inner coating or lining. Glass and stone linings are most effective, but often they are prohibitively expensive, and usually they are readily available only for tank sizes of 120 gal and smaller. Large numbers of alternative coatings and linings have been proposed, such as phenolic epoxy, butyl rubber, and coal tar, as well as vinyl, polyethylene, and polyvinylchloride (PVC). Adherence of the coating to the tank walls is important not only because corro-

sion protection depends on it but also because peeling can cause severe clogging problems, even when a strainer is used. It should also be noted that it is difficult to find coatings that stand up to the temperature and humidity conditions in this application.

2. Increased steel thickness. Extra wall thickness may be worth the extra expense, and it can easily be combined with other methods of protection.
3. Chemical corrosion inhibitors. Corrosion inhibitors protect the tank either by forming chemical complexes on the surface of the metal or by adjusting the acidity of the solution to a range at which corrosion takes place very slowly. In systems protected by chemical corrosion inhibitors, the solutions must be tested periodically, typically every 6 to 24 months, in order to maintain the effectiveness of the inhibitor. Corrosion inhibitors are available to be added to plain water in a tank, or they can be mixed with an antifreeze solution to protect the collector loop and heat exchangers. Because most corrosion inhibitors are toxic, their use in solar systems should be considered carefully, particularly when potable water is to be heated. Chromate-type inhibitors should not be used in systems involving potable water because they are carcinogenic and highly toxic. Moreover, proper disposal of used chromate solutions is quite troublesome.
4. Cathodic protection. A bar of electronegative metal, such as zinc, aluminum, or magnesium, can be attached inside the tank to provide cathodic protection against galvanic corrosion. The bar must make electrical contact with the tank and must be submerged in the water. The metal chosen for the protective bar should be more reactive[49] than the most reactive metal in the solar system. This method of protection works well in combination with tank coatings. Because protection for the solar system ends when the metal bar is completely dissolved, the bar must be replaceable.
5. Sealing the system against oxygen. If the system is completely sealed so that no air or water can enter it, the only oxygen available for the rusting process is the oxygen trapped inside the system. When the oxygen in the system is used up, no further rusting can occur. (Sealed systems usually must be designed to the standards of the ASME pressure-vessel code, and an expansion tank may be required to limit the pressure created as the liquid in the system expands.) This method of preventing rust does not protect against galvanic corrosion[50] or acid attack.
6. Stainless-steel tanks. Stainless-steel tanks do not require interior or exterior coatings or other methods of corrosion protection. When the cost of maintaining corrosion inhibition over the service life of a tank is considered, a stainless-steel tank may be more cost-effective than a carbon-steel tank, in spite of its substantially higher initial cost.

Concrete Tanks The major advantages of concrete tanks are that they are relatively inexpensive as long as the shape is kept simple, the mass of concrete becomes part of the storage system, and concrete is a readily available construction material. Concrete also has considerable resistance to underground loads. Because concrete can be cast in almost any shape, it is ideal for retrofit installations. Concrete is also fireproof and corrosion-resistant.

Concrete does have certain disadvantages. It is subject to capillary leakage; water can seep through cracks and joints unless the tank is lined. Concrete tanks require good design and workmanship. Because they are heavy, they often require special foundations. Often it is difficult to make leakproof connections through the tank walls, and substances leaching from unlined concrete tanks can cause corrosion of metals (especially aluminum) in the system.

Fiberglass-reinforced Plastic Tanks We recommend the use of factory-insulated fiberglass-reinforced plastic (FRP) tanks that are designed specifically for solar energy systems and are available in convenient sizes and shapes. A typical factory-insulated FRP tank consists of an inner FRP shell covered with 2 to 4 in. of urethane insulation. An outer FRP shell protects the insulation. Factory-insulated FRP tanks can be used outdoors above or below ground, as well as in a garage.

Nearly all FRP tanks have two severe limitations. First, they can withstand only little pressure. Unless specifically designed for some low-pressure application, they should never be pressurized, and a vent must be provided in the tank to ensure this condition. Second, the tank temperature must never exceed the limit specified by the manufacturer. If the tank temperature limit is below the boiling point of water (this is nearly always the case), the solar control system must stop adding heat to the tank before the temperature limit is reached.[51] This requires careful selection of the control system. In order to avoid stagnation of the collectors, one should consider appropriately programmed night-sky radiation or the use of a heat-purging unit (see Fig. 3.1) as an alternative solution to the problem of tank overheating.

The temperature limitation of an FRP tank is determined by the type of resin used to bond the glass fibers. Ordinary polyester resins have a limitation of 160°F, which is suitable for average solar space-heating applications with the appropriate protection against occasional overheating. With premium-quality resins, the temperature limitation can be raised to 180°F to 200°F.

Wooden Tanks with Plastic Liners Wooden tanks of the type generally used for storage of potable water can be used only in conjunction with a plastic liner that keeps the wood dry. The liner can be made of vinyl, butyl rubber, PVC, or another suitable material. The tank temperature must be strictly limited to the specifications for the liner. PVC liners have a limit of 160°F; more expensive EPDM liners are useful up to 212°F, but they are more difficult to fabricate than PVC. Other liners made of polyethylene, polypropylene, butyl rubber, chlorosulfonated polyethylene, chlorinated polyethylene, or chlorinated PVC (CPVC) have various temperature limitations up to and beyond the boiling point.

Tank Insulation

It is good practice to design for storage heat losses less than 2% of the daily collected solar heat. A typical cylindrical 1000-gal tank 5 ft in diameter and 7 ft in height, holding 300,000 Btu of solar heat at 60°F above ambient temperature, will have a daily heat loss of less than $0.02 \times 300{,}000 = 6000$ Btu. It will have 150 ft^2 of surface area and thus will require an R-36 insulation. This indicates that the tank insulation requirements are stringent and are higher than is commonly believed. If the tank is buried in the ground, the requirements become even more stringent. Seepage of moisture will, in time, occur even into ''closed-cell'' insulation materials such as polyurethane foams[52] and will lower the insulation value. Twice the normal thickness of the insulation layer is recommended for underground applications [14]. Further precautions in below-ground installations, such as a coal-tar coat on the outside of the sprayed-on closed-cell insulation and a plastic envelope for the entire insulated tank, are usually worth the extra expense.[53]

Mounting and Anchoring Tanks

If a tank is placed in a garage (e.g., in a retrofit application), one must be sure that the concrete floor can hold the significantly increased load. In earthquake-prone areas, top-heavy tanks should be properly anchored to the building. In below-ground applications, the buoyancy of an empty tank should be taken into consideration. Figure 1.22 shows an example of how an underground tank can be installed [14]. However, one must check the local building codes and carefully evaluate the possibility of an above-ground installation before selecting a tank installation method as costly as that shown in Fig. 1.22.

Specifications for Pressurized Tanks

It is good practice to specify steel tanks whenever pressurization is required. The only exception to this rule is a design calling strictly for static pressure only, such as is shown in Fig. 2.5. Pressurized tanks should be rated according to Section VIII of the ASME pressure-vessel code.

1.3.2 Rock-Bed Heat-Storage Systems

If air is used as the collector heat-transport medium, a rock bed is the most appropriate medium for heat storage. Figure 1.23 shows a design for a typical pebble-bed heat-storage unit [38]. The following requirements must be met in the design of the bin: (a) Plenum sizes must be adequate. If a plenum is too small, the friction loss for the air is too high within the plenum volume, resulting in uneven airflow through the rock bed, as shown in Fig. 1.24. As a rule of thumb, each vertical plenum area (for the top plenum and the bottom plenum) should be $>10\%$ of the horizontal cross-sectional area of the rock bed [14] (Fig. 1.25). (b) The pressure drop through the rock bed must be adequate. A total pressure drop of 0.15 to 0.30 in. water across the rock bed is considered desirable [14]. (c) The rock bed depth must be at least 20 times the nominal rock diameter in order to assure optimum heat transfer to and from the rocks at the correct height within the bin.

There is a distinct interrelationship among the average pebble size, the dimensions of the rock bed, and the pressure drop across the rock bed (Fig. 1.26). The following design example illustrates the use of this figure, as well as a practical method of meeting the preceding requirements.

Example. Already predetermined are the following: 400 ft^2 collector area (air type); airflow 2 ft^3/min per square foot of collector area; rock-bed vertical clearance 5.5 ft; volume of rock 200–300 ft^3. The cross-sectional area of the rock bed is between 200 ft^3/5.5 ft = 36.4 ft^2 and 300 ft^3/5.5 ft = 54.5 ft^2. Choosing the length and width of the rock bed as 6 ft 3 in. and 6 ft 10 in., respectively, gives a cross-sectional area of 42.7 ft^2, which is within the preceding limits. The *face velocity*, defined as the velocity of the air as it enters the rocks, is

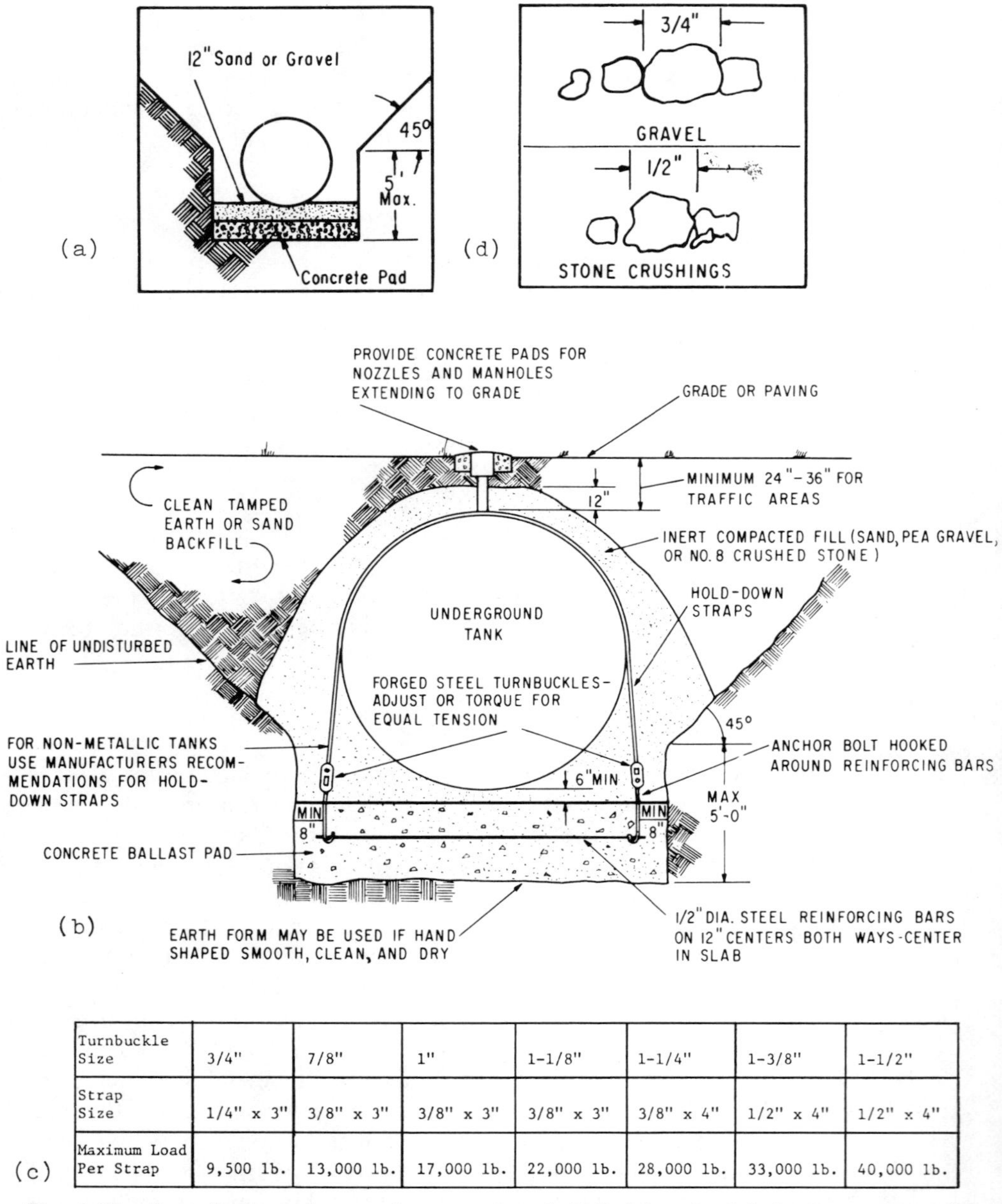

(c)

Turnbuckle Size	3/4"	7/8"	1"	1-1/8"	1-1/4"	1-3/8"	1-1/2"
Strap Size	1/4" x 3"	3/8" x 3"	3/8" x 3"	3/8" x 3"	3/8" x 4"	1/2" x 4"	1/2" x 4"
Maximum Load Per Strap	9,500 lb.	13,000 lb.	17,000 lb.	22,000 lb.	28,000 lb.	33,000 lb.	40,000 lb.

Fig. 1.22. Installation of an underground tank [14]: (a) typical hole dimensions; (b) installation details; (c) specifications for hold-down straps; (d) specifications for backfill material.

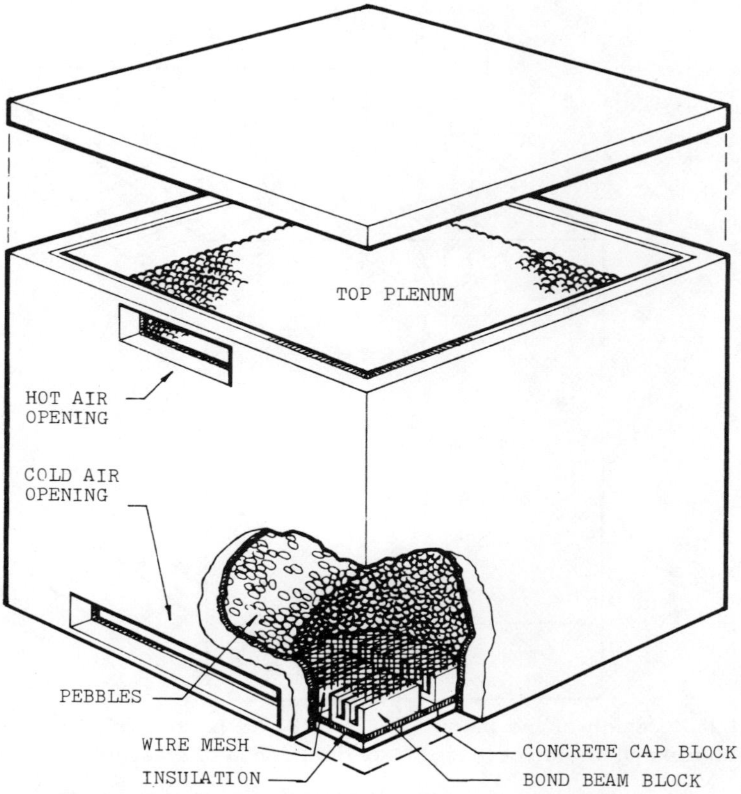

Fig. 1.23. Typical rock-bed heat-storage unit [38].

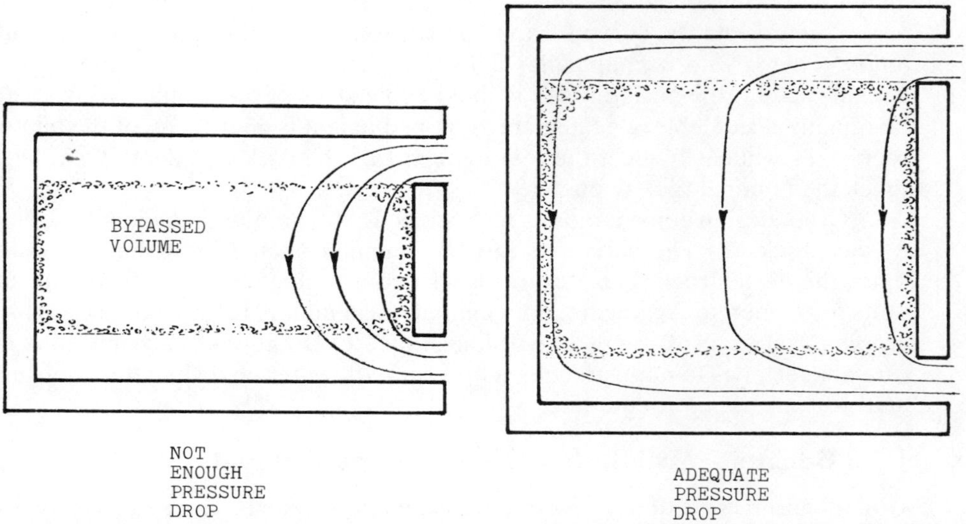

Fig. 1.24. Airflow patterns when the pressure drop is insufficient or when the plenum is too small [14].

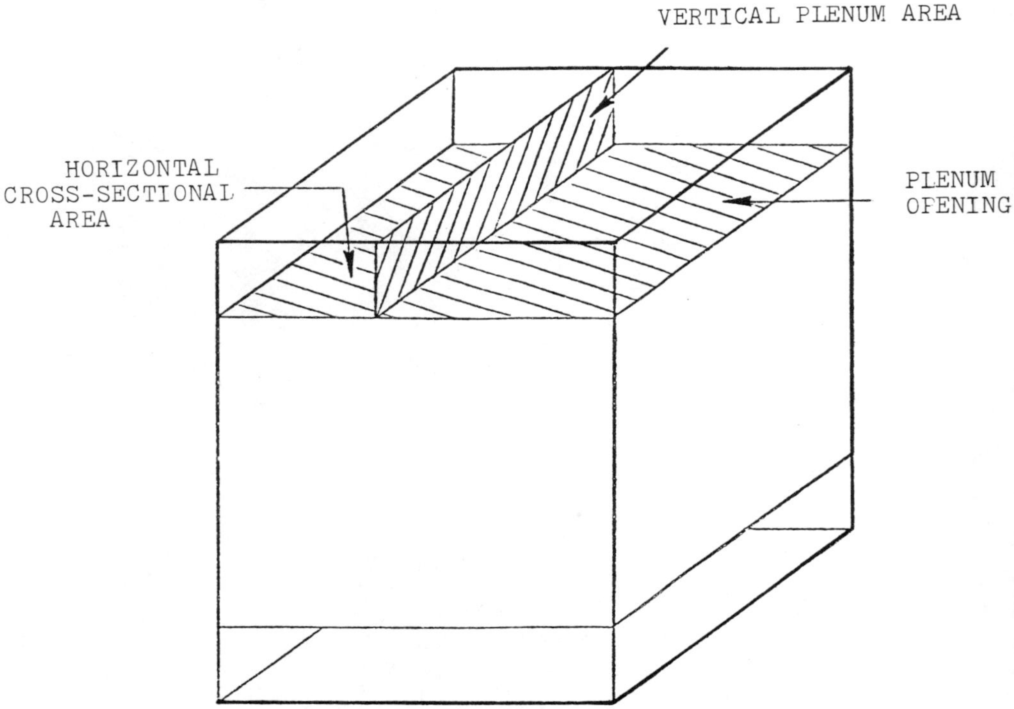

Fig. 1.25. Design of the plenum for a rock-bed heat-storage system. The vertical plenum area should be at least 10% of the horizontal cross-sectional area [14].

800 ft^3-min^{-1}/42.7 ft^2 = 18.7 ft/min. The *pressure gradient*, defined as pressure drop per unit length through the rock bed, must be between 0.15 in./5.5 ft = 0.027 in. water/ft and 0.30/5.5 = 0.055 in./ft. These limits are indicated in Fig. 1.26 by horizontal lines. A vertical line through the 18.7-ft/min face velocity, as determined earlier, intersects the curve for a 0.5-in. pebble size within the pressure-gradient limits. Taking into account allowable tolerances in the face velocity (approximately ±20%), the design result is that pebble sizes of 0.375 and 0.75 in. are also marginally acceptable. Because the rock-bed height is 66 in., any of the allowable rock sizes will easily meet the requirement that the rock-bed depth be at least 20 times the nominal rock diameter.

Other design considerations for rock beds [14] are the following: (a) Use only washed rock. (b) The variations in rock diameter should be within 0.75 and 1.5 times the design rock diameter. (c) Rocks conforming to ASTM C-33, Specifications for Concrete Aggregate, are generally acceptable. (d) Do not use rocks that crumble, such as soft sandstone or limestone. (e) Do not use carbonaceous rocks (fire danger). (f) Do not use rocks that react with water or carbon dioxide, such as marble, limestone, or dolomite.

1.3.3 Concrete Wall/Slab as Heat-Storage Elements

Under certain conditions, concrete walls and floors can be used effectively for the storage of actively collected solar heat.[54] The heat transfer is made through pipes that are embedded in the concrete. Copper pipes and plastic pipes have been used for such purposes. If plastic pipes (ABS, PVC, CPVC) are used, one must be sure

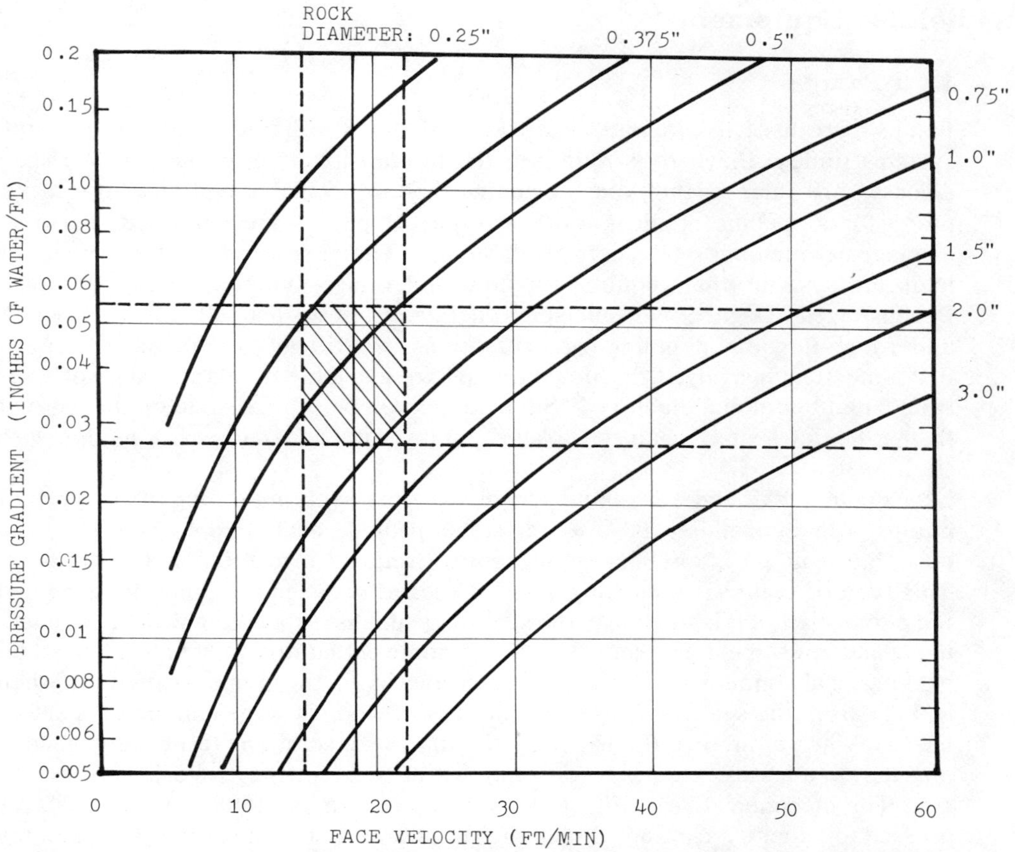

Fig. 1.26. Interrelationship among average pebble size, velocity of airflow through the rock bed (face velocity), and pressure gradient [14].

to plan for enough pipe length to allow heat transfer at a reasonably low temperature difference. The collector efficiency will suffer badly if there is an excessive difference between the water temperature and the concrete temperature. As a rule of thumb, one should figure about 4 ft of 0.75-in. pipe (nominal diameter, plastic pipe) per square foot of active solar-collector area.[55] This ratio can be lower if copper pipe is used, as long as the spacing of the pipes in the concrete does not exceed 12 in., which is adequate to establish even distribution of heat in the concrete.

An important criterion for the acceptability of concrete slabs or walls as heat-storage elements is that they must have sufficient thermal mass; their temperature increase during a sunny day should be only a few degrees. A concrete wall 16 ft wide, 8 ft high, and 18 in. thick will store approximately 20,000 Btu at a temperature increase of 5°F, the energy collected by one small (3 ft × 6 ft) solar collector during a sunny day. If the perimeter of a concrete slab is well insulated, the sand and the soil on which the slab is built can be used effectively as an additional heat-storage medium, making a slab an attractive heat-storage element (so long as one makes appropriate use of the collected solar heat during the summer season, in order not to increase the air-conditioning load).

1.4 Related Equipment

1.4.1 Pumps

Pumps[56] are used to circulate heat-transport fluids in liquid-based solar energy systems (unless they are working on the thermosiphoning principle). We shall discuss here some of the types of pumps that are available and how to select a pump for a specific application. Two types of pumps are marketed: positive-displacement pumps and centrifugal pumps. The first type is well suited for hydronic systems that require low flow under high-pumping-head conditions. Because well-designed hydronic solar energy systems practically always operate under high-flow/low-pumping-head conditions, centrifugal-type pumps are specified almost exclusively. Centrifugal pumps are available in a large range of sizes from a number of manufacturers. Such pumps can be subclassified on the basis of their types of sealing against leakage: pumps with dynamic seals; pumps with static seals.

Figures 1.27 and 1.28 show schematic cross sections through two types of pumps with dynamic seals. The adjustable-packing seal (Fig. 1.27) is the least desirable sealing system because it requires frequent inspection and adjustment. This type of seal uses a packing gland squeezed between the pump housing and the pump shaft; if it is too loose, the seal will leak. As the packing wears, the packing gland must be tightened. There are some situations in which adjustable-packing seals should not be used: when access to the pump is limited; when leakage from the seal can cause system failure; when leakage can create a safety hazard; when antifreeze fluids are used. Antifreeze solutions (especially silicone oils and, to a lesser extent, glycol solutions) have a tendency to leak.

The mechanical seal (Fig. 1.28) consists of two carefully polished surfaces pressed together by a spring. One of the surfaces is part of the pump shaft, and the other surface is sealed against the pump housing with an O-ring. A minute amount of leakage lubricates the polished surfaces, usually so little that it evaporates before it becomes visible.

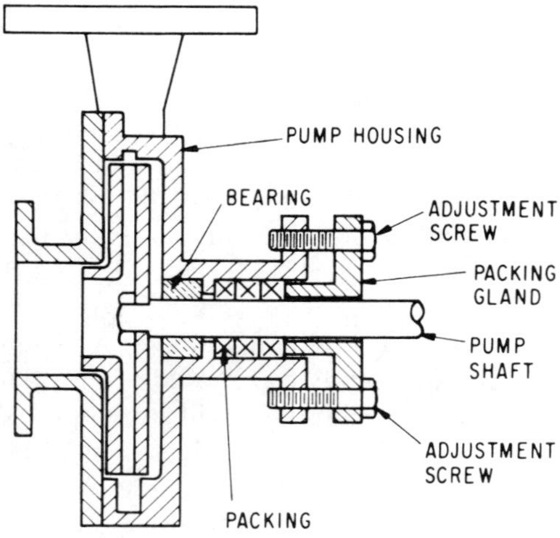

Fig. 1.27. Cross section through a dynamically sealed pump with an adjustable-packing seal [14].

Considerable experience with hydronic heating systems has shown that O-ring seals can last the lifetime of a system without requiring adjustment. However, the system must be kept clean, because sediment in the water can easily scratch the polished surfaces. Chromate-type corrosion inhibitors have caused failures when hard chromate crystals have developed between the polished surfaces. High temperatures and pressures can also cause premature failure. Some antifreeze fluids, such as silicone oils, tend to leak excessively through O-ring seals.

A pump that uses a magnetic coupling, as shown in Fig. 1.29, has no troublesome rotating seals. Instead, the rotor of the electric motor and its bearings are placed entirely inside an extention of the pump housing. The stator of the electric

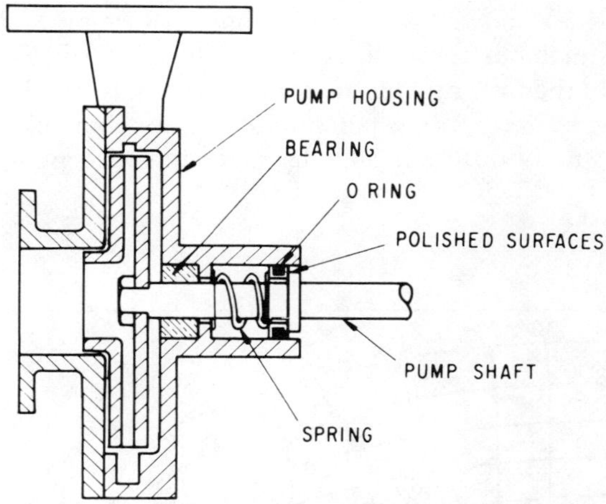

Fig. 1.28. Cross section through a dynamically sealed pump with an O-ring seal [14].

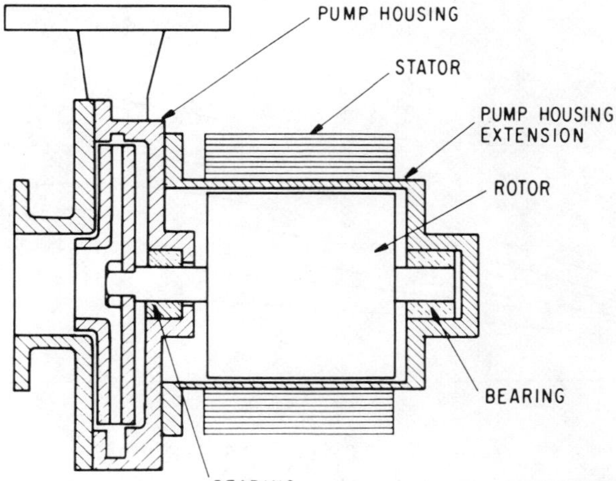

Fig. 1.29. Cross section through a statically sealed pump with a magnetic coupling [14].

motor fits outside the pump housing and drives the rotor with a rotating magnetic field. In some designs, a set of rotating magnets replaces the stator, and an external electric motor turns a shaft that rotates the magnets. Magnetically coupled pumps can be expected to last the lifetime of the system. However, they do tend to require service if they are installed in hydronic systems that are not used on a regular basis. Some solar space-heating systems have pumps in their DHW interfacing systems (see Fig. 2.23) that may never operate during the winter months, when the main heat-storage tank is continuously at relatively low operating temperatures because of incessant space-heating demands. During such long resting periods, the magnetically coupled impeller can become "frozen" by corrosion or sediment, and the pump must be dismantled to free the impeller. Although this usually does not involve the purchase of spare parts, the labor for this service must be considered.[57]

Figure 1.30 shows a typical pump performance characteristic (see also Fig. 2.16). It includes three different sets of curves.[58] The upper set shows the dependence of the flow rate on the friction loss in the hydronic system. The figure shows six such curves corresponding to six impeller diameters. Pump manufacturers often supply different impeller sizes for each pump model in order to be

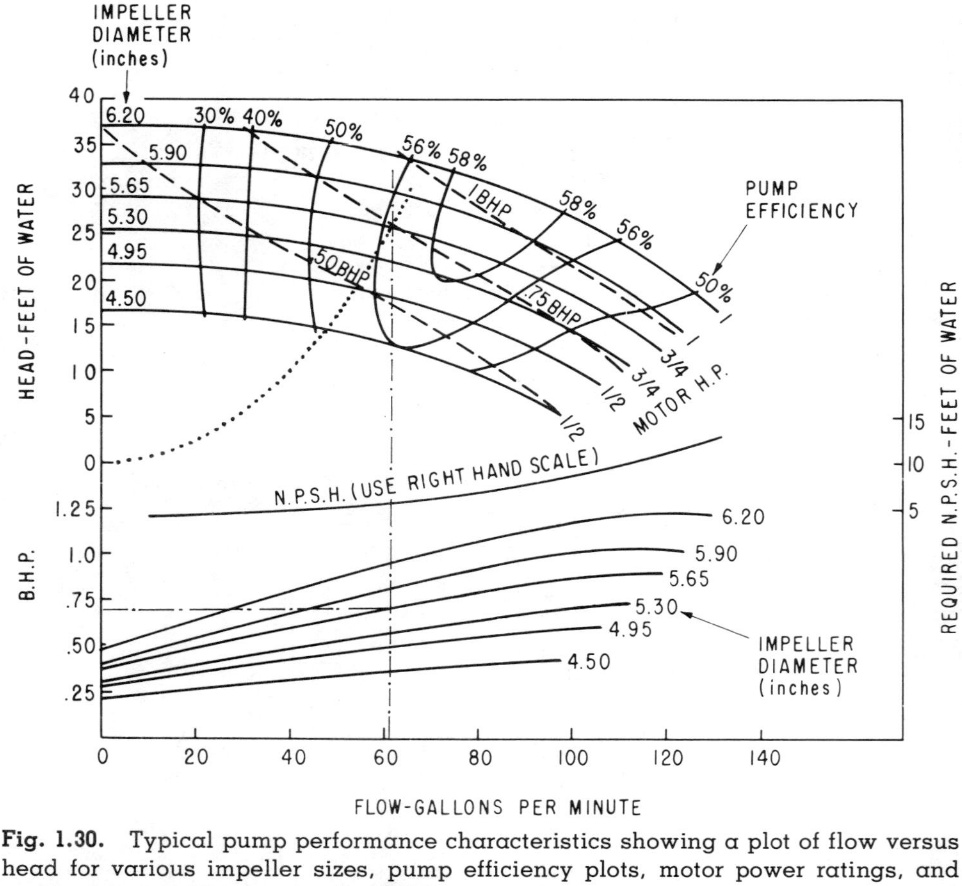

Fig. 1.30. Typical pump performance characteristics showing a plot of flow versus head for various impeller sizes, pump efficiency plots, motor power ratings, and required net positive-suction head [14].

able to closely match the pump characteristics to the system characteristics. Also shown by the upper set of curves are the pump efficiency and the required motor power.

In order to use the upper set of curves in selecting a pump, one must determine the system flow characteristic. A procedure to determine this important function is presented in Sections 2.3.2 and 2.3.3, and the result is indicated by a dash line in Fig. 2.16.[59] The system will operate at the head/flow (delivery) conditions given by the intersection of the system flow characteristic with the performance curve of the selected pump. The designer must select a pump (or an impeller size) that will give a flow rate that is within a few percent of the required flow rate. In situations in which more than one pump and impeller diameter combination will give acceptable flow rates, curves similar to those in Fig. 1.30 will help in the selection of the pump with the highest efficiency (note the efficiency curves in Fig. 1.30).

The lower set of curves in Fig. 1.30 shows the power consumed by the pump versus the flow rate for various impeller diameters. After selecting a pump, an impeller diameter, and a flow rate, one can use the lower set of curves to determine the motor size required for the pump.[60] For example, for the system flow characteristic shown by the dotted line in Fig. 1.30, a 5.65-in. impeller was selected. This pump will deliver 61 gpm at a water head of 26 ft. The dash-dot vertical line intersects the curve for a 5.65-in. impeller at a BHP value close to 0.75; i.e., a 0.75-hp motor is required.

The middle curve in Fig. 1.30 shows the net positive-suction head (NPSH) requirement. This is the absolute head at the pump inlet minus the vapor head of the liquid being pumped. If the NPSH does not exceed the pump's NPSH requirement, cavitation can occur and can destroy the pump in a short time. Most systems have sufficient NPSH if the pump is mounted below the water level of the storage tank of a nonpressurized system, or if the system is pressurized.

The NPSH at the pump inlet can be calculated as follows: (a) Find the atmospheric pressure (measured in feet of water) using Fig. 1.31. (b) Measure the difference in height (in feet) between the pump inlet and the surface of the water in the

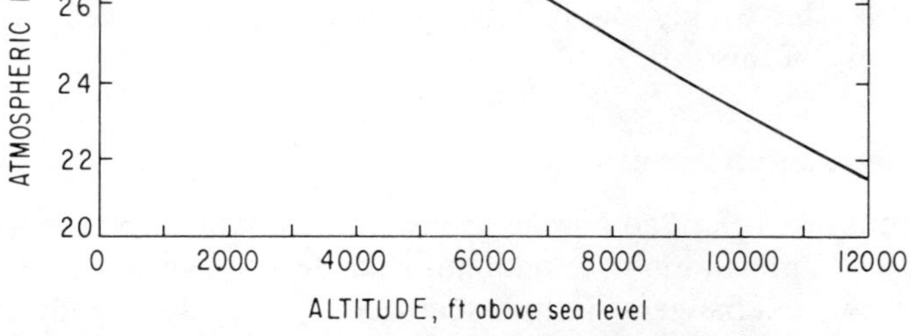

Fig. 1.31. Dependence of atmospheric pressure on altitude [14].

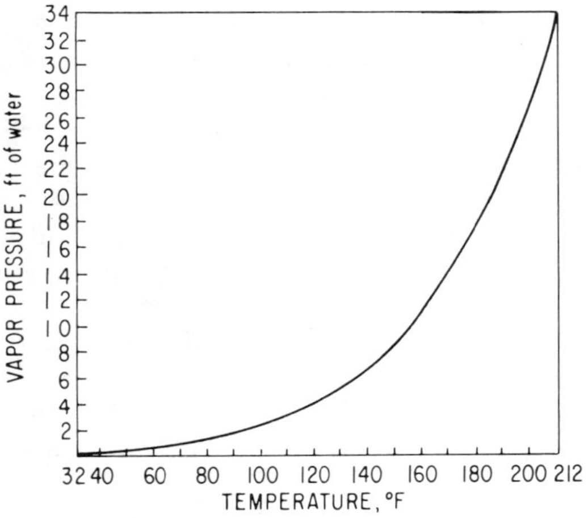

Fig. 1.32. Dependence of the vapor pressure of water on temperature [14].

tank. If the pump is located below the water surface, add this difference; otherwise, subtract it. (c) Subtract the head loss caused by friction in the pipe connecting the pump inlet to the storage tank. See Section 2.3.3 or the ASHRAE handbook [7] for the method of calculating this friction loss. (d) If the system is pressurized, add the tank gauge pressure (measured in feet of water). (e) Subtract the vapor pressure (measured in feet of water) at the highest temperature the system will reach (Fig. 1.32).

Pumps usually are made of iron, bronze, or stainless steel. Iron will quickly rust in systems that are open to the atmosphere, but it can be used in closed-loop systems that are treated with corrosion inhibitors. Both bronze and stainless-steel pumps can give long service life. Iron pumps are not recommended, and they do not conform to most building codes for use in DHW systems that operate without a heat exchanger.

1.4.2 Hand-operated Valves

Solar energy systems generally use the same types of hand-operated valves commonly found in residential water systems. There are two types of valves: globe valves and gate valves. Globe valves (Fig. 1.33) control the flow. Because globe valves offer more resistance to flow than gate valves, they should be used only where flow control is essential. Gate valves (Fig. 1.34) are not recommended for flow control, but they are suitable for opening and closing lines. When open, gate valves have only small flow resistance. Most valves used in solar energy systems are of this type.

1.4.3 Heat Exchangers

The discussion is limited here to the two types of heat exchangers that have found widespread application in residential solar-heating systems: convective liquid-to-liquid heat exchangers (immersed coils or plates) and small tube-in-shell exchangers for liquid-to-liquid exchange. In larger solar installations (such as in

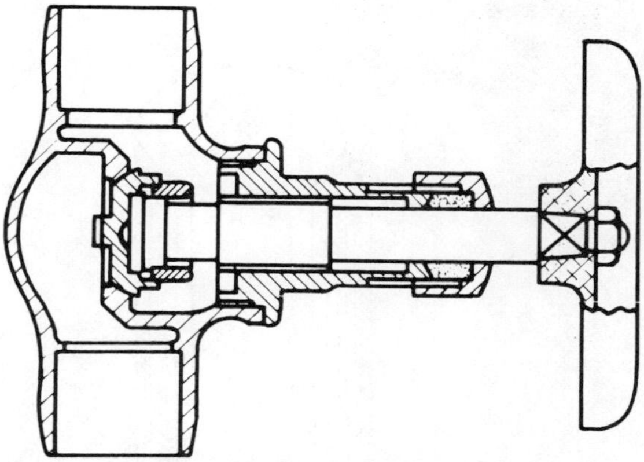

Fig. 1.33. Cross section through a typical globe valve [14].

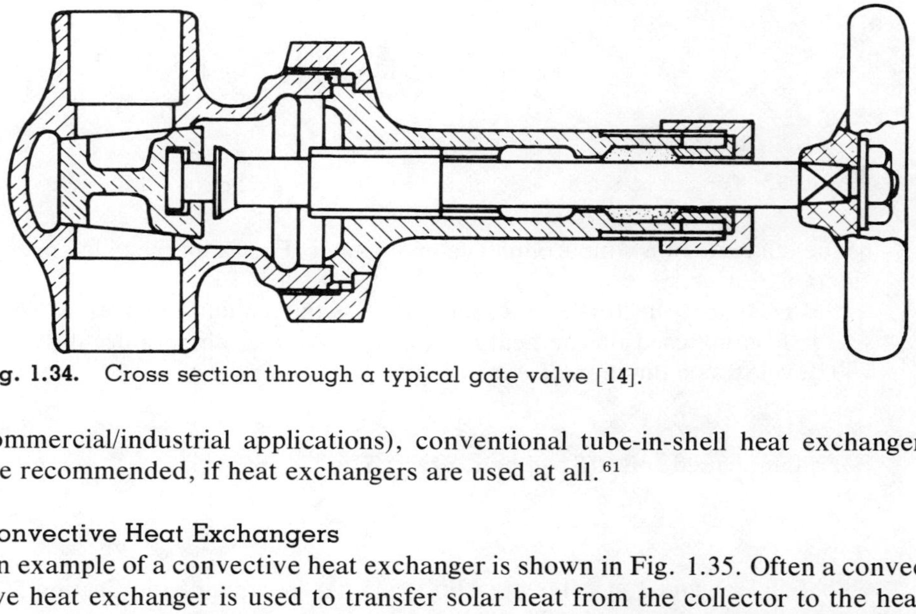

Fig. 1.34. Cross section through a typical gate valve [14].

commercial/industrial applications), conventional tube-in-shell heat exchangers are recommended, if heat exchangers are used at all.[61]

Convective Heat Exchangers

An example of a convective heat exchanger is shown in Fig. 1.35. Often a convective heat exchanger is used to transfer solar heat from the collector to the heat-storage tank (see Fig. 2.3); it also finds application in solar preheating of domestic hot water, as shown in Fig. 4.2b and Fig. 2.23. However, care must be taken to meet the strict safety standards described in Section 1.4.3 whenever the heat exchange involves potable water.

The amount of exchanged heat S in a fluid-to-fluid heat exchanger is proportional to the active heat-exchanger area A and to the mean temperature difference δT between the two fluids:

$$S = E_x A \delta T \tag{1.17}$$

The proportionality constant E_x represents the heat-exchange capacity (Btu-hr^{-1}-°F^{-1}-ft^{-2}). It depends strongly on the design and type of the heat exchanger. Generally, E_x is several times larger for tube-in-shell heat exchangers than for convective heat exchangers. A typical value of E_x for an immersed coil was found to be 80 Btu-hr^{-1}-°F^{-1}-ft^{-2} (see Appendix E). For an immersed-plate

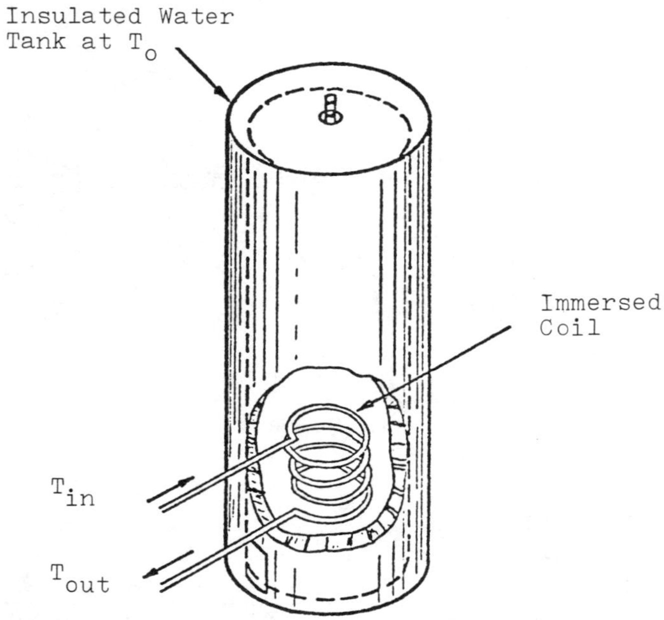

Fig. 1.35. Typical convective (immersed-coil) heat exchanger [14].

heat exchanger or a wraparound heat exchanger (Fig. 1.36) the value may be considerably lower.[62]

In a solar-collector-loop installation, S is the amount of collected solar heat, A_0q.[63] It is imposed on the heat exchanger; i.e., the temperature differential δT will develop as a function of A_0q:

$$\delta T = A_0q/E_xA \tag{1.18}$$

For an immersed coil, δT can be expressed as

$$\delta T = \frac{T_{in} + T_{out}}{2} - T_0 \tag{1.19}$$

with T_{in}, T_{out}, and T_0 as shown in Fig. 1.35. As is explained in Section 5.2.5, the value of δT is additive to ΔT, the operating collector temperature parameter, any increase in which decreases the system efficiency. Thus it is always advisable to keep δT as low as possible, if the use of a heat exchanger cannot be avoided altogether. From equation 1.18 it follows that δT decreases as the active heat-exchanger area A increases. Because this, in turn, increases heat-exchanger cost, a reasonable compromise must be found. It is generally believed that $\delta T \approx 20°$ is an acceptable value. (Note that, due to temperature stratification in the tank, the actual increase of the collector inlet temperature due to the heat exchanger is often less than this value; see Section 5.2.5). For $q \approx 200$ Btu-hr^{-1}-ft^{-2} and $E_x \approx 80$ Btu-hr^{-1}-ft^{-2}-°F^{-1}, this gives $A/A_0 \approx 1/8$. If the immersed coil is a 0.5-in. nominal copper pipe (5/8-in. OD), its surface area per foot of length is roughly 1/6 ft^2. This gives the convenient rule of thumb that an immersed-coil heat exchanger is approximately of the proper size if the numerical value of the length of the immersed 0.5-in. nominal pipe (measured in feet) equals the numerical value of the collector area (measured in square feet).

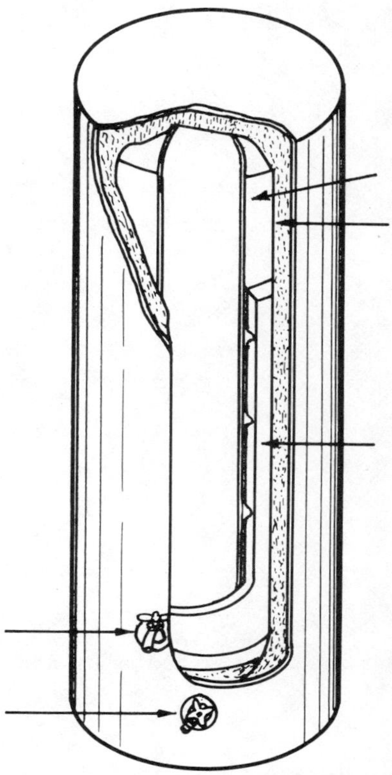

Fig. 1.36. Example of a wraparound heat exchanger. A pressure-bonded metal plate with integral fluid passages is clamped around the outside of the storage tank [14].

Some solar designers specify one or several immersed plates, such as the copper Rollbond absorber plates described earlier, as heat exchangers for the collector loop. The heat-exchange capacity E_x of such an arrangement can be estimated to be about 50 Btu-hr^{-1}-ft^{-2}-°F^{-1}.[64] In analogy to the development of the rule of thumb described earlier, we obtain $A/A_0 \approx q/E_x\delta T \approx 0.20$. Therefore, the surface area of the immersed plate should be about one-fifth of the collector area.[65]

For a wraparound heat exchanger, as shown in Fig. 1.36, the situation is often less favorable. This type of convective heat exchanger, which is frequently used in solar DHW systems, requires good thermal contact to the tank wall. For a 60-ft^2 collector system, the contacted area should be about 12 ft^2, which is a good fraction of the entire mantle area of the storage tank.

Tube-in-Shell Heat Exchangers

Generally, tube-in-shell heat exchangers work on the forced-counter flow principle, as illustrated in Fig. 1.37. The fluid that is receiving heat is separated from the fluid giving off heat by a highly conductive wall. The fluids may have opposite flow directions, imposed by mechanical pumps. It is important that the flow velocities in shell and tube be high enough to warrant turbulent flow.[66] Effective tube-in-shell heat exchangers for residential solar energy systems can be custom-built from standard copper pipe fittings.

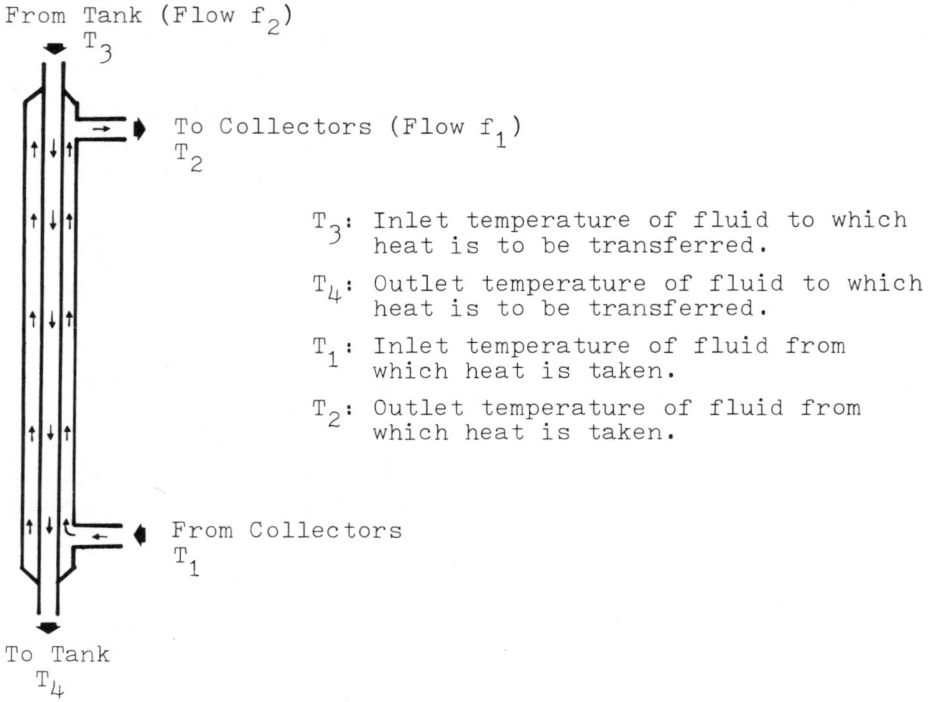

Fig. 1.37. Typical tube-in-shell heat exchanger used in solar installations.

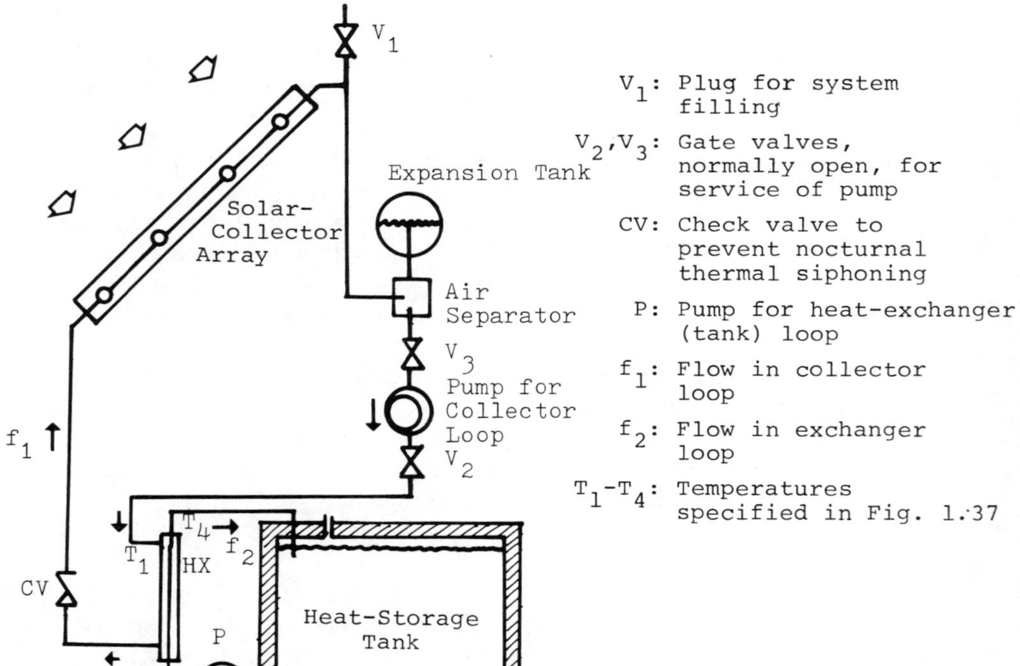

Fig. 1.38. Schematic diagram of a solar-collector system with tube-in-shell heat exchanger (HX). Compare Fig. 2.3 (an immersed convective-type heat exchanger).

A schematic diagram of a typical solar-collector system with a tube-in-shell heat exchanger is shown in Fig. 1.38. The mean temperature difference δT between the shell and the tube of the exchanger (i.e., between the collector loop and the tank loop) can be approximated as:

$$\delta T = \frac{\delta T_{in} - \delta T_{out}}{\ln(\delta T_{in}/\delta T_{out})} \tag{1.20}$$

where $\delta T_{in} = T_2 - T_3$ and $\delta T_{out} = T_1 - T_4$. It is significant to note that the collector operating temperature, which should always be as low as possible, is increased by $\delta T_{in} = T_2 - T_3$, not by δT. This difference is equivalent to the term $C_0 q/i$ in equation 5.1 in Section 5.2.5 and depends on the flows in both heat-exchanger sections,[67] as well as on the heat-exchanger area, as shown in Fig. 1.39. These curves

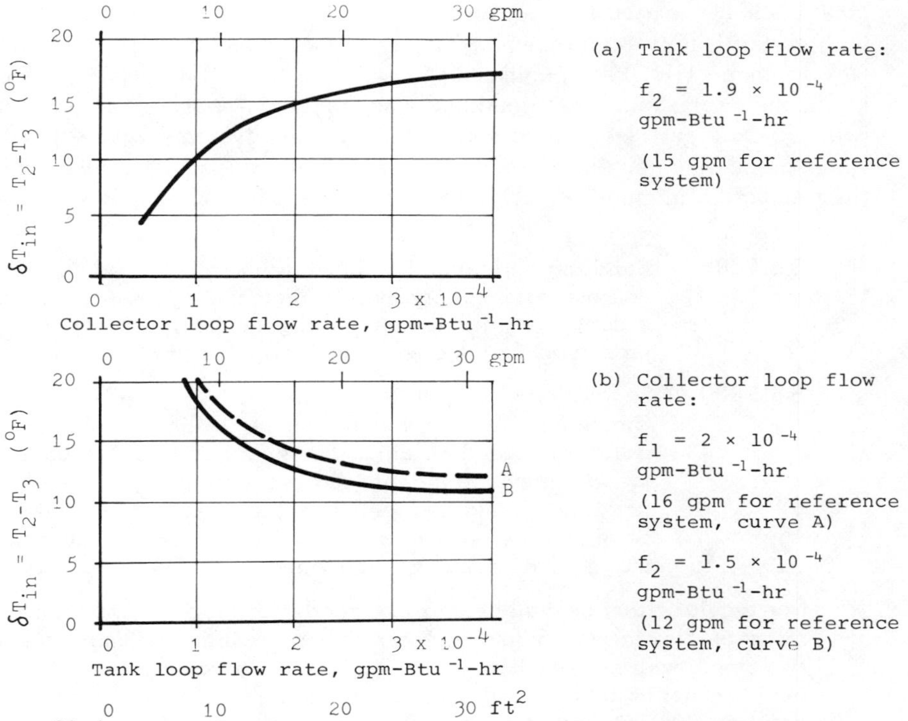

Fig. 1.39. Curves for heat-exchanger sizing. Calculations were made for water as the heat-transport fluid, using E_x = 350 Btu-hr^{-1}-ft^{-2}-°F^{-1} as heat-exchange capacity of the tube-in-shell exchanger. The reference system has a 440-ft^2 collector area operating at 180-Btu-hr^{-1}-ft^{-2} collection.

were calculated for $E_x = 350$ Btu-hr^{-1}-ft^{-2}-°F^{-1}. To facilitate interpretation of these curves, the general scale of the diagrams is expressed for the specific case of a 440-ft^2 collector system operating at 180 Btu-hr^{-1}-ft^{-2} collection (top abscissa), whereas the bottom abscissa indicates the values of interest reduced to 1 Btu-hr^{-1} collected with the collector system.

The curves of Fig. 1.39 suggest the following rules of thumb for sizing a tube-in-shell heat exchanger in the collector loop of a solar energy system: (a) The flow rate in the heat-exchanger (tank) loop should be $> 2 \times 10^{-4}$ gpm per Btu of hourly collected solar heat or > 0.04 gpm per square foot of collector area.[68] (b) The effective area of the heat exchanger should be $> 2 \times 10^{-4}$ft^2 per Btu of hourly collected solar heat or > 0.04 ft^2 per square foot of collector area.

Safety Aspects

Heat exchangers used to heat potable water must be constructed in compliance with certain safety requirements. The U.S. Department of Housing and Urban Development (HUD) has published standards for property receiving federal funding in their solar-initiative programs. The standards apply to all solar domestic water-heating systems and to solar space-heating systems with solar domestic water-preheating facilities, if nonpotable heat-transfer liquids are used. The HUD Intermediate Minimum Property Standards list the following requirements:

S-615-10 PLUMBING

S-615-10.1 Handling of Nonpotable Substances

Potable water supply shall be protected against contamination in accordance with the prevailing model plumbing code having jurisdiction in the area, as well as the requirements which follow.

S-615-10.1.1 Separation of Circulation Loops

Circulation loops of subsystems utilizing nonpotable heat transfer fluids shall either be separated from the potable water system in such a manner that a minimum of two walls or interfaces is maintained between the nonpotable liquid and the potable water supply or otherwise protected in such a manner that equivalent safety is provided.

Commentary: Double wall heat exchanger designs are one way of meeting the intent of this criterion. When double wall heat exchanger designs consisting of two single wall heat exchangers in combination with an intermediary potable heat transfer liquid are used, leakage through one of the walls would result in a single wall configuration. Although this design is considered to meet the intent of this criterion, there are several other designs that avoid this problem.

The use of single wall configurations which solely rely upon potable water pressure to prevent contamination is not considered to be an acceptable solution. Similarly, extra thick single walls are not considered to meet the intent of this criterion.

For approval of other than double wall designs, the procedures described in S-101 should be utilized.[69]

S-615-10.1.2 Identification of Nonpotable and Potable Water

In buildings where dual fluid systems, one potable water and the other nonpotable fluid, are installed, each system may be identified either by color marking or metal tags as required in ANSI

A13.1–1956 (Scheme for the Identification of Piping Systems, American National Standards Institute) or other appropriate method as may be approved by the local administrative code authority. Such identification may not be required in all cases.

These standards are fulfilled in the following heat-exchanger designs: (a) wraparound heat exchanger (Fig. 1.36); (b) two separate single-wall heat exchangers, one for the (nonpotable) collector fluid, one for domestic hot water, immersed in (potable) heat-storage water (see Figs. 2.3 and 2.23); (c) coil soldered to tank (similar to wraparound exchanger); (d) two tubes swaged together.[70] One tube is inside another tube in such a way that the outside of the inner tube is firmly pressed against the inside of the outer tube. Because this type of heat exchanger has heat-exchange properties almost as good as those of a single-wall exchanger, and because it is relatively inexpensive to manufacture, this double-wall heat exchanger is often used.

1.4.4 Pipe and Storage-Tank Insulation

Storage tanks should be insulated with materials whose insulating properties do not deteriorate significantly when they become wet, except when tanks are mounted above ground and when they are easily accessible. Closed-cell foam materials, such as urethane foam, have been used successfully. One must make sure that the specified insulation material is compatible with the maximum temperature the tank can reach. For most applications this requirement excludes, for example, styrofoam, a closed-cell insulating material that is inexpensive and has excellent insulating properties.

For easily accessible above-ground installations, building-grade fiberglass insulation has frequently been used (one or two layers of R-19 and/or R-11 material, depending on the necessary insulation value). Alternatively, construction of a simple enclosure around the tank has been proposed, with the space between the tank and the enclosure being filled with inexpensive cellulose insulation material. For recommended insulation values see Section 1.3.1.

The heat loss in pipes can be determined by the following equation:

$$q_0 = \frac{\pi}{6} k \Delta T_0 \frac{L}{\ln(r_1/r_2)} \tag{1.21}$$

k is the insulation constant (Btu-hr^{-1}-ft^{-2}-°F^{-1}-in.), ΔT is the temperature difference between the pipe and ambient, r_1 and r_2 are the outer and inner radii (inches) of the pipe insulation, respectively, and L is the length of the pipe (feet). For $r_1 \lesssim 2r_2$, which is the case for most pipe insulation, we can express $\ln(r_1/r_2)$ as

$$\ln \frac{r_1}{r_2} \approx 2 \frac{r_1 - r_2}{r_1 + r_2}$$

which gives

$$q_0 \approx \frac{\pi}{12} k \Delta T \frac{r_1 + r_2}{r_1 - r_2} \tag{1.22}$$

Insulation values q for various commercially available pipe insulation materials are listed in Table 1.6 for L = 100 ft and T_0 = 100°F.[71] Because the heat losses in

TABLE 1.6 Heat Losses[a] for Various Pipe Insulation Materials

Nominal Pipe Size (Copper) (in.)	Fiberglass (1 in. nominal)	Fiberglass (1.5 in. nominal)	Fiberglass (2 in. nominal)	Foam (1 in. exact)	Armaflex[b] (0.75 in. nominal)	Fiberglass (0.75 in. exact)
0.5	1.12	0.97	—	0.58	1.20	1.36
0.75	1.41	1.10	0.96	0.70	1.42	1.68
1	1.48	1.32	1.12	0.82	1.62	1.97
1.25	1.79	1.41	1.20	0.93	1.81	—
1.50	1.86	1.49	1.36	1.04	2.00	2.55
2	2.23	1.74	1.47	1.26	2.36	3.13
2.50	3.11	2.23	1.81	1.48	2.70	—
3	2.91	2.23	—	1.69	3.05	—

[a] Heat losses in kBtu per hour for a 100-ft pipe length and a 100°F temperature difference.
[b] Armaflex is a closed-cell flexible material produced by Armstrong, Lancaster, Pennsylvania. Some other insulation-material manufacturers produce similar products.

pipes can constitute a significant heat-loss factor for a solar energy system, equations 1.21 and 1.22, as well as Table 1.6, suggest that (a) the pipe runs should be as short as possible, (b) the pipe runs should be in sheltered locations, and (c) the superior insulation characteristics of foam, as compared with fiberglass or Armaflex (which has a k value similar to that of fiberglass), are accentuated by the logarithmic heat-loss law applicable to pipes, as compared with the linear law that applies to planar surfaces.

Notes

[1]See Fig. 1.7 and its accompanying text for discussion of the significance of double glazing versus single glazing.

[2]A rock bed is the best natural storage system for air-type collectors. See Section 1.3 for details.

[3]As a first-order approximation, a and b are reduced by a factor close to the concentration ratio (see Section 1.1.2).

[4]Realistic reflectivity values are 0.75 to 0.85. Glass mirrors have values up to 0.95, but they are prohibitively expensive; aluminized polymers can be equally good, but they lack longevity.

[5]The value of cos ϑ, as defined in equation 1.2 and in Appendix A, is always appreciably greater than for stationary flat-plate collectors, except for the time period near the noon hour, where the two types of systems have roughly the same geometric exposure toward the sun.

[6]This type of equation is used in official testing procedures for solar collectors (e.g., ASHRAE 93–77). A physically more nearly correct representation would have a fourth-power loss term that would more correctly describe the black-body radiation losses of the collector. However, for the temperatures at which flat-plate collectors are normally used, the radiation losses may be considered to be included in the second-power loss term (which is also sometimes omitted, see the topic Determination of Efficiency Parameters by Testing in Section 1.1.2).

[7]The reason for the introduction of the parameter c is to allow the reader to use tabulated ambient temperature values, which are indeed often lower than the ambient temperature values near the collectors, and to compensate explicitly (using the parameter c) rather than implicitly for this discrepancy.

[8]The sun's altitude (at any location within the continental United States) at noon is 47 deg higher in the summer than in the winter (see Appendix I, Fig. I.3).

[9]More details on the subject of collector tilt and corresponding solar collection expectance are presented in Section 5.2.2 and in Appendixes A, B, and I.

[10]See the topics Stationary Concentrating Collectors and Tracking Concentrating Collectors in Section 1.1.1.

[11]Usually < 50 to 100 Btu/hr-ft^2, as compared with ~ 300 Btu/hr-ft^2 during sunshine conditions (see Sections 5.2.1 and 5.2.8 and Appendix C).

[12]At $I = 60$ Btu/hr-ft^2 and $\Delta T - c = 60$°F, the parameter $(\Delta T - c)/I$ becomes 1.0, and the collector efficiency curve A will be below zero (Fig. 1.7).

[13]Compare Tables H.3 and H.4 in Appendix H.

[14]It is now becoming customary to use the collector's outer dimensions as a basis for determination of collector efficiency. Formerly, the "net absorber" area was preferred. The opening in the glazing (window or aperture) was also used in some cases. Care must be exercised in evaluating test results, because they may look quite different when related to different areas.

[15]In the British system of units, f is given in gpm, C_w in Btu per pound of fluid per degree F, I in Btu-hr^{-1}-ft^{-2}, and A_0 in ft^2. The constant factor a_0 then computes to 500.3 (min/hr × lb/gal).

[16]On the other hand, flow rates that are too high are also undesirable because they eliminate the possibility of effective temperature stacking and stratification in the heat-storage tank.

[17]The temperature parameter c (see equations 1.1 and 1.4) is not used explicitly in the official test procedure. However, the value of ΔT that is determined in the official tests is exactly equivalent to $\Delta T - c$ as we have defined it here (see footnote 7).

[18]Alten 100 G collector having 30.11 ft^2 gross area, using water as the heat-transfer fluid. Test performed 2/15/79 by DSET Laboratories, Phoenix, Arizona.

[19]The superscript notations (1) and (2) in equations 1.7 and 1.8 indicate constants applicable to linear fit and quadratic fit, respectively.

[20]For more details about these quality/workmanship-related aspects of solar collectors, see the topic Materials Aspects in Section 1.1.2.

[21]Such evaluation is equally important for the manufacturer, since testing according to ASHRAE 93–77 is not only costly but also time-consuming and therefore not practical as a design tool for new products.

[22]The method described here is most accurate for single-glazed collectors. However, in a later subsection a change in the number of glazings will be discussed.

[23]Note, in particular, that the glazing transmission is significantly reduced when going to double-glazed collectors. It should also be noted that $a_n < a_r$ if a second glazing is added.

[24]A second (inner-collector) glazing that falls into this category is Teflon. However, Teflon has the significant advantage of a very high transmissivity in the visible range, which certainly outweighs the advantages of glass for inner glazing in the medium-ΔT range (see the topic Glazings in the subsequent section Materials Aspects).

[25]The absorptivity of the coating itself is 97%; an additional increase is due to the upright fins (see Fig. 1.9) of the test collector. These fins intercept about 50% of the primarily nonabsorbed radiation, thus increasing the total absorptivity to 98.5%.

[26]The radiation from an absorber at temperature T_c is $R_B = \epsilon_0 \sigma (T_c^4 - T_{sky}^4)$, with ϵ_0 being the IR emissivity $(0 < \epsilon_0 < 1)$ and $\sigma = 0.173 \times 10^{-8}$ Btu-ft.$^{-2}$-hr^{-1}-°R^{-4} as the black-body radiation constant. Assuming a sky temperature of 0°F (459°R), the radiation loss of an absorber/emitter with $\epsilon_0 = 0.89$ is given in Table 1.2. It is therefore evident that the radiation loss is, in principle, substantial, even at lower-ΔT collector operating conditions. In this range, a 20% reduction in the a_n value may be advisable for selectively coated single-glass-glazed collectors.

[27]See Appendix D, which compares various glazed collectors under the same operating conditions. Although some selectively coated collectors have high ratings, many flat-black-coated collectors perform better.

[28]See the topic (5) Absorber Emissivity in Section 1.1.2.

[29]The physical meanings of "absorptivity" and "emissivity" are identical. It is for

reasons of ease of understanding that one often uses the word "absorptivity" if the absorbing function is addressed and "emissivity" if re-radiation is discussed.

[30]See the topic (5) Absorber Emissivity in Section 1.1.2.

[31]During the night, a typical collector will be exposed to nearly 100% humidity; during daytime operation, the relative humidity will be low.

[32]If the coating is a paint, one must carefully assess the possible forms of pretreatment of the metal base to which the paint will be applied.

[33]A coating of 20 mils thickness and 2 Btu-hr^{-1}-ft^{-2}-°F^{-1}-in. heat conductivity will establish (at 300 Btu-hr^{-1}-ft^{-2} insolation) a temperature difference between surface and metal of $\delta T = 300/2 \times 0.020 = 3$°F. This can be considered a significant disadvantage.

[34]For further details on this subject, see the subsection Efficiency Characteristic in Section 1.1.2, particularly Fig. 1.7, the topics (2) Glazing Transmission and (3) Number of Glazings in Section 1.1.2, Section 5.4.1, and Section 5.5.

[35]This includes the reflection losses, which alone are 4% at each surface for glass panes.

[36]See the topic (3) Number of Glazings in Section 1.1.2.

[37]See the topic (8) Insulation in Section 1.1.2, which indicates only a slight increase (by 0.045 Btu-hr^{-1}-ft^{-2}-°F^{-1}) in the linear temperature-loss coefficient when the back insulation is decreased from R-12 to R-8.

[38]In some instances these thin-film condensates can increase solar transmission. Such an increase has been observed during the testing of solar collectors. Their heat-transfer characteristics occasionally are better after a severe stagnation test. The reason for this behavior could be either a reduced reflection loss at the interior glazing surface due to the coating (having a low refractive index) or a reflection of IR radiation inside the collector back to the absorber due to optical-interference phenomena at the condensed film. Whatever the reason for this (abnormal) improvement might be, it is usually of little significance, because it is hardly possible to quantitatively and reproducibly create such a thin-film layer.

[39]This is true only for ideal collectors. As is the case for glazed collectors, the ϵ value for tested collectors is found to be reduced because of fin-efficiency influences [see topic (7) Fin Efficiency in Section 1.1.2] and various other, usually less significant, operating parameters.

[40]Because this type of radiation loss occurs in the presence of very low differences between emitter (collector) and receiver (surroundings), it is permissible to develop the term $T_c^4 \times T_a^4$ ($T_c \approx 90$°F = 549°R, collector temperature; $T_a \approx 70$°F = 529°R, ambient temperature) to $T_c^4 - T_a^4 \approx 4T_a^3(T_c - T_a) = 4T_a^3\Delta T$. The Value of $\epsilon_0\sigma \times 4T_a^3$ for $\epsilon_0 \approx 0.9$ (absorber IR emissivity) and $\sigma = 0.173 \times 10^{-8}$ Btu-ft^{-2}-°R^{-4}-hr^{-1} happens to be almost exactly 1.0 Btu-ft^{-2}-hr^{-1}-°F^{-1}.

[41]See Section 3.6 for nocturnal radiation with solar collectors.

[42]Some tanks cannot be exposed to boiling-water temperatures (see Section 1.3.1 for details).

[43]The control system in this example is somewhat less sophisticated than the one discussed as most desirable in section 2.4 which also takes into account the outside ambient temperature and leaves the furnace pump running whenever space heating is demanded, regardless of the temperature of the heat-storage medium (see Section 2.4.1 and Fig. 2.17).

[44]It should be noted that if the house thermostat also controls an air conditioner, the fan center must turn on the furnace fan some way other than by using terminal G.

[45]This will also be slightly advantageous in forced-air heating applications, as is discussed in Section 2.4 (Fig. 2.17).

[46]The control of the valves bypassing the storage tank when auxiliary heat is being delivered is not indicated in Fig. 1.20.

[47]Not indicated in Fig. 1.21 is the circuit that activates the auxiliary heater. Auxiliary heat will have to be used whenever the storage-tank temperature is below the preset DHW minimum temperature level and after the DHW has been solar-preheated optimally. However, such an additional circuit is relatively expensive and may not be cost-effective in many instances.

[48]An excellent detailed report by the Solar Energy Group of Argonne National Laboratory [14] discusses the characteristics of liquid-based and solid-based heat-storage systems. Some of the material used in this chapter is taken from that report.

[49]The metals commonly used in solar installations, ranging from least reactive (most noble) to most reactive (least noble), are copper, iron, aluminum, and magnesium.

[50]Thus it is still necessary to use galvanically protecting couplings at the transitions between dissimilar metals.

[51]A control setting 5°F below the limit is recommended. One should note that this can introduce stagnation conditions for the collectors, conditions that are generally undesirable and should be avoided in the interest of long service life.

[52]Fiberglass will soak full of moisture and must not be used in below-ground applications.

[53]See Section 1.4.4 for insulation materials.

[54]The use of large thermal masses is, of course, quite customary in "passive" solar designs. Especially in California, tens of thousands of homes have radiant floor heating systems (Eichler homes, for example). These floors generally lend themselves for storage.

[55]Assuming 200 Btu-hr^{-1}-ft^{-2} collection, 0.1-in. wall thickness of the plastic pipe, and 2 Btu-hr^{-1}-ft^{-2}-in. as the heat-transfer coefficient of the pipe material, this rule of thumb gives a 10°F temperature difference, which is still quite tolerable as a heat-exchanger temperature difference (see Sections 1.4.3 and 5.2.5).

[56]Parts of this subsection are adapted from the recent work of Cole et al. [14].

[57]The pumps marketed by Grundfos [23] are not known to have this type of problem. They are some of the most widely used and most reliable pumps for solar installations.

[58]Such curves can be obtained for any pump from the manufacturer or distributor.

[59]In addition, data on head losses through collectors and heat exchangers are available from the various manufacturers or distributors. The *ASHRAE Handbook of Fundamentals* also gives a method for calculating the head loss caused by friction in pipes [7].

[60]In a well-designed system, the total pumping power should not exceed 1.5% of the solar power being collected (1 hp = 2546 Btu, 1 kW-hr = 3412 Btu).

[61]Their sizing is common practice for the hydronic-system design engineer.

[62]The tank fluid, having received thermal energy from a coil and thus having a locally increased temperature, will separate faster (when rising because of its lower specific gravity) from a coil than it will from a plate, thus more rapidly making room for colder fluid to reach the pipe to pick up further energy.

[63]A_0 is the collector area; q is the collected heat per square foot of collector area.

[64]This value still depends on a number of variables that are difficult to control; it should not be taken as an accurate value but rather as an approximate value.

[65]Note that both outside surfaces of the immersed plate count.

[66]For water, the flow velocity should be well above 1 ft/sec. However, note that flow velocities that are too high can cause erosion problems and can unduly increase the friction losses. A flow velocity of 2 ± 0.5 ft/sec is recommended in solar-system applications (see also Section 2.3.3).

[67]Actually, the dependence on the flow in the collector loop, f_1, as indicated in Fig. 1.39 (top), may be misleading. The graph indicates that the best conditions are obtained at low flow rates f_1. However, the temperature of the fluid increases strongly with decreasing f_1 inside the collector (see Section 2.3.2), and for that reason the total system collection efficiency is lowered. These two effects nearly cancel each other, making the efficiency of a solar energy system with a tube-in-shell heat exchanger nearly independent of the collector-loop flow rate.

[68]This is about the same value recommended for the collector-loop flow rate (see Section 2.3.2).

[69]S-101, "Variations to Standards," is an earlier section of the HUD Intermediate Minimum Property Standards. It lists some general goals and refers to Section 101–4 of

MPS 4900.1, which contains details for approval of one-time-only and multiple-occurrence variations.

[70]This type of heat exchanger should actually be considered a single-wall heat exchanger with an extra-thick wall. It then will not fulfill the HUD Intermediate Minimum Property Standards. However, many local building-code authorities have allowed this design when it is presented in conjunction with a "low-toxicity" heat-transfer fluid such as food-grade propylene glycol or a silicone oil. Codes may become more definite in this respect in the near future.

[71]Nominal insulation thicknesses of commercial materials may differ from the exact thicknesses.

2

Solar Systems for Space Heating

A solar heating system consists of three interconnected subsystems: the collector system, the storage system, and the terminal subsystem by which the collected solar heat is transferred into the building. The terminal subsystem normally includes domestic water heating.

The most important design goal is that the solar-collector system operate at the lowest possible collector temperatures. Low collector temperatures increase collector efficiency[1] and therefore increase the overall system cost-effectiveness. The importance of this general design criterion cannot be overemphasized. The utmost effort should be made, for example, to select a building-heating system that requires a low-temperature (solar) heated-water input, such as a radiant floor heating system or an oversize forced-air heating system. A baseboard (convector) heating system, which functions only at relatively high water temperatures (170°F–190°F), is not suited for solar heat (Section 2.4). A heat exchanger between the collector loop and the heat-storage medium increases the operating temperature of the collectors and is therefore equally undesirable (see Section 5.2.5 and Appendix E). An undersize heat-storage tank also reduces the system efficiency (Section 2.1.2).

A typical solar space-heating system is shown in Fig. 2.1. The various components were discussed in Chapter 1 of this handbook, and the overall solar heat collection that can be expected with any complete, integrated solar heating system is the subject of Chapter 5 (see also Appendixes I and J). In this chapter the emphasis is on (a) sizing of the collector system and the storage system, (b) some design aspects of collector mounting and system piping, including specifications for pipe and pump sizes, (c) the freezing and overheating protection, and (d) the interface to the space-heating system and domestic water-heating system of the building.

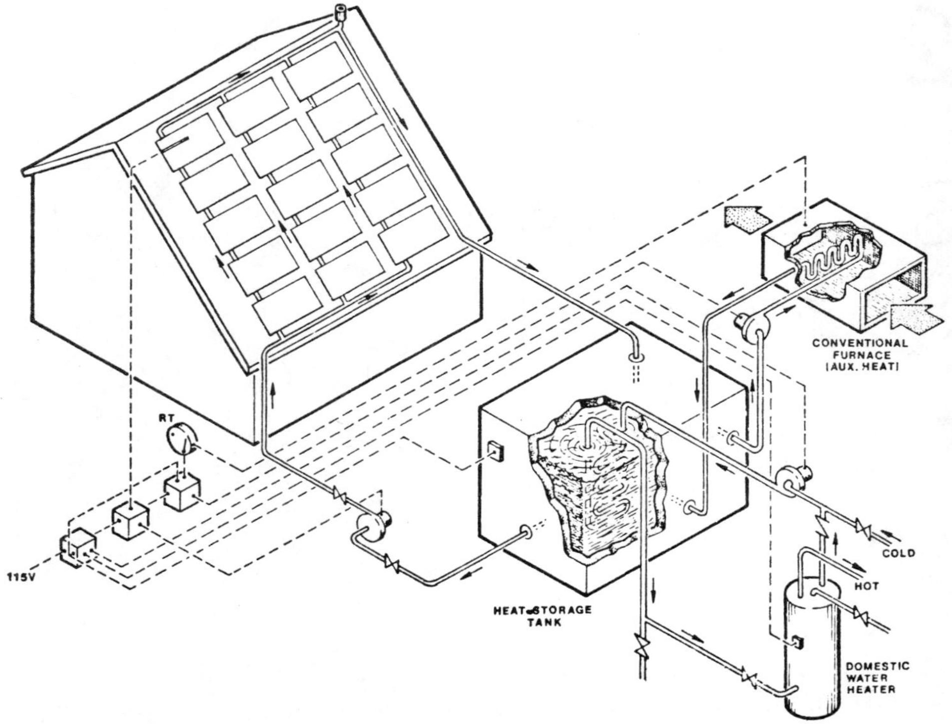

Fig. 2.1. Typical solar space-heating system with solar domestic water-heating loop.

2.1 Sizing

2.1.1 Collector Area and Tilt

The sizing of the collector area usually is quite simple. It is often determined by indirect factors such as cost limitations, limitations on space available for the collectors, and existing roof tilt and orientation. Generally, a very important design goal is to use all collected solar heat throughout the year. In most instances the following sequence can be followed in sizing a solar space-heating system:[2]

1. Compute the building heat-loss coefficient[3] (in Btu-hr^{-1}-°F^{-1}). The degree of accuracy required for this calculation is not high. Usually an accuracy within 10% is quite sufficient.
2. Compute the direct solar gain through the windows for a sunny day.[3] This calculation should also have an accuracy of 10%. It should be performed for one representative day of each month (usually the 21st day).[4] One should always carefully take into account to what extent the east-, south-, and west-facing window areas are shaded by roof overhangs and landscaping. For the collector sizing procedure, the direct solar gain through the windows is used to partially offset the space-heating requirements. Therefore one must carefully determine

if it is realistic to count all direct solar gain through the windows or if (because of insufficient heat-distribution facilities in the building or because of a low thermal capacity of the building itself) only a portion of the direct solar gain through the windows can be used to offset the space-heating needs.

3. Consult applicable degree-day tables,[5] and calculate for each month the effective space-heating requirement: the result from step 1 × 24 × degree-days per day, minus the direct solar window gain.
4. Select an appropriate desirable solar coverage for a representative sunny winter day. This will normally be somewhat less than 100%, which is favorable in cases in which full utilization of all collected solar heat in the off-season (in the summer) is questionable.[6] Economic considerations demand that excess in heat collection be avoided: 100% solar coverage on a sunny winter day will mean that the average solar coverage equals the sunshine probability during the winter months; more than 100% coverage on a sunny winter day will require excessive heat-storage facilities and is not recommended for (regular) solar space-heating systems, where cost-effectiveness is a design criterion. Multiply the result from step 3 by the selected solar coverage.
5. Add an allowance of 15,000 Btu per person for solar domestic water heating. Subtract a reasonable amount for internally generated heat, such as heat generated by lights, appliances, and people (200 Btu/hr per person). The result is the target solar supply value for a sunny winter day.
6. Assume an average solar system collection value of 1000 Btu per square foot of collector area per day (full sunshine conditions)[7] if the location is in a mild climate zone (2500–3500 degree-days per season), if single-glazed collectors of medium quality are used, if the lowest useable storage temperature is 100°F, if the storage capacity is at least 2.5 gal of water per square foot of collector area, and if no heat exchanger is used in the collector loop.[8] Divide the result from step 5 by this assumed average solar system collection per square foot of collector area, add a 10% allowance for heat losses in pipes and storage tank,[9] and obtain the desired collector area.
7. Determine the number of collectors that most closely fits the area established in step 6. Quite often, architectural considerations will suggest an even number of collectors. The availability of space for the collectors may be another limiting factor.

The collector tilt angle should be determined with great care. As explained in Section 5.4.2,[10] the solar system collection will require quite steep collector tilt angles for winter space-heating applications in most of the continental United States. A compromise toward lower tilt angles, which often are architecturally more pleasing, not only reduces the solar collection expectance in the winter but also shifts the optimum collection season toward the summer, when excessive solar heat may actually become a problem for the collectors and the storage-tank materials.[11] Therefore, it is strongly advised that the building architecture be adapted to the necessary tilt angle. For strong emphasis on space heating during the winter months, a collector tilt angle equal to the numerical value of the geographic latitude plus 20 to 25 deg is recommended. If there is reasonable utilization of the solar heat collected during the summer months (such as swimming-pool heating for private residences), a tilt angle equal to the value of the geographic latitude plus 5 to 10 deg is most desirable. In this case, an easy-to-construct 45-deg tilt is acceptable throughout most of the continental United States.[12]

After the system has been sized, all design parameters that are needed for a complete solar system collection analysis (see Chapter 5) should be determined, and the system collection computation should be performed.[13] Depending on the discrepancy that is found between the result of this computation and the targeted solar supply value determined in step 5 of this sizing procedure, appropriate adjustments in collector size may be made.

Some solar designers specify a lower angle, such as the value of the geographic latitude plus 10 to 15 deg, as the optimum collector tilt angle for space heating, and the value of the latitude itself for domestic water heating (e.g., Beckmann et al. [8], p. 45). This rule applies only in those cases in which maximum annual solar heat collection is desired, irrespective of the season during which the collection peaks. However, most solar installations, and virtually all domestic space-heating systems, have a heating demand season that does not at all fit with this rule. Furthermore, it can be shown that the maximum annual collection peaks only very slightly at this tilt angle. A variation in the tilt angle by as much as ~ 20 deg from this rule results in annual collection values that are within a few percent of the optimum annual cumulative collection.

2.1.2 Heat-Storage Capacity

The storage capacity must be determined with respect to the temperature that the storage medium reaches at the end of the solar collection day (T_0 at 5:00 P.M.). An excessively high temperature means that the collector system will be operating at a high-ΔT condition (i.e., at low efficiency) during the afternoon collection hours. A low value of T_0 (5:00 P.M.) may mean, instead, that the solar heating system was designed with an oversized (i.e., unduly expensive) heat-storage facility.

The storage temperature at the end of the collection day (5:00 P.M.) is, for any given storage volume, primarily a function of (a) the collector area and (b) the heating demand during the collection hours. A building with a substantial amount of direct solar heat gain through the windows, and located in a relatively mild climate, may require little additional space heating during the daytime. Consequently, a large fraction of the solar heat collected with the collector system will have to be stored for use during the evening and night hours, and the storage capacity will have to be relatively high. On the other hand, if a substantial fraction of the collected solar heat can be used during the daytime hours, only a relatively low storage capacity is required. This may, for example, be the case for commercial buildings requiring space heating only during the daytime office hours, as well as for substantially undersized systems in colder climates.

Figure 2.2 shows an example of the dependence of solar system collection on the storage capacity and the heating demand during the collection hours. The amount of solar heat collected during one sunny day (Jan. 21) is plotted versus the storage capacity, which is expressed in terms of gallons of water per square foot of collector area. Curve *A* is calculated for an average, properly sized domestic solar space-heating system.[14] The building has substantial solar heat gain through its windows and requires virtually no additional heat during the solar collection hours. As curve *A* indicates, the solar system effectiveness (i.e., collection per square foot of collector area) is quite low for low storage capacities (< 1 gal/ft²) and increases to an acceptable value at about 3 gal/ft². A further increase in storage capacity (although it would further improve collection effectiveness) usually is not cost-effective.[15] Curve *B* was computed for the same system, but the building was assumed to be located in a colder climate (18°F colder during the solar collection hours). The dwelling now uses almost all of the heat collected with the solar collectors during the daytime. Consequently the storage requirements are significantly reduced, which is clearly indicated by the flatness of curve *B* for storage capacities greater than 1 gal/ft². Thus, for this application, a considerably smaller storage volume is acceptable. However, the curve for system *B* can be expected to be less flat during the fall and spring transition months, and for that reason the selection of a storage-capacity size that is marginal may be undesirable. The results of the calculations presented in Fig. 2.2 are in agreement with the rule of thumb for storage-capacity sizing presented in Table 2.1.

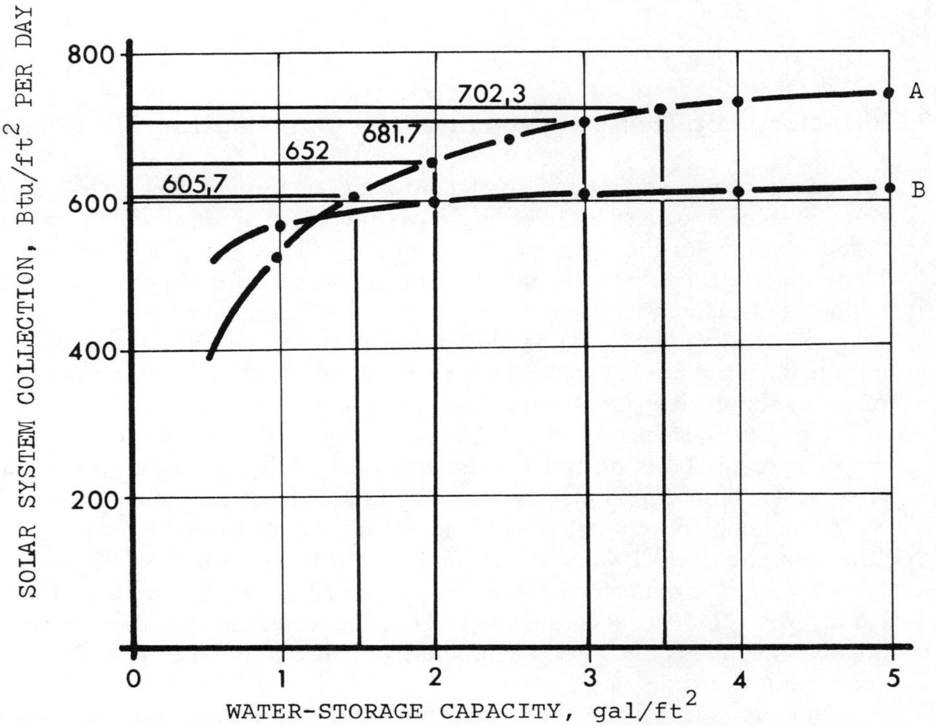

Fig. 2.2. Dependence of solar system collection on heat-storage capacity. Curve *A* is for a system using no solar heat during the collection hours; curve *B* is for a system using almost all the solar heat during the daytime.

TABLE 2.1 Suggested Storage Capacity for Various Solar Space-heating Applications

Storage Capacity (gal/ft²)	Suggested Space-heating Application
<1	Not recommended
1–1.5	Systems supplying only daytime space heating at relatively constant rates of use per hour
1.5–2	Systems supplying a substantial amount of daytime heating and some small fraction of evening and nighttime heating
2–2.5	Systems supplying some fraction of daytime heating and a substantial amount of heating after sunset
2.5–3	Systems supplying virtually no space heating during the solar collection hours
3–3.5	Systems supplying no space heating during the daytime collection hours and operating at unusually high efficiency (in mild climates, under low minimum storage-tank temperature conditions, no heat exchangers)
>4	Not recommended[a]

[a] A storage capacity greater than 4 gal/ft² may be desirable in special circumstances (e.g., in systems that do not require heat during several consecutive days, such as weekend homes). The solar heat then must be cumulatively stored in a large storage tank.

2.2 Collector-Loop Design: Freezing and Overheating Protection

The most common and potentially the most serious form of damage to a solar collector system is caused by freezing. It is therefore appropriate to discuss the various options for the hydronic design of the collector loop in conjunction with the method selected for freezing protection. The need for effective freezing protection can be increased because of the black-body radiation characteristics of the solar collectors.[16] Because of radiation, the absorber reaches a temperature equilibrium that can be well below the ambient temperature on clear nights. The following simplistic example may illustrate this influence of black-body radiation:

Using the formula in footnote 26 of Chapter 1, and assuming an effective sky temperature of −60°F, an ambient temperature of 36°F, and an absorber emissivity of 89%, the absorber will radiate 59 $\text{Btu-hr}^{-1}\text{-ft}^{-2}$ at a temperature of 36°F, i.e., at the ambient temperature. If the absorber is glazed with a single sheet of glass, approximately 10% of this radiation, or 6 $\text{Btu-hr}^{-1}\text{-ft}^{-2}$, will actually leave the collector.[17] A collector having a linear temperature-loss coefficient of $a = 1$ $\text{Btu-hr}^{-1}\text{-°F}^{-1}\text{-ft}^{-2}$ (see equation 1.6) will reach a temperature equilibrium between radiative loss and convective gain of approximately $36°\text{F} - 6/a = 30°\text{F}$; i.e., it will cause water to freeze.

Numerous methods of providing protection have been proposed, including (a) closed-loop systems with a nonfreezing heat-transport fluid (antifreeze), (b) drain-down, (c) "hot-shot" freezing protection, and (d) natural siphoning. Some of the design intricacies, advantages, and disadvantages of these methods of freezing protection will be discussed in the following sections.

2.2.1 Closed-Loop System with Nonfreezing Heat-Transport Fluid

This is the safest method of providing freezing protection. The fluid in the collectors is usually a solution of 50% water and 50% antifreeze, based on glycol as the agent to reduce the freezing point of the solution.[18] Other non-water-based fluids are used in some applications. When using these fluids, one must determine that the nonmetal materials used in the system (e.g., rubber hoses) are compatible with the type of fluid used. Most non-glycol-based solutions have substantially increased boiling points that are beyond the stagnation temperatures of flat-plate collectors. Glycol solutions can be harmed under conditions of collector stagnation.[19] Differences in viscosity and heat capacity of heat-transport fluids other than water may require some adjustments in the design of the collector system flow rates.[14].

The substantial cost of an antifreeze solution and the necessity for periodic exchange of that solution (every other year) call for the design of a closed-loop collector system with a heat exchanger between this loop and the heat-storage tank. Figure 2.3 shows a typical piping diagram for such a system. The necessity of a heat exchanger[20] is the most significant disadvantage of a closed-loop system. The heat exchanger increases the operating temperature of the collectors, which in turn decreases the collection efficiency.[21] The heat exchanger, the air separator, the expansion tank, and the pressure regulator/pressure relief valve (components that are required for closed-loop systems) increase system costs, as compared with an open-loop system, although such increases are often overestimated. By comparison, the intricacy of the plumbing design to allow adequate drain-down in a drain-down system may represent a substantial additional cost for that frequently specified type of collector fluid system.

A closed-loop system is an absolute requirement if collectors with aluminum or iron water channels are used.[22] This is necessary to be able to control the amount of oxygen that reaches the interior of the fluid channels and to be able to apply an effective corrosion inhibition in the heat-transport fluid.

2.2.2 Drain-down

The collector plumbing system must be designed to allow complete drain-down. Water pockets must be avoided. For many types of collectors this means that special care must be used to install the collectors with the fluid channels exactly in the horizontal position. Figure 2.4 shows the piping schematic for a typical drain-down system.

Two drawbacks must be considered with respect to drain-down designs. First, the probability of internal corrosion of the water ducts is increased because of periodic admission of air. If a corrosion-inhibiting water additive is used, this may not be an appreciable disadvantage for copper or stainless-steel water channels. However, periodic (usually biannual) checking and correcting of the inhibitor concentration is advised. The potential for corrosion problems in drain-down systems that use aluminum or steel water ducts in the collectors is so high that it is strongly advised that a closed loop be designed for these electronegative (corrosion-prone) collectors.[23] The potential for corrosion in open systems can be decreased considerably if the design incorporates an additional feature to ensure

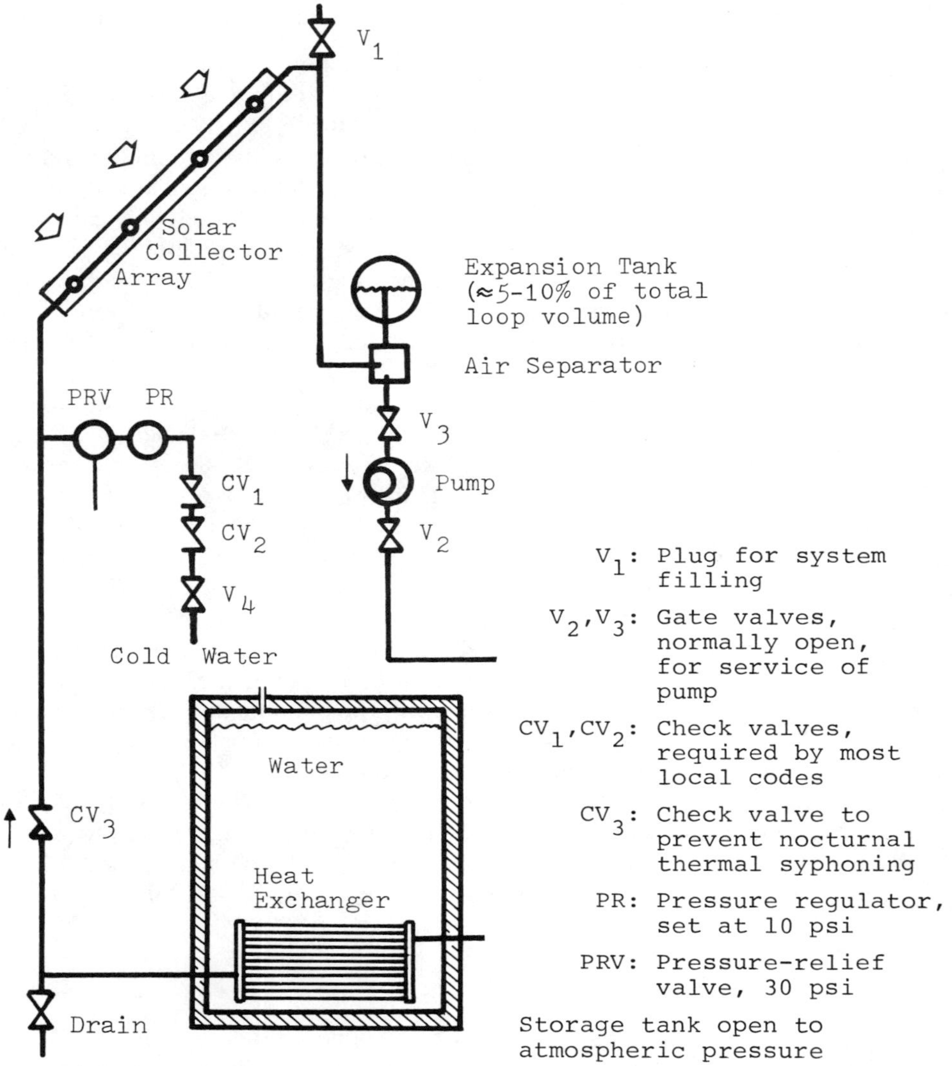

Fig. 2.3. Typical piping diagram for a closed-loop collector system with heat exchanger. V_4 should be closed except for intentional refilling of the collector-loop fluid. If a nonaqueous heat-transport fluid is used, delete *PR*, CV_1, CV_2, and V_4.

that drain-down occurs only when freezing protection is required, not every day, ie., whenever the pump stops. This can be achieved with a solenoid valve used in place of the vacuum-breaker valve V_4 (Fig. 2.4). The solenoid must be operated from a freezing-temperature-sensing circuit, which is usually part of a standard solar control system.[24] However, one should use a normally open valve (open whenever it is not energized) in order to provide freezing protection during periods of electricity outage.

The second drawback is that after each drain-down, the system must be re-primed. The solar pump must have enough pumping capacity (flow and head) to overcome the initial static pumping head between the top of the storage tank and

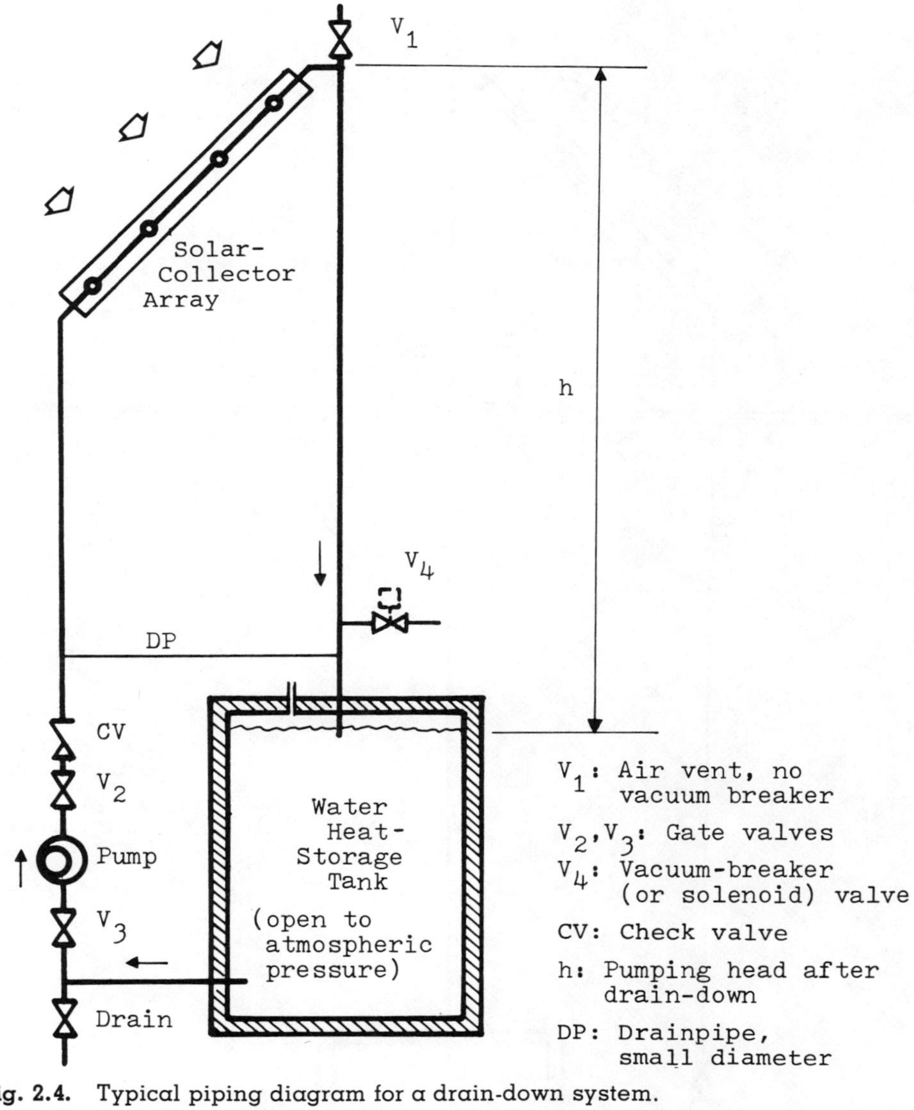

Fig. 2.4. Typical piping diagram for a drain-down system.

the top collector header. If this height difference h (Fig. 2.4) is appreciable (i.e., if $h > 10$ ft in residential space-heating systems), an uneconomically high powered pump will be required. Once the system is filled, the normal hydronic principle applies,[25] and then a lower pumping capacity will not only suffice but actually be preferable.[26] One possible and quite inexpensive solution to this problem is shown in Fig. 2.5. The collectors are drained into a separate water reservoir (auxiliary tank) that is close to the collector level and has a solid piping connection to the main heat-storage tank. This tank must be completely sealed; however, because the pressure never exceeds the static head of the water column between the auxiliary tank and the main tank, this pressure is very well defined, and catastrophic overpressures cannot occur (the system is open to the atmosphere). That

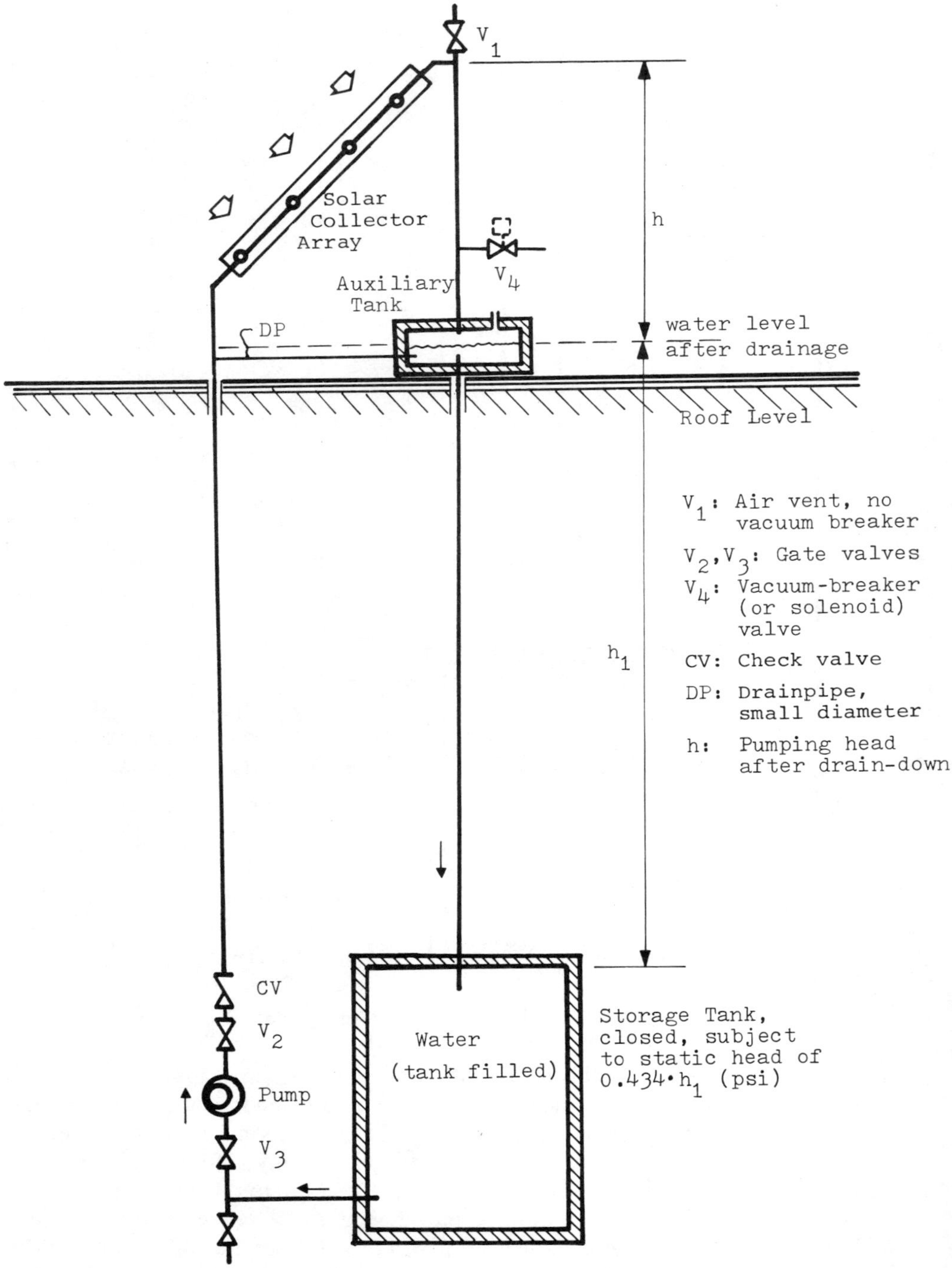

Fig. 2.5. Piping schematic for a drain-down system with auxiliary tank.

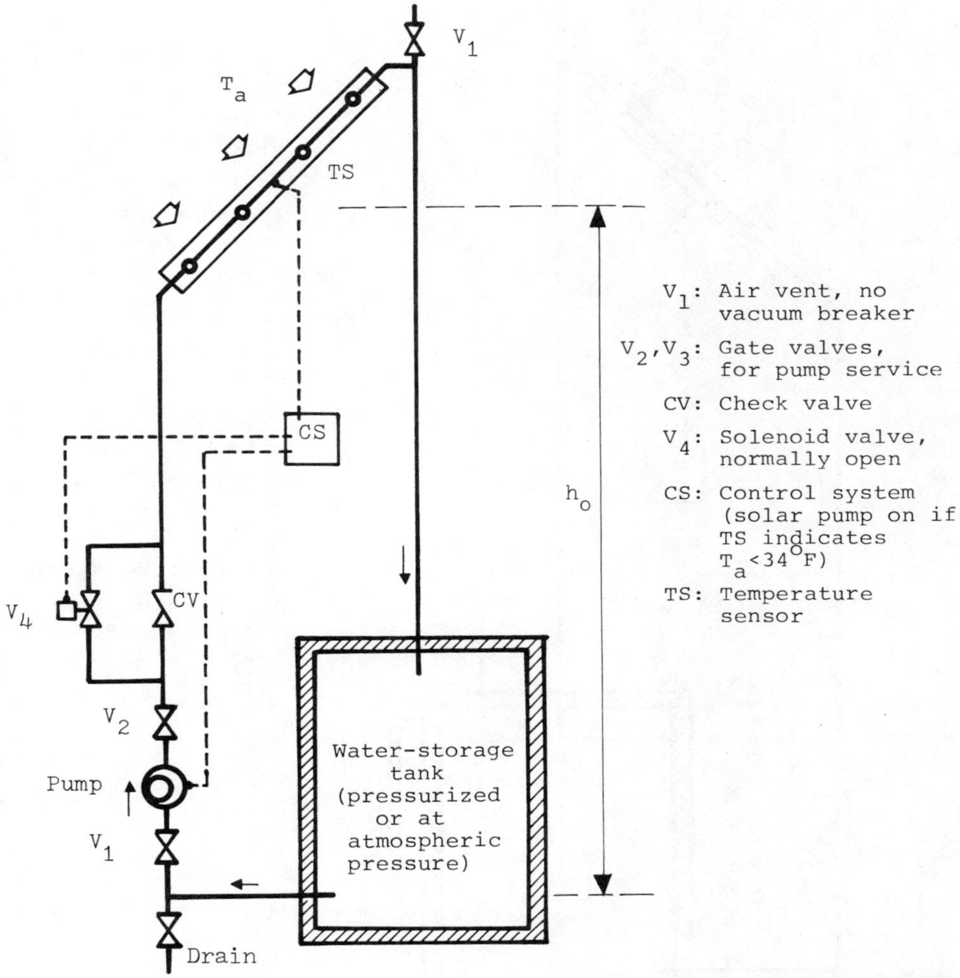

Fig. 2.6. Piping schematic for "hot-shot" freezing protection with automatic siphoning as backup.

is, the safety margin in the specification of the main tank is almost as uncritical as in the case of the tank in Fig. 2.4, and the extra expense for the tank is marginal. Most fiberglass tanks will work well in this type of pressurized application, although generally they are not advisable for higher-pressure applications (see Section 1.3.1).

2.2.3 "Hot-Shot" Freezing Protection

This method is recommended only for mild climates[27] where freezing conditions occur infrequently. Basically, the system is a closed loop, but no heat exchanger is used between the storage tank and the collectors (Fig. 2.6). Whenever the need for freezing protection is sensed by a temperature sensor mounted in the center of one of the absorber plates, the panel pump is turned on to pump hot water from

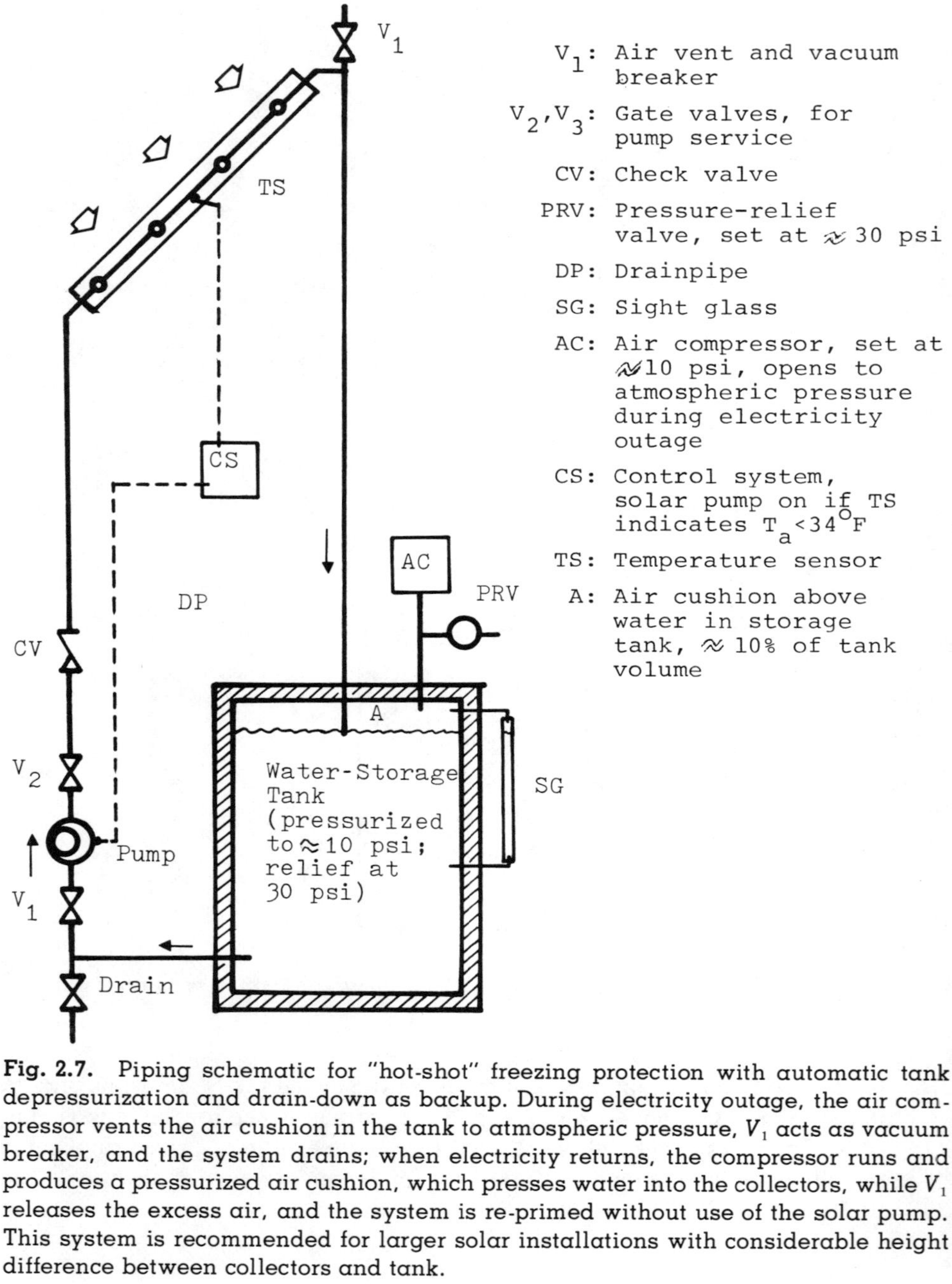

Fig. 2.7. Piping schematic for "hot-shot" freezing protection with automatic tank depressurization and drain-down as backup. During electricity outage, the air compressor vents the air cushion in the tank to atmospheric pressure, V_1 acts as vacuum breaker, and the system drains; when electricity returns, the compressor runs and produces a pressurized air cushion, which presses water into the collectors, while V_1 releases the excess air, and the system is re-primed without use of the solar pump. This system is recommended for larger solar installations with considerable height difference between collectors and tank.

the storage tank into the collectors until the sensor detects a certain higher absorber temperature (usually about 45°F) and turns the pump off. Because the electronic part of this system is very inexpensive and is actually incorporated in some commercially marketed solar control systems,[28] this can be an inexpensive method of providing freezing protection. It can be shown that for mild climates (<3000 heating degree-days per year at 65°F base temperature) the annually accumulated heat loss for all "hot shots" in a typical system is far less than the annual sacrifice of heat collection that would occur if a heat exchanger were used between the collector loop and the heat-storage tank.

The major drawback of the "hot-shot" method of freezing protection is that it depends on electricity and on a functioning collector pump. When the pump fails, or when there is an electricity outage, no freezing protection is available. This risk may warrant (at least for larger solar installations) backup freezing protection, such as that provided by a solenoid valve bypassing the antisiphoning check valve (Fig. 2.6) or by depressurization (Fig. 2.7).

The backup freezing protection shown in Fig. 2.7 seems quite sophisticated. Yet it has been shown that in larger solar systems (e.g., commercial installations with more than 1000 ft^2 of collector area) this method is quite attractive. Draindown is activated by depressurization of a normally pressurized tank.

2.2.4 Natural Siphoning Freezing Protection

If the collectors are located substantially above the water level of the heat-storage tank, freezing protection by natural siphoning may be considered. This technique works well as backup for the "hot-shot" method (Fig. 2.6). To obtain siphoning, the check valve that usually prevents reverse siphoning is bypassed with a water line containing a normally open solenoid valve that is energized (closing the valve) whenever the pump is off. Only if freezing protection (or backup protection) is required is the valve opened to allow a thermosiphoning current to pass by the check valve. The valve should be electrically connected in such a way that it is open (not energized) whenever the pump is running. This reduces electricity consumption and, more important, causes the valve to cycle on a daily basis, which helps prevent sticking of the valve plunger and makes the system more reliable.

In order to obtain adequate siphoning, several conditions must be fulfilled. First, the collectors must be mounted and interconnected such that a natural siphoning current opposite to the normal flow direction can develop when the bypass valve V_4 is open. For this to occur, hot water must rise in the return pipe to the collectors, replacing heavier water that has lost temperature in the collectors because of radiation. Collector systems with complicated routing of the supply and return pipes or of the collector-bank interconnecting pipes will not necessarily fulfill this siphoning condition.[29] They may actually develop a siphoning flow in the normal flow direction.[30] The second condition is that the flow restriction must be small enough for an apprecible siphoning current to develop to offset the heat losses that would cause freezing.

The flow can be estimated by the following procedure: First, the density of water at a typical storage-tank temperature should be determined using appropriate tables (the density of water at 100°F is 0.99299 g/cm^3). Assuming an exit temperature of 39°F (at which the density of water is 1.00000 g/cm^3), the density difference (0.00701 g/cm^3 in our example) can be computed. If this value is multiplied by the height difference between the heat-storage tank and the collectors, h_0, one obtains the approximate pressure sustaining the siphoning flow (for $h_0 = 15$ ft = 456 cm we obtain a pressure of 3.2 g/cm^2 or 0.045 psi, which is equivalent to 1.25 in. of head). With this result, the flow can be computed in accordance with standard hydronic flow computation practices. It depends, as usual, on pipe length and pipe diameter (for 100 ft of 1-in. copper pipe, the standard hydronic calculations give a flow rate at 1.25 in. of head of 1.0 gpm). This flow rate must be used to calculate the rate of heat delivery to the collectors [$1.0 \times 60 \times 8.34 \times (100 - 39) \approx 30{,}000$ Btu/hr]. This result must then be compared with the rate of heat supply required for freezing protection under adverse conditions. The system in our example may have 400 ft^2 of collector area and a linear heat-loss coefficient of $a = 1.0$ Btu-hr^{-1}-°F^{-1}-ft^{-2}; i.e., it loses $1.0 \times 400 = 400$ Btu-hr^{-1}-°F^{-1}. The preceding rate then will be sufficient to sustain a ΔT of 75°F; i.e., it will protect to an ambient temperature of 32° − 75°F = −43°F. The orifice of the solenoid valve must be large in order not to restrict the flow excessively (in the preceding example, a 1-in.-diameter valve opening was assumed).

2.2.5 Overheating

There are two sets of circumstances that can produce overheating. The first is pump failure or electricity outage. If the failure occurs during high collector insolation conditions, the collector and the heat-transfer fluid can be exposed to stagnation temperatures as high as 300°F for single-glazed collectors and 400°F for double-glazed collectors. Most collectors are designed to withstand these stagnation conditions without damage.[31] A water-based heat-transfer fluid can develop steam, which necessitates a pressure-relief valve in a closed-loop piping arrangement (Figs. 2.3 and 2.7). In pressurized systems, the boiling point is increased; this may actually prevent the generation of steam under most stagnation conditions for single-glazed collectors.[32] On the other hand, the pressure inside the collectors of an open system may actually be slightly below the atmospheric pressure, which lowers the boiling point of water to less than 212°F and increases the possibility of steaming in such systems. Stagnation overheating can damage glycol-based heat-transfer fluids. The acidity and the concentration of the corrosion inhibitor in such fluids must be tested after each prolonged stagnation period. Biannual exchange of fluid is recommended regardless of whether or not stagnation has occurred.

The second set of circumstances that can cause overheating involves extended periods of solar heat storage without adequate periodic use of the collected heat. This condition, which can arise during vacation periods or during the summer season (where it specifically applies to systems with low collector tilt angles and double-glazed collectors), can raise the temperature of the heat-storage tank to a level that can damage some fiberglass tank materials. To avoid this danger, night-sky radiation or heat purging into a cooling tower should be provided as automatically controlled preventive measures.

2.3 Collector System Design

2.3.1 Mounting of Collectors

The geometric factors that must be considered for appropriate collector mounting (tilt, azimuth) are discussed in Chapter 5. It should be pointed out that departure from the desired tilt angle and, to a lesser degree, departure from the required collector azimuth orientation (20 deg west of south is usually optimum) can have very undesirable consequences, such as poor solar heat collection in the winter months, overheating problems during the summer months (if the collector tilt angle is too low), or generally low collection efficiency (if the collector azimuth is wrong). Rather than adapting the collectors to a particular desired architecture for a building (and sacrificing effectiveness of the solar installation), one should adapt the architecture to the requirements of the solar-collector system. Many architecturally pleasing solar homes have been built during the past few years, and ample literature is available on this subject.

A solar-collector system can be designed as the primary watershed; i.e., it can replace the conventional roofing material. However, the designer of such a system must ensure (a) that the collectors are waterproof and are built of high-quality materials (glass glazings are definitely advisable for this application) and (b) that the waterproofing between the collectors is maintained with proper (usually custom-designed) flashings. A second waterproof membrane below the collectors is advisable as a precautionary measure; however, this can be an inexpensive single layer of felt. An example of a collector installation in which the collectors are used as the watershed is shown in Fig. 2.8.

Fig. 2.8. Solar space-heating retrofit installation on a residence in a moderate climate (3000 degree-days per heating season). The 350 ft^2 of single-glazed aluminum Rollbond collectors supply 70% of the heating requirements of the 2300-ft^2 house. This is a closed-loop system with antifreeze heat-transport fluid and a 1000-gal storage capacity; collectors were built in place by the owner and are the new watershed. The 38-deg collector tilt (roof tilt) is below optimum for space heating (should be 60 deg), but it was chosen for architectural reasons.

Fig. 2.9. Solar space-heating installation in a conventional tract house in a moderate climate (3000 degree-days per heating season). The system is designed for low cost and high efficiency. The use of more collectors flat on the roof would have been architecturally more desirable but much less cost-effective. Specifications: 1500 ft^2; 300 ft^2 of collector area, single-glazed; 800 gal of storage above ground in garage; drain-down on freezing protection demand. (Courtesy of Alten Corporation [2].)

The most common and easiest method of installing solar collectors is to install them above existing roofing material.[33] Typically, collectors are mounted on a series of stringers (2 × 4 in. or 4 × 4 in.) that are securely fastened into the rafters, using long lag bolts and elastic mastic for sealing. This type of installation offers the possibility of mounting the collectors easily at a tilt angle other than the roof angle (which is usually too low to be practical for winter space heating). However, in designs of new homes this mounting method has the disadvantage that no cost offset is gained because no costly roofing materials are replaced by the solar collectors. An example of installation of solar collectors on the conventional roof of a new tract house is shown in Fig. 2.9. If the orientation of the roof does not lend itself to the mounting of solar collectors, a location on the ground may be considered.

The anchoring method chosen for a collector system must make adequate provision for wind loads. Wind loads can be significant if the collectors are not mounted flat on the roof surface; in such circumstances it is strongly advised that wind-load calculations be performed using locally applicable wind-load factors. The static weight of solar collectors (typically 4–7 lb/ft^2) usually can be handled by existing roof structures without the addition of rafters or support structures. However, local codes may call for special measures.

Fig. 2.10. Solar space-heating installation on a residence in Woodside, California. Collectors are mounted on racks at 55 deg tilt on flat roof which is painted white to augment insolation into collectors by reflection. Collector mounting racks are vertically offset to reduce collector-to-collector shading and encased to blend with architecture of building. (Courtesy of Alten Corporation [2]).

2.3.2 System Flow Rates and Hydronic Design of Collector Banks

The choice of the collector-loop design depends on the method of freezing protection chosen for the particular system, as discussed in Section 2.2. The most important functional parameter of the hydronic system is the collector-loop flow rate. It depends on the desired temperature increase δT during a single pass of the heat-transport fluid through the collectors. Because the collector efficiency decreases with increasing δT (i.e., with decreasing flow), and because the pipe diameters and the pump size must be increased for increased flow rates, one must find the optimum compromise between the costs for plumbing materials and electricity on the one hand and the sacrifice in system efficiency on the other hand. Such a compromise is usually found at a flow rate between 0.03 and 0.05 gpm per square foot of collector area.[34] However, the value chosen must be consistent with the following three collector flow conditions: (a) The fluid velocity inside the collector water passages must be greater than 1 ft/sec, preferably about 2–3 ft/sec, in order to assure turbulent flow, which is required for optimum heat transfer from the channel walls to the fluid.[35] (b) The fluid velocity should be less than 6 ft/sec, preferably less than 4 ft/sec, in order to avoid erosion damage of the collector fluid channels.[36] (c) The friction loss at the chosen flow rate must be reasonable.

These three conditions are not trivial, and in order to meet them, different plumbing methods are required for different collectors. Various commercially

Fig. 2.11. Flow pattern in parallel-flow collector.

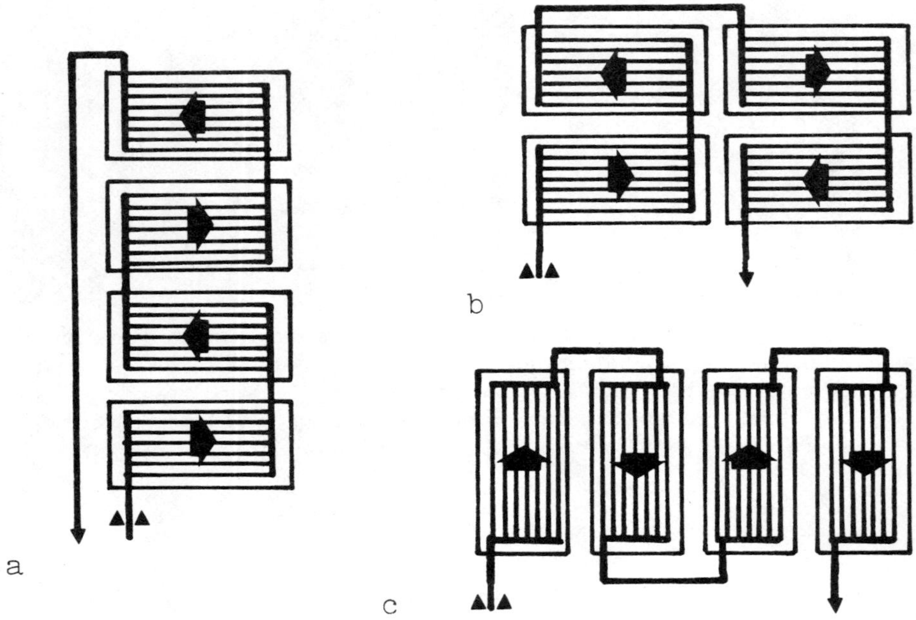

Fig. 2.12. Piping for collectors that require high flow rates because of parallel-tube construction in space-heating applications.

available collectors of the copper-tube/aluminum-fin design have the flow pattern shown in Fig. 2.11; i.e., there is an inlet header and an outlet header and an array of pipes in which the water moves in parallel-flow fashion. If, for example, eight 0.5-in. pipes are used in a 4-ft × 8-ft collector (32 ft^2), a minimum flow rate of 8 × 0.7 gpm = 5.6 gpm is required to keep the fluid velocity above 1 ft/sec. This computes to 0.18 gpm/ft^2 collector area, which is three to four times the recommended flow rate per collector. If this type of collector is used in space-heating applications, it is necessary to connect at least four of the collectors in series if the effective heat-removal rate per collector is to be equivalent to the preceding rule of 0.03 to 0.05 gpm/ft^2 for water as the heat-transport fluid.[37] Figure 2.12 shows some examples of connections for such collectors.[38]

Whereas the parallel-flow collector type shown in Fig. 2.11 requires particularly high flow rates, a serpentine-type collector requires particularly low flow rates. Such a collector flow pattern is shown in Fig. 2.13. A serpentine collector based on 0.5-in. type L copper pipes requires a flow rate of 1.4 gpm to maintain an

Fig. 2.13. Flow pattern in serpentine-type collector.

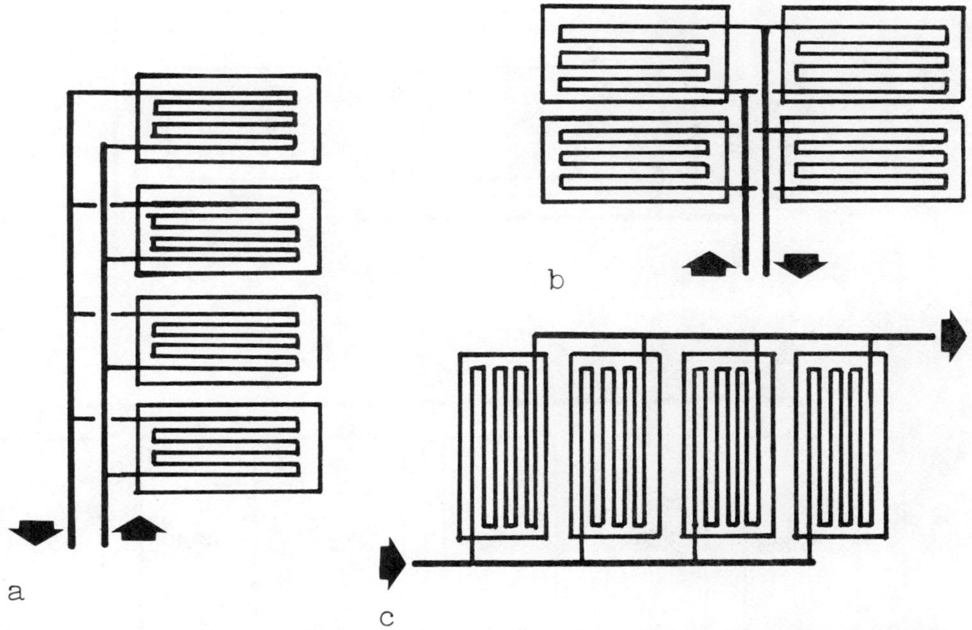

Fig. 2.14. Piping for collectors that require low flow rates, such as serpentine-type collectors, in space-heating applications.

optimum flow velocity of 2 ft/sec in the fluid channels. Obviously, this required flow rate is independent of the size of the collector. If the gross collector area is 32 ft^2, the flow rate per square foot of collector area will compute to 0.04 gpm, which coincides well with the desirable single-collector flow rate (0.03–0.05 gpm/ft^2). It is therefore suggested that collectors of the serpentine type be connected in a parallel arrangement, such as that shown in Fig. 2.14.[39] The series arrangement shown in Fig. 2.12 would not be adequate for this collector type.

It should be emphasized that there are no intrinsic advantages or disadvantages connected with either of the extremes of collector types shown in Figs. 2.11 and 2.13, as long as they are plumbed in their proper series/parallel arrangements.[40] It may become desirable, however, in any particular collector layout to select one or the other type of collector flow. Most commercial collectors fall between the two extremes described in Figs. 2.11 and 2.13 and allow the formation of groups of two or three collectors in series. Some examples of such

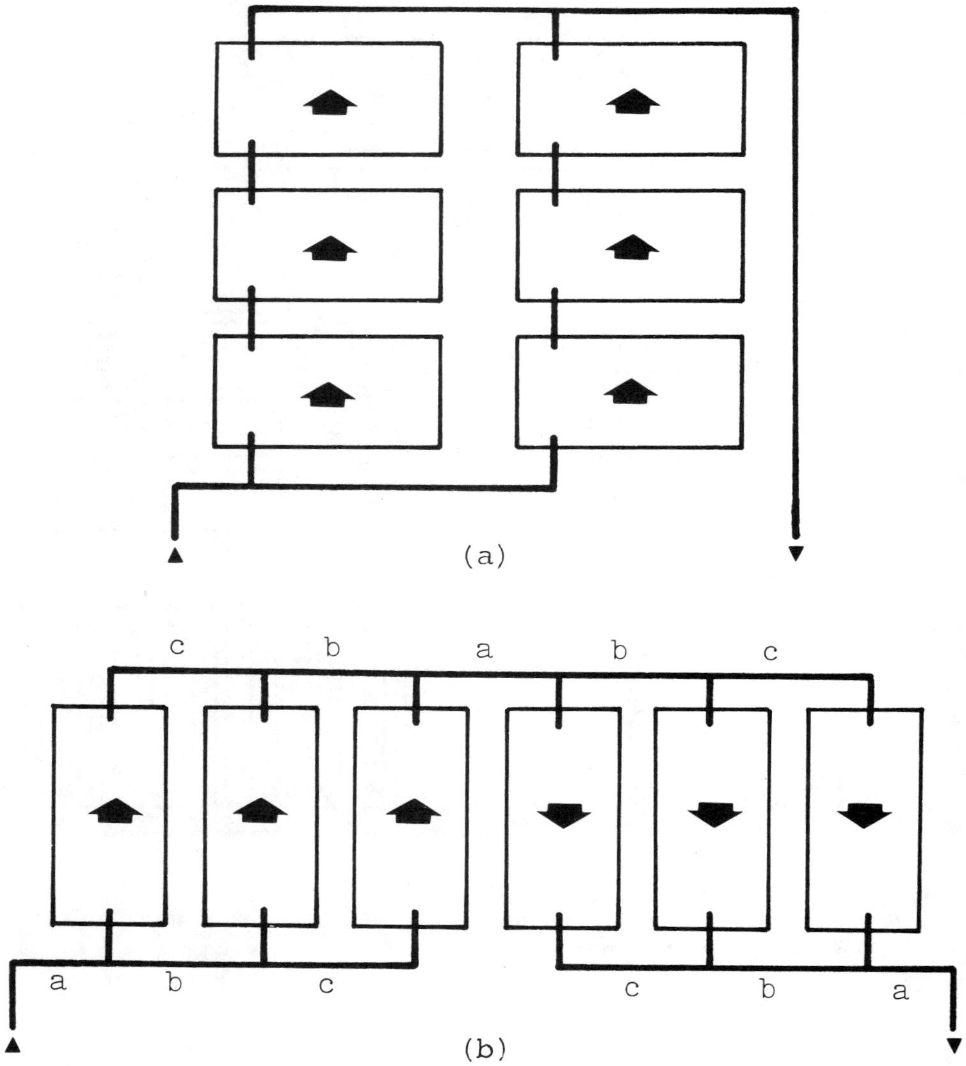

Fig. 2.15. Examples of collector groupings of three collectors (a) and two collectors (b) each in series.

arrangements are shown in Fig. 2.15. If a heat-transfer fluid other than water is used, adjustments in the flow design may be required to compensate for the differences in viscosity and heat capacity of that fluid as compared with water.

2.3.3 System Friction Loss, Pipe Sizes, and Pump Size

The total friction loss of the collector system can be determined from the friction-loss characteristic of the individual collectors, the chosen flow rate, and the number of collectors connected in series. For example, the friction loss for the arrangement shown in Fig. 2.15b for a system flow rate of 6 gpm and a collector friction-loss characteristic as shown in Fig. 2.16 amounts to 12.5 ft of head (two collectors in series, i.e., 2 × 1 gpm flow through each collector). This loss will, of course, be the same if additional collector groups are added, as long as the total system flow rate is increased by 2 gpm per added collector group.

The letters in Fig. 2.15b indicate different pipe sizes, reflecting the sequentially lower flows through pipes *b* and *c* as compared with pipe *a*. The sizing of the pipes should be in accordance with common design practice for hydronic systems. A design goal of 2 ft/sec flow velocity in all pipes is recommended. For quick reference, Table 2.2 may be used for sizing copper pipes (type L; the loss is given in feet of head per 100 ft of pipe length, for water).

The pump for the collector loop should be selected after the sizing and layout of the collectors and the pipes have been completed and the resulting system friction loss has been established. The desired working point *M* (gallons of flow, feet of head of friction loss) should fall at or slightly below the performance characteristic of the selected pump. An example is given in Fig. 2.17, which shows the performance characteristics of various booster pumps. The system working point *M* is assumed to be 20 gpm at 13 ft of head. The friction-loss characteristic of the solar-collector system can be represented by a second-order curve through *M* with zero slope at the origin of the coordinate system.[41] The performance characteristics for the pumps Series A and B intersect the system friction-loss characteristic (dash curve) at approximately 19 gpm and 21 gpm, respectively. Thus the larger pump (B) will normally be the first choice (see Section 1.4.1).

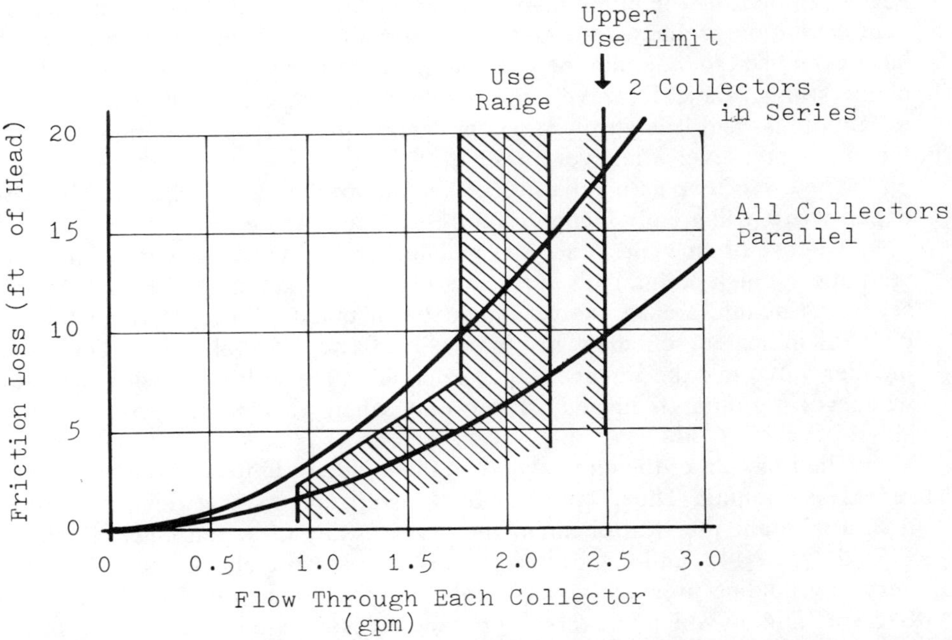

Fig. 2.16. Typical friction-loss characteristic of a serpentine-type collector. The graph is specifically valid for the reference collector discussed in Section 1.1.2.

TABLE 2.2 Friction Loss and Flow Velocity in Type L Copper Pipes[a]

	Flow Rate (gpm)											
	1		2		5		10		20		40	
Pipe Size (in.)	Velocity (fps)	Loss (ft)	Velocity (fps)	Loss (ft)	Velocity (fps)	Loss (ft)	Velocity (fps)	Loss (ft)	Velocity (fps)	Loss (ft)	Velocity (fps)	Loss (ft)
0.50	1.4	2.7	2.8	9.6								
0.75	0.7	0.4	1.3	1.5	3.3	8.2						
1.00			0.8	0.4	1.9	2.0	3.8	7.3				
1.25					1.2	0.7	2.6	2.6	5.0	9.5		
1.50					0.9	0.3	1.8	1.1	3.6	4.0	7.2	14.6
2.00							1.0	0.3	2.0	1.0	4.0	3.5
2.50									1.3	0.3	2.6	1.2

[a] Friction loss in feet of head per 100 linear feet of pipe. Flow velocity in feet per second.

2.3.4 Pump Location and Valves

In an open-loop system (Fig. 2.4) there are two reasons that the solar pump should always be mounted well below the water level in the heat-storage tank.[42] First, most of these pumps are of the centrifugal type; i.e., they are not self-priming. If such a pump loses its priming, which can happen if it is mounted above the drain-down level, the system will stagnate, and the pump can overheat and burn out. Second, mounting the pump below the water level in the storage tank helps prevent cavitation problems during the start-up after drain-down. Various methods have been used to maintain the priming in a pump mounted above the water level of the storage tank. However, the reliability of these methods has been below expectations, and it is strongly recommended that the pump be mounted below the tank water level whenever possible.

The use of two gate valves on either side of the pump[43] is an easily justified expense that will greatly facilitate future service of the pump.

The use of air vents requires special attention. Open systems require an air vent at each high point. These vents should not be vacuum breakers.[44] Instead, a separate vacuum-breaker valve should be installed farther down in the return pipe, as indicated schematically in Figs. 2.4 and 2.5 (valve V_4). The vacuum-breaker valve may be replaced by a solenoid valve if drain-down is not desired whenever the pump is turned off, but only when freezing protection is required [14].

The function of the check valve in the collector loop is to prevent nocturnal reverse siphoning. Those collector-bank arrangements that do not ensure that nocturnal siphoning will occur in the reverse-flow direction (such as shown in Figs. 2.14a, 2.14b, and 2.15b) need a positive-pressure check valve. However, a very low holding pressure is sufficient, as was demonstrated in the example in Section 2.2.4, in which a reverse siphoning pressure equal to 1.25 in. of head was calculated.

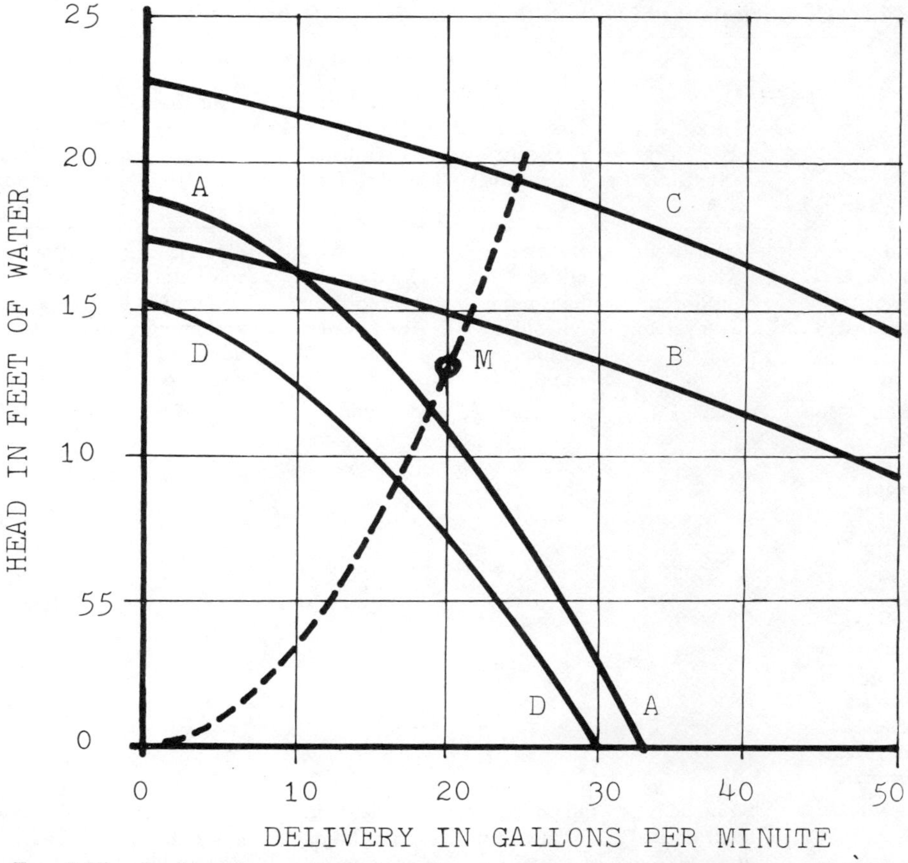

Fig. 2.17. Performance characteristics for several booster pumps, with sample system friction-loss characteristic (dash curve) indicating assumed desired working point *M*.

2.4 Interface with Space-heating Systems

In principle, a solar heating system always interfaces with a fully sized and fully functioning conventional heating system that can be used as a backup heating system during periods of cloudiness. It is not advisable to undersize the backup heating system; the cost savings would be relatively low, and most building codes require a full-size heater based on conventional fuel anyway. Because of these factors, a solar heating installation can be considered a retrofit to an existing heating system, even in a new building. Solar heating can be adapted easily to forced-air-type heating systems (including heat pumps) and to radiant slab heating systems.[45] In some cases the conventional heating system can be left unchanged, with a separate solar heat-distribution system being added. This is usually done for small solar-collector systems designed to heat only one room or a section of a house.

TABLE 2.3 Characteristics of Various Solar Heat-Supply Subsystems

	Solar Heat Supply to:				
Heater	Forced-Air Heating System	Heat Pump (Solar-assisted)	Heat Pump (direct Solar Space Heat)	Radiant Floor Heating System	Unit Ventilator Room Heater
Main interfacing element with solar heat-storage tank (water)	Water-to-air heat exchanger in duct system	Heat exchanger in heat pump (or solar storage tank)	Water-to-air heat exchanger in duct system	Pipes to solar storage tank, with motor-operated valves	Unit ventilator with pipes to DHW
Minimum practical storage-tank temperature at mild ambient temperature	100°F	<50°F	100°F	90°F	100°F
Minimum practical storage-tank temperature at severe ambient temperature (use of some auxiliary heat is assumed)	100°F	50°F	100°F	140°F	100°F
Pressure in solar heat-supply subsystem	Same as in solar heat-storage tank	Imposed by heat pump	Same as in solar heat-storage tank	Same as in solar storage tank; change pressure in floor subsystem	City water pressure
Feasibility and suggested use	Feasible in all climates at efficient collector operation	Feasible only in severe climates and if no low-cost gas or oil available	Feasible in all climates (check if heat pump itself is cost-effective)	Feasible in mild climates, less in severe climates	Feasible in all climates at efficient collector operation
Ease of design and installation	Easy	Moderate	Easy	Change in hydronic system of floor may be complicated	Easy

In the following sections, some details of solar heat-supply systems for forced-air heating and radiant floor heating systems are discussed. Furthermore, the subject of solar-assisted heat pumps is addressed, and the characteristics of separate solar heat-distribution systems (solar room heaters) are described. The main features of the different interfacing systems are summarized in Table 2.3.

2.4.1 Interface with Forced-Air Heating Systems

The typical subsystem for solar heat supply to a forced-air heating system (Fig. 2.18) consists of a water-to-air heat exchanger that is added in the plenum of the central heating system, a water-circulation pump, and the necessary piping connections to supply solar-heated water from the storage tank to the heat exchanger.

It is important to select a heat exchanger that is efficient even at low entering water temperatures. This will reduce the temperature requirement under which the solar collectors must operate. This, in turn, increases the overall solar system efficiency. It is advisable and possible to achieve effective transfer of solar heat to

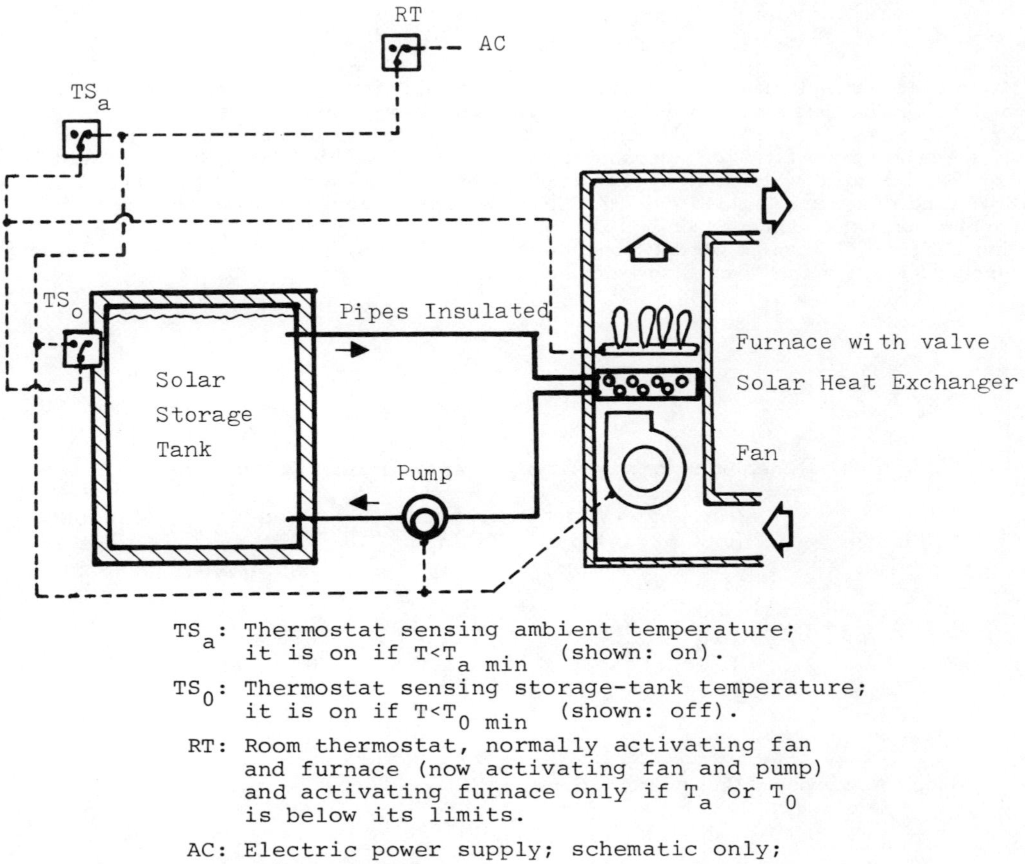

TS_a: Thermostat sensing ambient temperature; it is on if $T<T_{a\ min}$ (shown: on).

TS_0: Thermostat sensing storage-tank temperature; it is on if $T<T_{0\ min}$ (shown: off).

RT: Room thermostat, normally activating fan and furnace (now activating fan and pump) and activating furnace only if T_a or T_0 is below its limits.

AC: Electric power supply; schematic only; relays not shown.

Fig. 2.18. Subsystem for solar heat supply to a forced-air heating system.

TABLE 2.4A Heating Data for a Series of Water-to-Air Heat Exchangers[a]

Entering Water Temperature	Model No.	Air flow (cfm)	Water Flow (gpm)	Heating Capacity (Btu/hr)	Leaving Water Temperature	Pressure Drop (ft)
140°F	VHW-2	800	5	29,750	128.1	0.20
	VHW-3	1200	7	38,500	129.0	0.28
	VHW-4	1600	9	52,200	128.4	0.45
	VHW-5	2000	12	70,200	128.3	0.99
	VHW-7½	3000	16	110,400	126.2	1.37
	VHW-10	4000	18	135,900	124.9	3.70
120°F	VHW-2	800	5	22,250	111.1	0.20
	VHW-3	1200	7	29,400	111.6	0.28
	VHW-4	1600	9	40,050	111.1	0.45
	VHW-5	2000	12	52,800	111.2	0.99
	VHW-7½	3000	16	83,200	109.6	1.37
	VHW-10	4000	18	102,600	108.6	3.70
100°F[b]	VHW-2	800	5	14,750	94.1	0.20
	VHW-3	1200	7	20,300	94.2	0.28
	VHW-4	1600	9	27,900	93.8	0.45
	VHW-5	2000	12	35,400	94.1	0.99
	VHW-7½	3000	16	56,000	93.0	1.37
	VHW-10	4000	18	69,300	92.3	3.70

[a] Magic Aire Division, United Electric Company [32]. Magic Aire Division now offers a series of water-to-air heat exchangers especially designed for solar space heating applications. These products, the SHW-2000, SHW-3000, and SHW-4000 series, perform even better and are offered in a larger variety of sizes. The new model SHW-2208 compares approximately with the VHW-2 and VHW-3 models at 800 and 1200 CFM air flow, respectively. The newer model SHW-2347 is similar in performance to the VHW-4 and VHW-5 models at 1400 and 2000 CFM, respectively. We advise you to contact the manufacturer for more detailed information.

[b] The data for 100°F entering water temperature were obtained by linear extrapolation of the data supplied for 120°F and 140°F entering water temperatures.

TABLE 2.4B Dimensions for a Series of Water-to-Air Heat Exchangers[a]

Model No.	Air flow (cfm)	Water Flow (gpm)	Face Area (ft^2)	Coil Size (in.)	Manifold Size (nom. in.)
VHW-2	800	5	2.0	20 × 14½	⅞
VHW-3	1200	7	2.0	20 × 14½	⅞
VHW-4	1600	9	2.4	20 × 17½	1⅛
VHW-5	2000	12	3.2	20 × 23	1⅜
VHW-7½	3000	16	5.7	25 × 33	1⅝
VHW-10	4000	18	6.3	20 × 45½	1⅝

[a] Magic Aire Division, United Electric Company [32].

the heat exchanger for space heating at inlet temperatures as low as 100°F. One should size the ducts and the heat exchanger as if for air conditioning, i.e., larger than for normal space-heating applications.

Table 2.4A shows the heating capacity of a typical commercial heat exchanger that is suitable for this purpose. The sections for 140°F and 120°F entering water temperatures were supplied by the manufacturer; the values for 100°F were extrapolated. It is apparent that a coil 20 in. × 17.75 in. (VHW-4, Table 2.4B), costing less than $200, can supply a substantial amount of solar heat.

It should be noted that most solar heating systems, because they are combined with a conventional heating system, can simply and automatically be supplemented by conventional energy (e.g., gas or oil) whenever the ambient temperature is so low that the hourly rate of heating demand exceeds the capacity of the solar system heat exchanger. For example, the residence analyzed in column 2 of Table 5.2A will, at an average daytime ambient temperature of 28°F and an inside daytime design temperature of 68°F, require space heat at the rate of 800(68–28) = 32,000 Btu/hr. The solar system design was based on 120°F as the lowest usable temperature of the solar heat-storage tank. The Magic Aire VHW-3 heat exchanger delivers 29,400 Btu/hr at 120°F input temperature and will be an adequate choice, considering that the solar storage tank will normally be somewhat hotter, thus increasing the heat-exchange capacity (38,500 Btu/hr at 140°F). Under an extreme ambient temperature condition of −12°F, the building heating demand will be increased to 800[68 − (−12)] = 64,000 Btu-hr. Rather than sizing the solar heat exchanger for this extreme demand, it is good practice to rely on an auxiliary heat for the difference between what a moderate-size solar heat exchanger can deliver and what might be needed if it is unusually cold outside.[46]

The control for the subsystem for solar heat supply to a forced-air heating system can have various degrees of sophistication. A method that is adequate in most cases is to activate the pump whenever the conventional room thermostat calls for heat and then to activate auxiliary power (gas, oil, electricity) only if the storage-tank temperature (T_0) is below the design limit and if the ambient temperature is below a preset limit at which the calculated heating demand exceeds the rating of the heat exchanger (Fig. 2.18). This requires two adjustable thermostats. The heat exchanger may be installed before or after the conventional furnace, with respect to the airflow direction.[47]

2.4.2 Radiant Floor Heating Systems

Interfacing with this type of conventional heating system is extremely simple, but there are disadvantages that will be considered shortly. Only minor changes in the existing piping are required to connect the solar storage tank in series with the conventional furnace.[48] One must be aware, however, that the floor/furnace loop is now subject to the hydronic pressure conditions of the solar storage tank. This means that in most cases the conventional automatic water makeup with pressure regulation (usually 10 psi) must be disconnected. If the solar storage tank is at atmospheric pressure, its level must be sufficiently above the pump to assure its priming as well as cavitation-free operation at all times.

The control for the subsystem for solar heat supply to a radiant floor heating system is very similar to the control for a solar-driven forced-air heating system, as is shown in Fig. 2.19. The only difference is that the control for the pump is deleted; instead, two valves are operated from the sensors TS_a and TS_0 to assure that either solar heat or auxiliary heat is used, not both simultaneously. Again, auxiliary heat is activated whenever T_a or T_0 is below its lower limit.

There is a disadvantage inherent in the principle of solar radiant floor heating, as compared with solar forced-air heating, as is illustrated in Fig. 2.20. In the case of a forced-air heating system, the return air is always close to room temperature, i.e., independent of the ambient temperature. Therefore, solar heat can always be used, even at very low ambient temperatures that require high supply temperatures (solar heat is then used for the initial temperature increase of the air, auxiliary heat for the rest). On the other hand, in the case of radiant floor heating

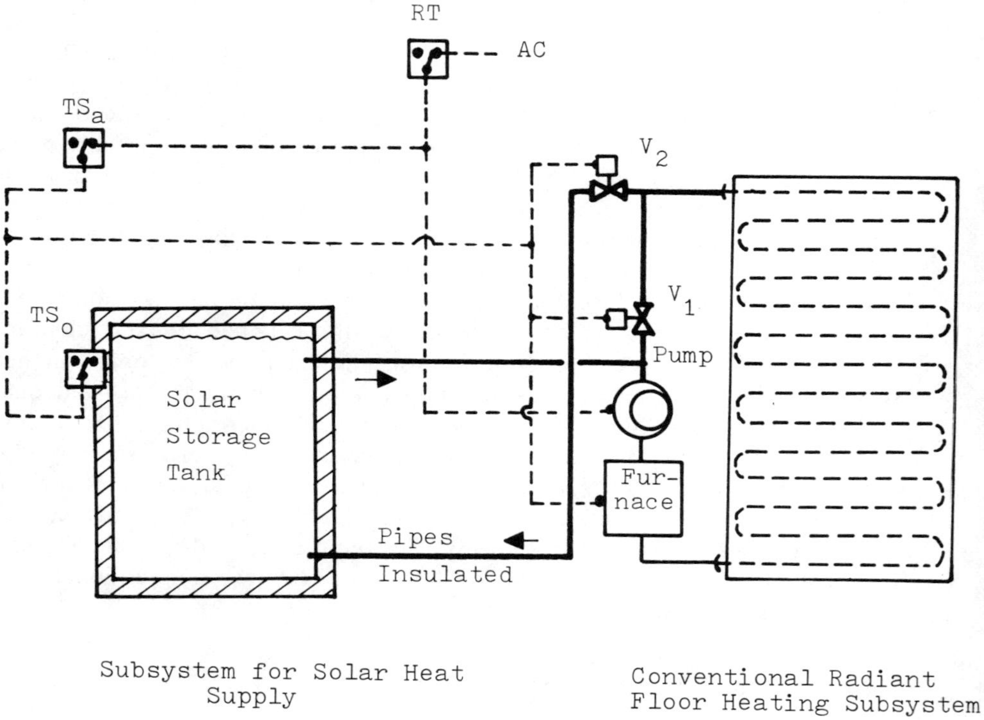

TS_a, TS_0, RT, AC: As in Fig. 2-18.

V_1: Valve, motor-driven or solenoid, open when energized.

V_2: Valve, motor-driven or solenoid, closed when energized.

Water makeup, expansion tank, pressure regulator, and pressure-relief valve for conventional radiant floor heating system are not indicated.

Fig. 2.19. Subsystem for solar heat supply to a radiant floor heating system.

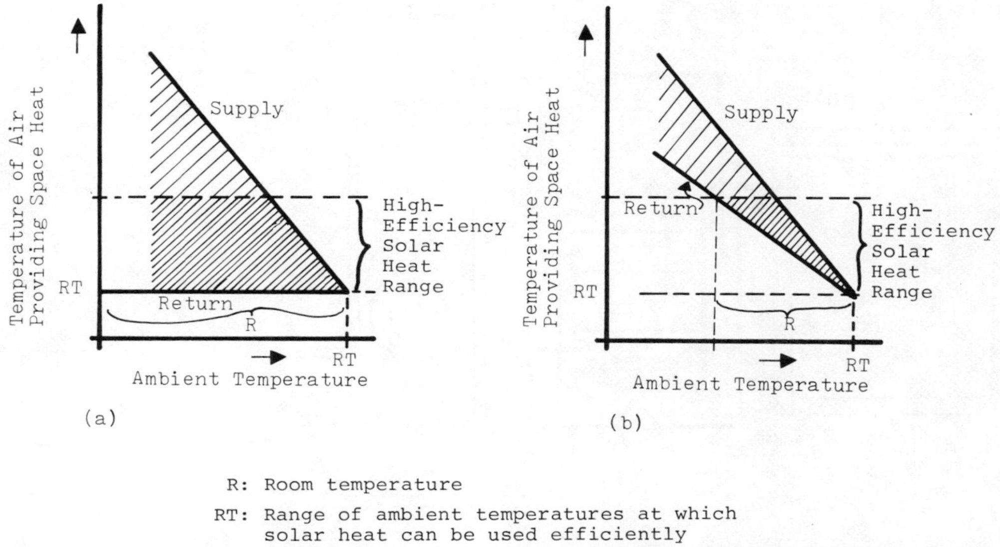

Fig. 2.20. Comparison of forced-air heating system (a) and radiant floor heating system (b) with respect to supply- and return-temperature dependence on ambient temperature.

systems, the return temperature increases in proportion to the supply temperature, which must be increased if the ambient temperature drops.[49] Consequently, the solar storage-tank temperature will have to be increased when the ambient temperature decreases, a fact that will lower the overall solar system efficiency.

2.4.3 Solar-assisted Heat-Pump Systems

A heat pump is solar-assisted [45] if its heat reservoir obtains energy from solar collectors (Fig. 2.21a). The principle is that the solar collectors increase the temperature of the heat reservoir to a level at which the heat pump can function more efficiently. Because this temperature usually is quite low (60°F–80°F), the solar collectors can also function very efficiently, and consequently the solar-collector installation is believed to be especially cost-effective. However, quite frequently there is a mistake in this chain of reasoning. It can be shown that solar assistance to a heat pump is advisable only if the ambient temperature (i.e., the heat reservoir from which the heat pump draws its heat supply if it is not solar-assisted) is consistently below freezing[50] for long periods of time.[51] In all other cases, a solar heat-distribution subsystem should be added to a heat-pump system (if a heat pump is chosen, for other reasons, as the means of space heating) in the same way as was described earlier for the case of a forced-air heating system; i.e., the solar heat should be supplied directly to the air-duct system (Fig. 2.21b). The fact that the solar collectors then naturally have to run hotter (> 100°F) does not generally make the overall solar system installation less cost-effective. Also see the work of Granryd [21] and Wurm and Kono [45] on the subject of solar-assisted heat pumps.

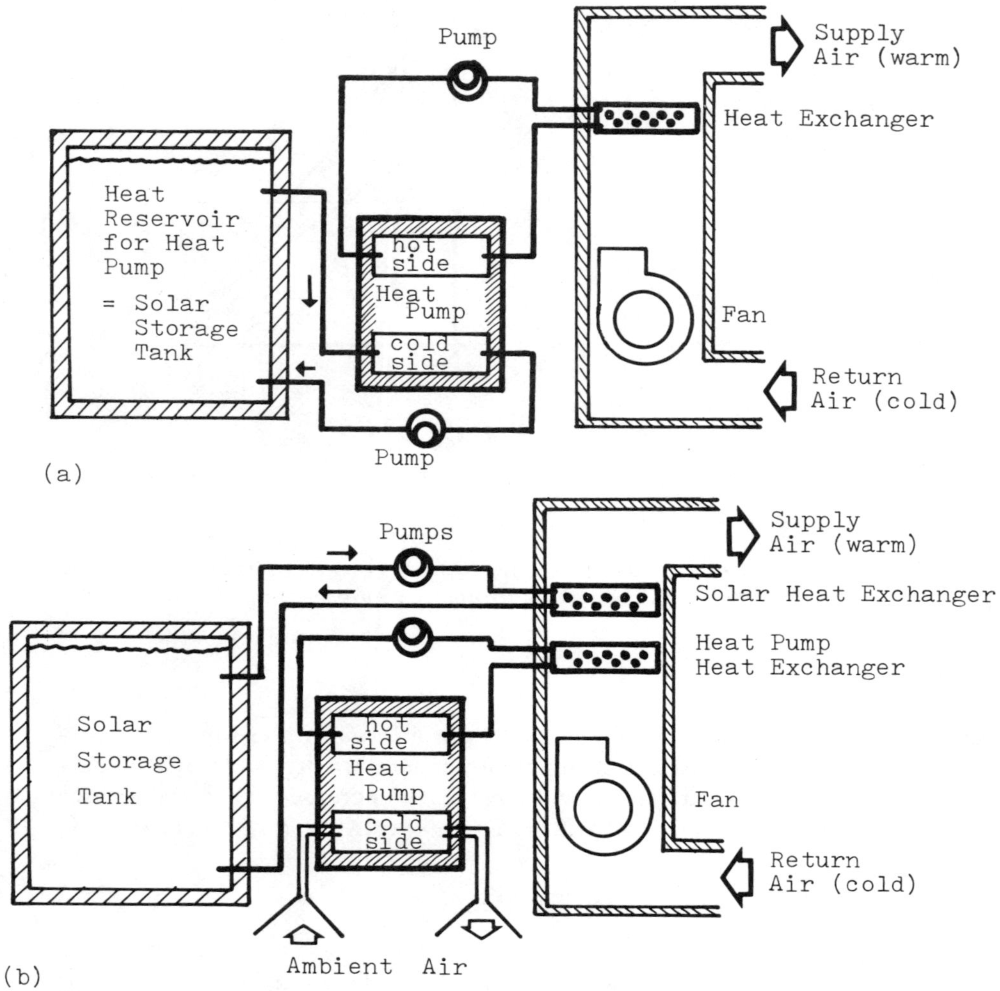

Fig. 2.21. Solar-assisted heat pump (a) and direct solar space-heating system with heat pump as auxiliary heat supply (b).

This important recommendation is explained with the following example, in which solar assistance for a heat pump with the efficiency characteristic shown in Fig. 2.22 is evaluated for two houses, one located in a severe climate zone (1800 degree-days per average winter month, house A), the other in a mild climate (600 degree-days per average winter month, house B). The solar collectors are used to supply a temperature of 50°F (house A) or 60°F (house B) to the heat reservoir. The solar assistance of the heat pump is compared with a regular solar-driven forced-air heating system operating from 110°F (house A) or 100°F (house B) storage-tank temperature. The electricity cost is assumed to be 8 cents per kilowatt-hour; natural gas for conventional forced-air heating is assumed to cost 50 cents per therm (100,000 Btu). A daily heat requirement of 600,000 Btu is assumed. However, the validity of the comparison does not depend on this assumption. The results of the comparison are summarized in Table 2.5. The most important

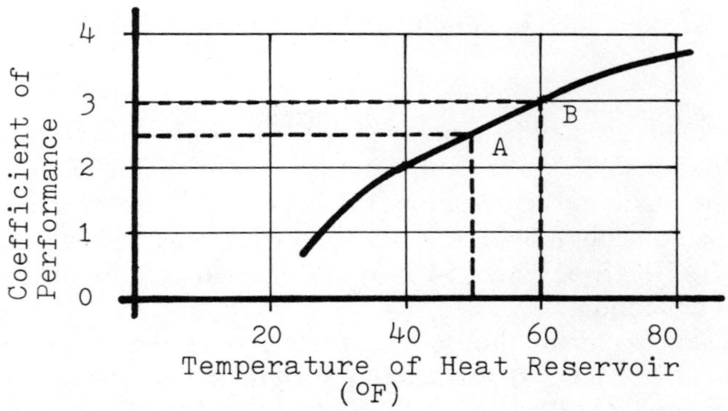

Fig. 2.22. Typical coefficient-of-performance (C.O.P.) diagram for a heat pump. For operating points marked with dash lines, see text.

TABLE 2.5 Feasibility Analysis for Solar-assisted Heat Pumps for Houses in Cold (A) and Mild (B) Climates[a]

		House A	House B
1	Average daytime ambient temperature	5°F	45°F
2	Average temperature at 8 A.M.	0°F	40°F
3	Collector operating temperature, for SAHP[b]	50°F	60°F
4	Collector operating ΔT, for SAHP	50°F	20°F
5	Collection in January, Btu-ft^{-2} per day[c]	1115	1415
6	C.O.P. of non-solar-assisted heat pump	1[d]	2.0
7	C.O.P. of solar-assisted heat pump	2.5	3.0
8	Square feet of collector area required[h]	367	310
9	Kilowatt-hours required to deliver 600 kBtu,[e] no SAHP	176	88
10	Kilowatt-hours required to deliver 600 kBtu, with SAHP	70	59
11	Cost to deliver 600 kBtu, no SAHP, $	14.06	7.04
12	Cost to deliver 600 kBtu, with SAHP, $	5.60	4.72
13	Savings in electricity due to SAHP, $	8.46	2.32
14	Storage-tank temperature for direct solar space heating (DSSH)	110°F	100°F
15	Collector operating ΔT for DSSH	110°F	60°F
16	Collection in January,[f] Btu-ft^{-2} per day	592	1015
17	Collection per day with system of size in step 8, for DSSH, kBtu	217	315
18	Auxiliary heat (kBtu) required to deliver 600 kBtu	383	285
19	Cost of auxiliary heat of step 18, $		
	(a) If makeup is with heat pump	8.97	3.34
	(b) If makeup is natural gas[g]	3.19	2.38

[a] Compare lines 12 and 19 for direct results; see text for details
[b] Solar-assisted heat pump.
[c] Single-glazed collector, from Table H.3, Appendix H, for 38° N. latitude, 60-deg. collector tilt angle, interpolated.
[d] C.O.P. = 1.
[e] 3414 Btu = 1 kW-hr at C.O.P. = 1.
[f] Double-glazed collector used for house A, value from Table H.4 interpolated.
[g] 60% gas-heater efficiency is assumed.
[h] {600,000 Btu/day required—step (10) × 3414 Btu/kW-hr × 0.8 (assured fraction of energy to drive heat pump actually contributing to delivered energy}/step (5).

results are contained in lines 12 and 19 of Table 2.5. They should be interpreted as follows:

1. The solar-assisted heat pump (SAHP) is not feasible for the house in a mild climate (house B). Even if no natural gas is available and a heat pump is selected as the auxiliary heat source, the cost to deliver the required 600,000 Btu per day is considerably less for the direct solar space-heating case ($3.34) than for the SAHP case, where $4.72 is still needed in spite of the more efficient solar heat collection.
2. In a severe climate (house A), a solar-assisted heat pump is (slightly) preferable to a direct solar space-heating system ($5.60 versus $8.97), but only if no natural gas (or oil) is available and all makeup heat is supplied by (very costly) electric-resistance heat. Otherwise, even in a severe climate a conventional direct solar space-heating system is more cost-effective than a solar-assisted heat pump[52] (only $3.19 would be required for auxiliary fuel, versus $5.60 for the case of the SAHP).

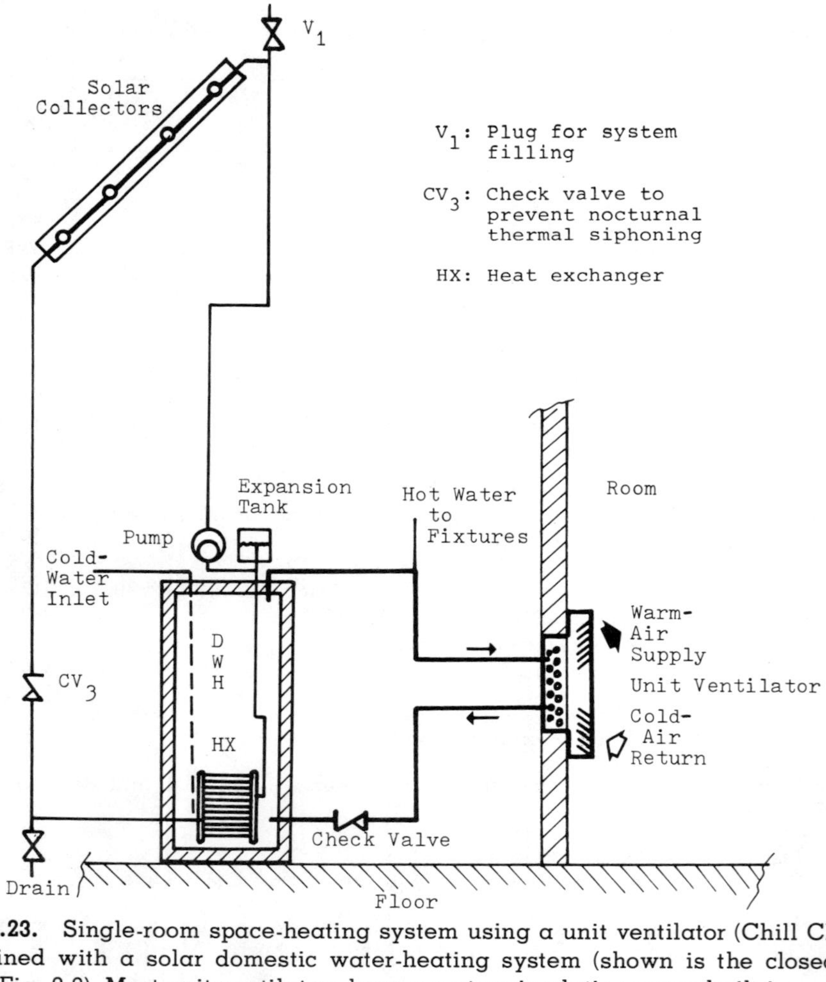

Fig. 2.23. Single-room space-heating system using a unit ventilator (Chill Chaser) combined with a solar domestic water-heating system (shown is the closed-loop type, Fig. 2.3). Most unit ventilators have a water-circulation pump built in.

2.4.4 Solar Room Heaters

The principle of heating only one room or a few individual rooms with solar energy, rather than supplying solar heat to the central heating system of a building, is very attractive because of the much lower system cost. Generally, such a system can be designed as an enlarged solar domestic water-heating system. One or more unit ventilators (such as the Chill Chaser by Turbonics, Inc., Cleveland, Ohio) are placed in strategic locations in rooms to be solar-heated and are directly connected to the solar-heated domestic water heater. A schematic drawing of such an arrangement is shown in Fig. 2.23. Because the water pump is part of the Chill Chaser, no extra pump need be installed, and the extra expense of a regular solar domestic water-heating system is limited to the unit ventilators (≈ $400 each), additional solar collectors (≈ $25/ft^2 installed), and some small-diameter piping, as well as installation costs.

Table 2.6 lists heat-delivery rates for a Chill Chaser. The sizing usually is much more approximate than for full-size central space-heating systems.

TABLE 2.6 Heating Data and Dimensions of the "Chill Chaser" by Turbonics, Inc. [42]

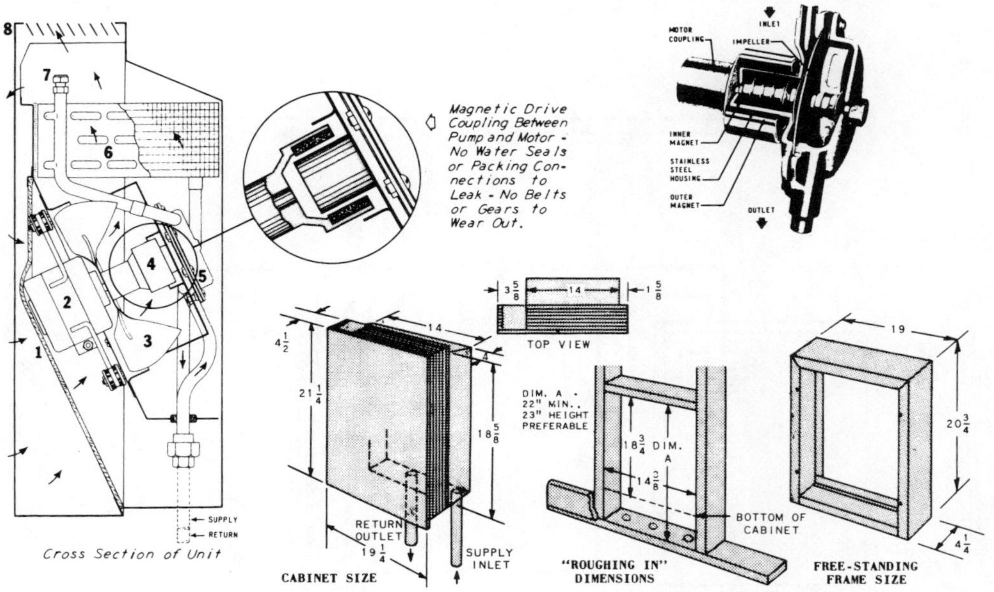

LEGEND

1. AIR FILTER
2. ELECTRIC MOTOR
3. FAN
4. MAGNETIC DRIVE
5. WATER PUMP
6. HEAT EXCHANGER
7. BLEED VALVE
8. OUTLET GRILLE

RATING DATA

GPM	UNIT OUTPUT - BTUH						TOTAL SUPPLY & RETURN TUBE MAXIMUM LENGTH, FEET *		
	ENTERING WATER TEMPERATURE						1/2" TUBE	1/2" TUBE	3/4" TUBE
	120 F	140 F	160 F	180 F	200 F	220 F	1/2" GLOBE VALVES	1/2" ANGLE VALVES	3/4" GLOBE VALVES
1000 RPM									
.8	3,600	5,200	6,700	8,200	9,800	11,400	80	100	100
.9	3,900	5,500	7,000	8,500	10,100	11,800	50	80	100
1.0	4,000	5,600	7,200	8,700	10,300	12,000	25	60	100
1.1	4,150	5,800	7,350	8,900	10,450	12,100	—	40	100
1.2	4,300	5,900	7,400	9,000	10,500	12,200	—	30	100
1500 RPM									
1.5	5,150	7,250	9,350	11,250	13,500	15,700	42	80	100
2.0	6,000	8,500	11,000	13,000	16,000	18,000	—	16	100

* Allowance of 8 elbows for fittings included in friction calculations. For 3/4" globe valves with 1/2" tube, use data above for 1/2" angle valves.

Typically, integrated domestic-water/single-room heating systems range in size from 90 to 150 ft² of collector area. The control of the room heater is strictly operated from a thermostat in the unit ventilator or is operated manually.

2.5 Interface with Domestic Water-heating System

A domestic water-heating (DWH) system being interfaced with a solar space-heating system is different in principle from a regular stand-alone solar domestic water-heating system. This is mainly because of the difference in the basic purposes of these two systems (i.e., space heating and DWH). Usually one uses a heat exchanger in the cold-water supply line to the regular domestic water heater and preheats the water entering the domestic water heater, using heat from the main solar storage tank.

Many solar heating systems have a more elaborate DWH interface that does not rely only on a one-time temperature pickup of cold water during its path through the heat exchanger but provides the entire domestic water heat whenever

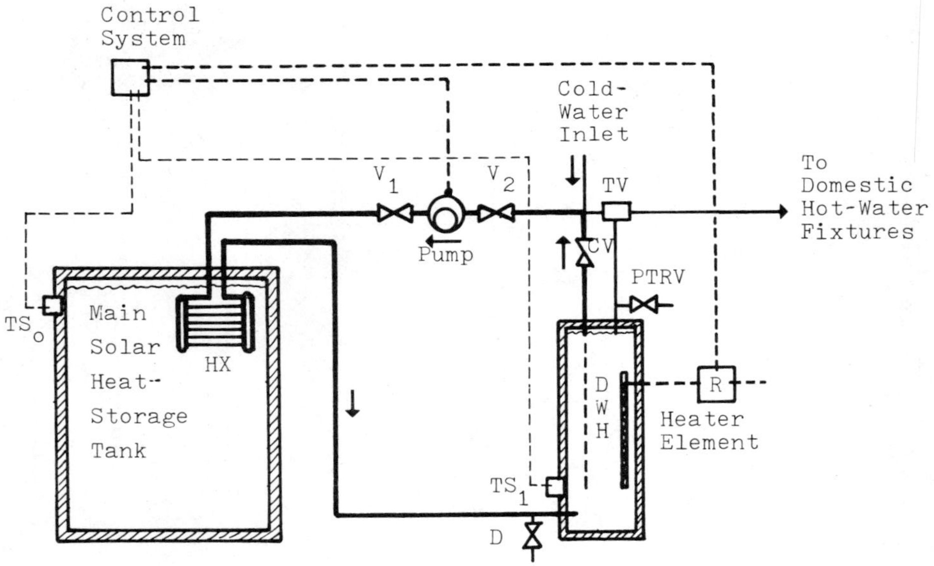

HX: DWH heat exchanger
V_1, V_2: Gate valves, for pump service
TV: 160°F tempering valve
CV: Check valve; note direction!
PTRV: Standard temperature/pressure relief valve
TS_0, TS_1: Temperature sensors
R: Relay for heater element
D: Drain

Fig. 2.24. Interface of solar space-heating system with electric domestic water-heating system.

the temperature in the main storage tank is sufficiently high (130°F–140°F), which is usually the case during the summer months and the late spring and early fall months, when little space heat is required. A schematic diagram of such a system is shown in Fig. 2.24.

For single-family residences, the heat exchanger may be a copper coil immersed in the main storage tank (60 ft of 0.5-in. coil). A tube-in-shell type of exchanger is preferable for larger dwellings. The pump may be small (1/20 to 1/200 hp) for single-family DWH interfaces. A 0.75-in. pipe diameter should be maintained throughout the piping system[53] to meet most code requirements.

The arrangement shown in Fig. 2.24 needs a control system that activates the conventional heating element of the domestic water heater only if TS_1 senses that makeup heat is required and TS_0 indicates that the temperature in the heat-storage tank is not high enough to supply the makeup heat. Otherwise (i.e., in the solar heat-supply mode), the control system activates the small circulation pump and leaves the conventional heating element off. Most gas boilers have a mechanically operated gas valve and therefore are not suitable for this mode of solar heat supply. A larger-size heat exchanger and the simple arrangement for solar preheating only are suggested for existing DWH system with gas-operated boilers.

Notes

[1]See equation 1.1; see also Sections 5.2.5 and 5.2.6 for further details on collection efficiency.

[2]Various computer programs have been proposed for use in sizing. A list of some of the programs available for public use is given in Appendix H (Table H.2). However, because of the practical limitations mentioned earlier, we suggest the use of the method of sizing presented here, with performance of accurate collection computations after the system has been sized, using the correct specified system components.

[3]See Appendix J for further details on heat-loss-coefficient calculations and solar heat-gain calculations through windows.

[4]See Section 5.1 and Fig. I.3 in Appendix I.

[5]See Appendix H (Table H.1) for monthly average degree-days for various locations in the United States.

[6]The ideal would be a combination of solar space heating and solar pool heating or solar air conditioning during the summer months.

[7]Alternatively (for more accurate results), consult Table H.3 or H.4 in Appendix H, using an average operating ΔT value, which can be assumed to be the difference between the lowest storage-tank temperature T_0 at 8:00 A.M. and the ambient temperature T_a at 8:00 A.M.

[8]Decrease this assumed collection value by 100 Btu/ft^2-day (a) for every 10°F increase in the lowest usable storage temperature, (b) if a heat exchanger is used, and (c) if the storage capacity is considerably less than specified above. Increase the assumed collection by 100 Btu/ft^2-day if an exceptionally good collector is used, such as the reference collector discussed in Section 1.1.2. Treat a double-glazed collector in a 6000-degree-day climate identically.

[9]This applies to a typical system (see Sections 1.3 and 1.4.2).

[10]See, in particular, Figs. 5.5. and 5.6a.

[11]Excessive heat collection will increase the storage-tank temperature to levels that can be dangerous to fiberglass containers and some types of insulation materials (see Section 1.3 for details).

[12]See Fig. 5.5 for the effect of tilt angle on solar collection.

[13]Using, for example, the program presented in Appendix I.

[14]The calculations were performed with the computer program described in Appendix I (program A) using the same building data and solar system parameters as shown in Table I.1, except for the storage volume (No. 11), which was varied.

[15]The improvement in collection effectiveness that is achieved by increasing the storage capacity from 1.5 gal/ft^2 to 2 gal/ft^2 is 7.6% (from 605.7 to 652 Btu/ft^2-day). At an installed collector cost of \$15/ft^2, this improvement will cost \$1.15 if done by increasing the collector area. This amount certainly exceeds the incremental cost for increase of the storage capacity, which typically amounts to \$1.00 per gallon, or \$0.50 for the 0.5-gal increase in our example. It is therefore more cost-effective to increase the storage capacity from 1.5 to 2 gal/ft^2 than to stay at 1.5 gal/ft^2 while increasing the collector area correspondingly. On the other hand, an increase in storage capacity from 3 gal/ft^2 to 3.5 gal/ft^2 amounts to only 1.9% improvement in the collector system effectiveness, or \$0.28/ft^2, which is less than the incremental cost for increased storage capacity. Thus the economics do not favor this further increase in storage capacity.

[16]Collectors with a flat-black coating are worst of all, but even selectively coated absorbers are subject to increased danger of freezing because of black-body radiation.

[17]This is an approximation, as was explained under the topic (5) Absorber Emissivity in the subsection Determination of Efficiency Parameters by Design in Section 1.1.2.

[18]Prestone II is a frequently used solution. This antifreeze liquid, which is designed for automotive use, contains effective corrosion inhibitors to protect copper, iron, and even aluminum in the wet system. However, Prestone II is based on toxic ethylene glycol. Toxicity is less of a problem if a propylene glycol solution is used.

[19]The solution can become acidic, and therefore it must be checked after each accidental stagnation event.

[20]See Section 5.2.5 and Appendix E for more details about the use of heat exchangers.

[21]See Section 1.4.3 for details about the design of heat exchangers for solar applications.

[22]See the subsection Materials Aspects in Section 1.1.2, in particular Table 1.3.

[23]See the topic Absorber under the subsection Materials Aspects in Section 1.1.2, in particular Table 1.1. Note that collectors of the copper-tube/aluminum-fin type of construction can be treated as copper collectors with respect to corrosion considerations. They can be used in open systems just as well as all-copper collectors.

[24]See Section 1.2. For example, the Delta T series DTT 200 by Heliotrope General has this feature.

[25]The down-flowing water "pulls" the up-flowing water (the valve V_1 must not be a vacuum breaker), and the pump has to overcome only the friction losses, once all pipes are filled.

[26]The high pumping power may, once the system is primed and functioning in the "hydronic flow mode," possibly cause excessive fluid velocities, which are undesirable because of material erosion from the collector and piping channels.

[27]Less than 600 heating degree-days during the coldest month (base 65°F).

[28]See Section 1.2. For example, the Delta T series DTT 200 by Heliotrope General has this feature.

[29]The collector arrangements shown in Figs. 2.12a, 2.14c, and 2.15a warrant nocturnal siphoning in the direction opposite to the normal daytime flow. The arrangements shown in Figs. 2.12b, 2.12c, 2.14a, 2.14b, and 2.15b will not siphon well.

[30]This siphoning pattern is actually a common design problem. If the siphoning and regular pumping flow directions are the same, the check valve (Fig. 2.6) will not fulfill its purpose. A spring-loaded check valve is a necessity in such a case.

[31]See the topic Materials Aspects in Section 1.1.2.

[32]At a relief-valve setting of 30 psi, the boiling point of water is raised to 275°F.

[33]In this application the building can be considered retrofitted with solar collectors. In fact, most solar homes (even if designed for solar heating) are actually retrofitted. This is particularly true for the heat-distribution subsystem, because most codes require a complete conventional backup heating system.

[34]At an effective solar collection of 200 Btu-hr^{-1}-ft^{-2}, a flow rate of 0.04 gpm/ft^2 results in δT = 20°F (for water as fluid).

[35]For aqueous solutions, the transition between (undesirable) laminar flow and turbulent flow is at approximately 1 ft/sec.

[36]According to Popplewell [37], 2 ft/sec is considered a safe maximum flow velocity inside the fluid channels of collectors.

[37]The number of collectors per group depends on the individual collector design. Additional collectors should be grouped accordingly, and all groups of a system should have equal numbers of collectors. The groups should be connected with each other in a parallel fashion.

[38]It should be noted that this collector type is particularly useful for swimming-pool-heating applications, where high collector flow rates are desirable because of the filtering requirements for pools. In a swimming pool arrangement, such parallel-flow collectors are connected in parallel arrangements that will assure high flow rates at low friction losses (Fig. 4.12).

[39]Any number of additional collectors can be connected in a parallel fashion.

[40]The serpentine-type collector becomes advantageous if the flow is underrated. Parallel-flow collectors require a substantial flow rate to keep the flow velocities in the parallel tubes the same. If the flow rate is low, some pipes within any one collector may have significantly higher flow velocities than others, which might then operate under laminar-flow conditions. Serpentine arrangements, on the other hand, impose a constant flow rate on all sections of the absorber plate, irrespective of the flow rate.

[41]The friction loss is assumed to be proportional to the square of the flow rate.

[42]In a system with an auxiliary tank (Fig. 2.5), the solar pump should be located below the auxiliary tank.

[43]V_2 and V_3 in Figs. 2.3 through 2.7 (see also Section 1.4.2).

[44]Because a pressure lower than atmospheric pressure usually is found in the return pipe of an open system, a vacuum-breaker valve too far toward the top of the system will let air in during normal operation. This is undesirable because of noise and the excess oxygen supply that might increase the system corrosion rate.

[45]It should be noted that radiant baseboard heating systems are not suitable for solar retrofit. The water temperatures required for baseboard heaters (convectors) are so high (170°F–200°F) that the solar heat supply would be uneconomical; i.e., the collectors would have to operate under excessively high ΔT conditions (too inefficient).

[46]An even better practice would be to reduce the minimum solar storage-tank temperature to 100°F and then increase the size of the heat exchanger (to VHW-4 or even VHW-5). This would raise the average solar heat collection by approximately 15%.

[47]An even simpler control schematic that does not include an ambient-air temperature sensor at all is described in Section 1.2.2.

[48]In some cases the storage tank is actually replaced by the slab, which acts as the heat-storage element; see Section 1.3.3 for details.

[49]Ideally, a logic control circuit, operating from the continuously measured ambient temperature, will be required to adapt the supply water temperature to the requirements at any given ambient temperature. Most radiant floor heating systems do not have such an elaborate control facility. Instead, they are always set at a fixed supply temperature that is high enough to meet the demand even at extremely low ambient temperatures. This will, of course, further decrease the solar system efficiency for this type of heat-distribution subsystem.

[50]More generally, below that point at which its coefficient of performance (C.O.P) falls below unity (see Fig. 2.22), which is usually near 30°F.

[51]Even then, only if no natural gas is available at relatively low cost as compared with electricity.

[52]The fact that for direct solar space heating in case A double-glazed collectors must be used (usually more costly than single-glazed collectors) will slant the results slightly more in favor of the SAHP, but it will not alter the interpretation of the results. It should also be noted that even in a severe climate zone the conditions during the transi-

tion months in spring and fall approach the winter conditions in a mild climate zone, which will slant the results further against the SAHP principle.

[53]Some small pumps have only 0.5-in. connections. In this case, a 0.75-in. bypass line (containing a check valve in the flow direction) should be provided around the pump. The cross section of a 0.75-in. pipe can be approximated with 0.5-in. coil by cutting the coil into two equally long sections and connecting them in a parallel flow pattern.

3

Active Solar Cooling Systems

Solar collectors can be used for air conditioning in two basically different ways. In the first approach, the collectors are used during the daytime to heat a heat-storage medium, such as water. The solar-heated fluid is then used to drive air-conditioning equipment, such as an absorption chiller or a Rankine-cycle engine. We define this approach as genuine active solar cooling. In the second approach, solar collectors are used during the night as radiators to directly cool a heat-storage medium, such as water.

3.1 General Description of Active Solar Cooling Systems

The most thoroughly developed method of active solar air conditioning involves solar-driven absorption chilling; its details will be discussed in the following section. Other methods of active cooling differ mainly in terms of the apparatus used to produce chilled air or chilled water;[1] but because they also operate from a solar-generated hot-water input, most of the aspects of solar-driven absorption-chilling air-conditioning systems also apply to these other types of chillers.

An absorption-chilling air-conditioning system consists, in addition to the elements that it has in common with a solar space-heating system (i.e., collectors, heat-storage tank, heat-distribution subsystem, auxiliary furnace, interface to DWH system), of a heat-rejection unit (cooling tower), a cold-storage tank, and, of course, the absorption chiller itself. A typical system is shown schematically in Fig. 3.1. Its details are discussed in the following sections. However, here we shall discuss only those aspects of solar air conditioning that are different from space-heating systems and that were not discussed in Chapter 2. In particular, we refer to Section 2.2 for collector-loop design details (freezing and overheating protection) and to Section 2.3 for the mounting of the collectors, the determination of flow rates, and the hydronic design of the collector banks, as well as the determination of pipe and pump sizes. In Section 2.4, space-heating interfacing aspects

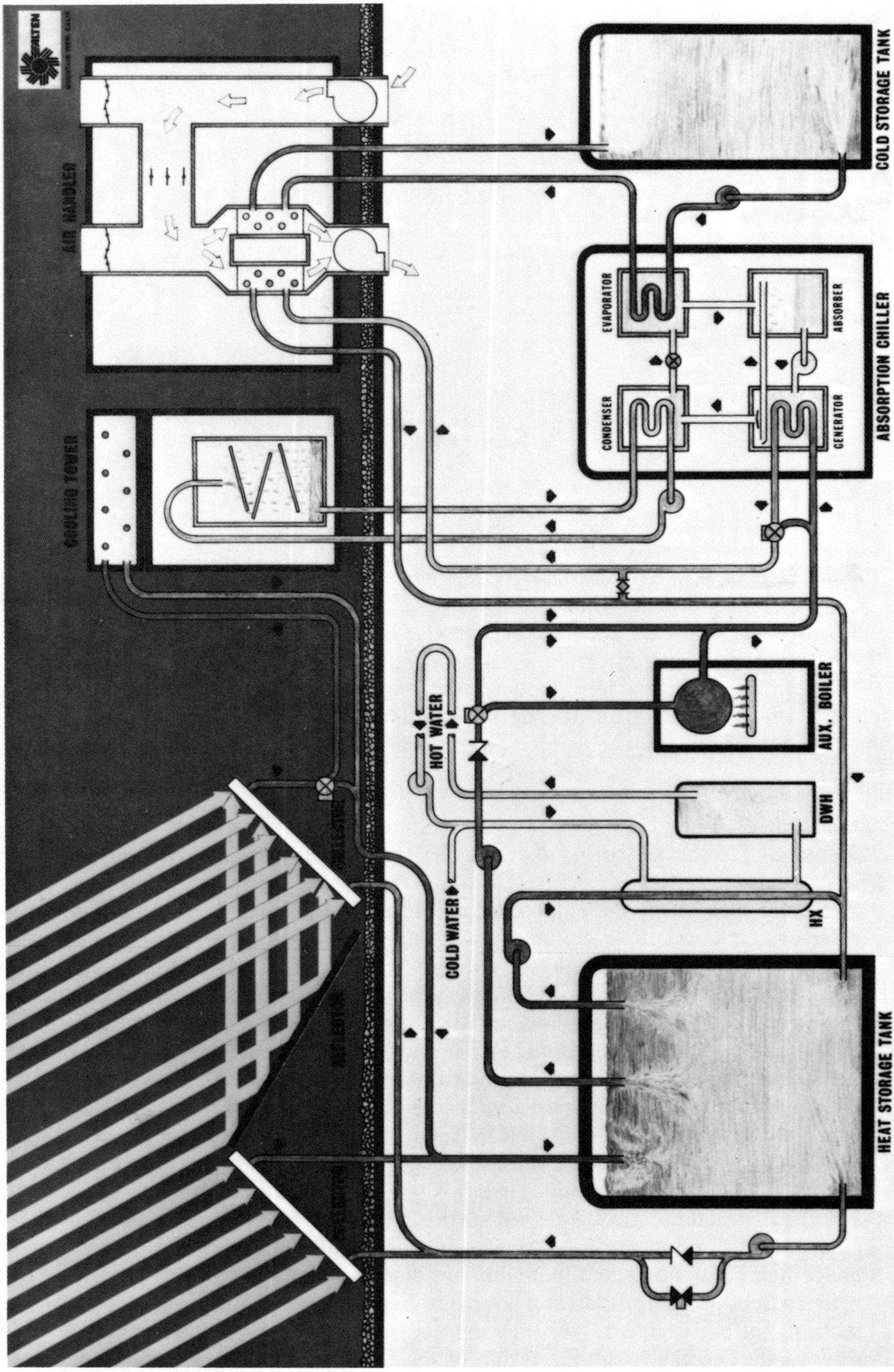

Fig. 3.1. Typical solar-driven absorption-chilling air-conditioning system. (Courtesy of Alten Corporation [1].)

were discussed, some of which also apply to solar absorption-chilling air-conditioning systems. The interface with the domestic water-heating system was discussed in Section 2.5.

3.2 Absorption Chillers

Absorption chilling is based on the principle of a two-stage vaporization–condensation cycle in which the heat of vaporization is supplied by solar energy and is extracted from the air (or water reservoir) to be cooled, respectively. A cooling tower is required to reject the sum of both energy components.[2] The coefficient of performance (C.O.P.) for a typical absorption chiller (i.e., the ratio of heat-removal capacity to required heat input) is approximately 0.7.[3]

A commercially available absorption chiller will reach its design cooling capacity at approximately 195°F input water temperature. However, an important feature should be noted: its efficiency (which should not be confused with cooling capacity) remains essentially constant down to 170°F to 180°F input temperature (see Section 3.5 and Fig. 3.2); i.e., it will also function well at these lower temperatures if the reduction in cooling capacity (about two-thirds of the rated capacity at 175°F input temperature) is of no concern. Tables 3.1 and 3.2 give specifications and performance data for two commercially available absorption chillers:[3] a 3-ton chiller and a 25-ton chiller.[4]

3.3 Sizing

Conventional air-conditioning systems are sized according to peak load conditions. For example, a building that will require an average heat removal of 240,000 Btu per day (20 ton-hr) might be equipped with a 6-ton chiller that is capable of removing 72,000 Btu per hour. This is needed because conventional air conditioners do not work from a cold storage as buffer medium and must therefore be capable of handling momentary greatest-load conditions. The fact that conventional air-conditioning tonnage is relatively inexpensive has contributed to such conservative design practices.

In the case of active solar air conditioning with absorption chillers, a heat-storage tank and a cold-storage tank are part of the system (cold storage is needed to decrease the number of on–off cycles of the chiller). Thus the chiller can work from a hot storage tank to a cold storage tank. It is activated whenever the cold tank reaches a present maximum temperature (≈55°F), and it is turned off if a preset minimum temperature (≈42°F) is reached. When the building demands cooling, the cold water is circulated through the fan coil units and is made available for cooling of the building, irrespective of the status (and of the hourly capacity) of the chiller. Thus a much smaller chiller can be selected. In the preceding example, a 3-ton chiller would be quite sufficient. It would have to operate only $20/3 \approx 7$ hr during an average day and therefore would still have ample reserves for hot spells.

The actual sizing of a solar-driven absorption-chilling air-conditioning system is more complicated than the sizing of a solar space-heating system. The parameters that must be considered in the sizing procedure include, in addition to

TABLE 3.1A Specifications for Arkla WF 36 3-Ton Absorption Chiller [5].

DESIGN DELIVERED CAPACITY, Btu/h	36,000[1]
DESIGN DELIVERED CAPACITY, Tons I.M.E.	3.0[1]
ENERGY REQUIREMENTS	
Design Hot Water Input, Btu/h	50,000
Design Hot Water Inlet Temperature, °F	195
Design Hot Water Outlet Temperature, °F	185.9
Permissible Range of Inlet Temp.	170 to 205
Design Hot Water Flow, gpm	11.0
Pressure Drop, Feet of Water, at 11 gpm	9.8
Permissible Range of Flow, gpm	5 to 22
Pressure Drop, Feet of Water, at 22 gpm	29.9
Maximum Working Pressure, psig	100
Unit Water Volume, Gallons, Approx.	3.0
Electrical Voltage, 60 Hz, 1 Phase	115[2]
Maximum Wattage Draw	250
CHILLED WATER DATA	
Design Inlet Temperature, °F	55
Design Outlet Temperature, °F	45
Design Flow, gpm	7.2
Pressure Drop, Feet of Water, at 7.2 gpm	4.6
Premissible Range of Flow, gpm	4 to 13
Pressure Drop, Feet of Water, at 13 gpm	12.5
Maximum Working Pressure, psig	100
Unit Water Volume, Gallons, Approx.	1.5
CONDENSING WATER DATA	
Design Heat Rejection, Btu/h	86,000
Design Inlet Temperature, °F	85
Design Outlet Temperature, °F	99.3
Permissible Range of Inlet Temp.	75 to 90
Design Flow, gpm	12.0
Pressure Drop, Feet of Water, at 12 gpm	9.6
Permissible Range of Flow, gpm	9 to 25
Pressure Drop, Feet of Water, at 25 gpm	33.9
Maximum Working Pressure, psig	100
Unit Water Volume, Gallons, Approx.	3.0
FOR COOLING TOWER SELECTION	
Maximum Heat Rejection, Btu/h	106,000
Range, °F	14 to 17
Minimum Permissible Sump Temperature, °F	75[3]
SERVICE CONNECTIONS	
Hot Water Inlet and Outlet	1" FPT
Chilled Water Inlet and Outlet	1" FPT
Condensing Water Inlet and Outlet	1" FPT
PHYSICAL DATA, APPROXIMATES	
Operating Weight, Pounds	675[4]
Shipping Weight, Pounds	680[5]
Crated Size, Inches	36W, 34D, 75H

NOTES:
1. Capacity at design conditions.
2. Units equipped for operation on 230V-50Hz-1Ph available on special order.
3. Thermostatic switch to control tower fan MUST be used. Set to "cut out" at 75°F.
4. Includes circulating water weights.
5. Units as shipped contain Lithium Bromide charge.

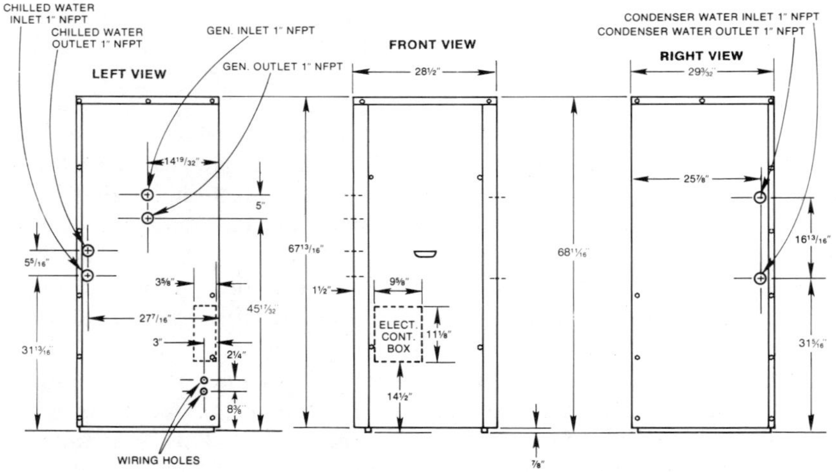

TABLE 3.1B Performance of Arkla WF 36 3-Ton Absorption Chiller as a Function of Hot-Water Inlet Temperature and Condenser Inlet Temperature [5]

Hot Water Inlet Temp. (°F)	Hot Water Outlet Temp. (°F)	Energy Input (Btu/hr)	Condenser Water Inlet Temp. (°F)	Delivered Capacity Btu/hr	Delivered Capacity Tons	Rejected Heat (Btu/hr)	C.O.P.
	167.0	16,400	80	9,700	0.81	26,100	0.59
170	167.4	14,500	85	6,400	0.53	20,900	0.44
	—	—	90	—	—	—	—
	170.7	23,800	80	17,300	1.44	41,100	0.73
175	171.1	21,600	85	13,100	1.09	34,700	0.61
	—	—	90	—	—	—	—
	174.3	31,200	80	24,400	2.03	55,600	0.78
180	174.8	28,800	85	19,400	1.62	48,200	0.67
	175.7	23,800	90	14,200	1.18	38,000	0.60
	178.0	38,400	80	31,100	2.59	69,500	0.81
185	178.5	35,900	85	25,600	2.13	61,500	0.71
	179.4	30,600	90	19,300	1.61	49,900	0.63
	181.7	45,800	80	36,800	3.07	82,600	0.80
190	182.2	42,900	85	31,300	2.61	74,200	0.73
	183.2	37,500	90	23,800	1.98	61,300	0.63
	185.3	53,100	80	40,600	3.38	93,700	0.76
195	185.9	50,000	85	36,000	3.00	86,000	0.72
	186.9	44,300	90	27,600	2.30	71,900	0.62
	189.3	58,800	80	41,800	3.48	100,600	0.71
200	189.8	56,000	85	40,200	3.35	96,200	0.72
	190.7	51,000	90	30,500	2.54	80,500	0.60
	193.4	63,800	80	42,000	3.50	105,800	0.66
205	193.9	60,800	85	42,000	3.50	102,800	0.69
	194.8	56,200	90	32,500	2.71	88.700	0.58

TABLE 3.2A Specifications for Arkla WFB 300 25-Ton Absorption Chiller [5].

DESIGN DELIVERED CAPACITY, Btu/h	306,000[1]
DESIGN DELIVERED CAPACITY, Tons I.M.E.	25.5[1]
ENERGY REQUIREMENTS	
Design Hot Water Input, Btu/h	447,000
Design Hot Water Inlet Temperature, °F	195
Design Hot Water Outlet Temperature, °F	184.8
Permissible Range of Inlet Temp.	160 to 200
Design Hot Water Flow, gpm	90
Pressure Drop, Feet of Water, at 90 gpm	20.7
Permissible Range of Flow, gpm	50 to 100
Pressure Drop, Feet of Water, at 100 gpm	25.6
Maximum Working Pressure, psig	100
Electrical Voltage, 60 Hz, 1 Phase	115[2]
Maximum Wattage Draw	150
CHILLED WATER DATA	
Design Inlet Temperature, °F	55
Design Outlet Temperature, °F	45
Design Flow, gpm	60
Pressure Drop, Feet of Water, at 60 gpm	9.8
Permissible Range of Flow, gpm	30 to 100
Pressure Drop, Feet of Water, at 100 gpm	26.9
Maximum Working Pressure, psig	100
Unit Water Volume, Gallons, Approx.	12
Fouling Factor	.0005
CONDENSING WATER DATA	
Design Heat Rejection, Btu/h	753,000
Design Inlet Temperature, °F	85
Design Outlet Temperature, °F	101.7
Permissible Range of Inlet Temp.	75 to 90
Design Flow, gpm	90
Pressure Drop, Feet of Water, at 90 gpm	22.9
Permissible Range of Flow, gpm	50 to 110
Pressure Drop, Feet of Water, at 110 gpm	33.5
Maximum Working Pressure, psig	100
Unit Water Volume, Gallons, Approx.	20
Fouling Factor	.001
FOR COOLING TOWER SELECTION	
Maximum Heat Rejection, Btu/h	853,000
Range, °F	16 to 17
Minimum Permissible Sump Temperature, °F	75[3]
SERVICE CONNECTIONS	
Hot Water Inlet and Outlet	2" FPT
Chilled Water Inlet and Outlet	2½" FPT
Condensing Water Inlet and Outlet	2½" FPT
PHYSICAL DATA, APPROXIMATES	
Operating Weight, Pounds	3,420[4]
Shipping Weight, Pounds	3,145[5]
Crated Size, Inches	114 W, 45 D, 69 H

NOTES: 1. Capacity at design conditions.
2. Units equipped for operation on 230V-50Hz-1Ph available on special order.
3. Thermostatic switch to control tower fan MUST be used. Set to "cut out" at 75°F.
4. Includes circulating water weights.
5. Units as shipped contain Lithium Bromide charge.

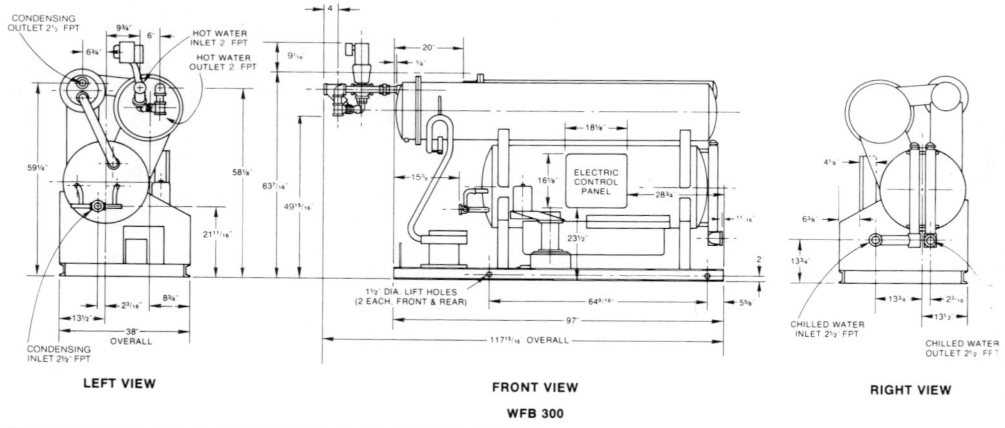

TABLE 3.2B Performance of Arkla WFB 300 25-Ton Absorption Chiller as a Function of Hot-Water Inlet, Chilled-Water, and Condensing-Water Inlet Temperatures [5]

Hot Water		Chilled Water	Condensing Water		Delivered Capacity			
Inlet Temp. (°F)	Outlet Temp. (°F)	Leaving Temp. (°F)	Inlet Temp. (°F)	Energy Input (Btu/hr)	Btu/hr	Tons	Rejected Heat (Btu/hr)	C.O.P.
165	160.3	40	80	210,000	138,000	11.5	248,000	0.66
	159.9	45	80	227,500	152,400	12.7	279,900	0.67
	161.1	40	85	172,600	98,400	8.2	271,000	0.57
	160.5	45	85	197,900	129,600	10.8	347,500	0.65
	159.9	50	85	227,600	145,000	12.1	372,600	0.64
	162.8	45	90	96,800	60,000	5.0	156,800	0.62
170	164.1	40	80	259,000	174,000	14.5	432,900	0.67
	163.8	45	80	270,900	189,600	15.8	460,500	0.70
	165.1	40	85	214,300	132,000	11.0	346,300	0.62
	164.6	45	85	240,700	166,800	13.9	407,500	0.69
	164.0	50	85	262,400	180,000	15.0	442,400	0.69
	166.9	45	90	139,100	96,000	8.0	235,100	0.69
175	168.0	40	80	306,700	206,400	17.2	513,100	0.67
	167.7	45	80	321,100	228,000	19.0	549,100	0.71
	169.1	40	85	257,000	163,200	13.6	420,200	0.64
	168.7	45	85	277,600	196,800	16.4	474,400	0.71
	168.2	50	85	297,500	213,600	17.8	511,100	0.72
	170.7	45	90	188,000	134,000	11.2	322.400	0.71
180	172.0	40	80	351,800	236,400	19.7	588,200	0.67
	171.8	45	80	357,500	259,200	21.6	616,700	0.73
	173.1	40	85	302,400	192,000	16.0	494,400	0.63
	172.9	45	85	313,800	224,400	18.7	538,200	0.72
	172.2	50	85	341,200	246,000	20.5	587,200	0.72
	174.7	45	90	234,000	168,000	14.0	402,000	0.72
185	176.0	40	80	395,700	260,400	21.7	656,100	0.66
	175.7	45	80	405,000	291,600	24.3	696,600	0.72
	176.9	40	85	352,300	218,400	18.2	570,700	0.62
	176.8	45	85	354,900	252,000	21.0	606,900	0.71
	176.3	50	85	380,700	276,000	23.0	656,700	0.72
	178.7	45	90	274,000	195,600	16.3	469,600	0.71
190	179.8	40	80	446,200	282,000	23.5	728,200	0.63
	179.7	45	80	449,800	321,600	26.8	771,400	0.71
	180.9	40	85	396,000	237,600	19.8	633,600	0.60
	180.8	45	85	400,000	276,000	23.4	676,000	0.69
	180.4	50	85	421,500	306,000	25.6	727,500	0.73
	182.5	45	90	327,900	225,600	18.8	553,400	0.69
195	183.6	40	80	500,000	300,000	25.0	800,000	0.60
	183.7	45	80	492,000	344,400	28.7	836,400	0.70
	184.7	40	85	448,700	258,000	21.5	706.700	0.57
	184.8	45	85	446,700	306,000	25.5	752,700	0.69
	184.6	50	85	451,600	328,800	27.4	780,400	0.73
	186.6	45	90	365,500	237,600	19.8	603,100	0.65

those that are relevant for solar space-heating systems (see Section 2.1), the size of the chiller and the volume of the cold-storage tank. Furthermore, the type of collectors, the necessity of augmentation of insolation by reflectors, the collector tilt angle, and the type of backup energy-supply system must be determined with more care than is usually required for stand-alone solar space-heating systems.

3.3.1 Collector Area and Tilt

One should follow the following sequence of steps:

1. Compute the building heat-gain coefficient.[5]
2. Compute the direct solar heat gain through the windows.[5]
3. Consult tables listing the average ambient temperatures at various times during the day (or "cooling degree-days"), and compute, for one day of each summer month, the conductive heat gain (Btu) per day, using a conservation-oriented desired inside temperature (e.g., 78°F–80°F).
4. Assume an average number of people occupying the building, and compute the daily heat gain due to people, using 300 Btu/hr per person.
5. Assume an average amount of electricity consumed inside the building, and compute the average daily heat gain due to lighting and appliances (1 kW-hr = 3414 Btu).
6. Make a reasonable assumption of the cooling load due to ventilation of ambient air.
7. Add the results from steps 2 to 6, divide by the chiller efficiency[6] ($\approx$0.7), and add an allowance of 15,000 Btu per day per person occupying the building for domestic hot-water energy consumption (see Section 4.1). The result is the target supply value for summer air conditioning. Note that in contrast to the procedure described in Section 2.1 for space heating, the full average air-conditioning load is the target, not some fraction of it. This is in order not to have to supplement an appreciable amount of air-conditioning energy by gas[7] and to avoid the additional installation of compressor-type auxiliary chillers.
8. Go through steps 1–5 of the sizing procedure for winter space-heating systems (Section 2.1.1) to determine the approximate winter space-heating demand.
9. Using Table H.4 of Appendix H, determine a collector area[8] that, at a given tilt, best matches the requirements established in steps 7 and 8. In case of an imbalance (between summer and winter coverage) that is not obviously eliminated by a particular choice of collector area and/or tilt, try to use reflective surfaces in front of the collectors.[9] As a rough estimate, assume that each square foot of reflector area will augment the solar collection during the proper season by 40% of the collector collection. Note that the collector tilt angle must be relatively steep when reflectors are used.[9]
10. Determine the number of collectors that will most closely approach the area established in step 9, and verify that they fit in an architecturally pleasing fashion at their proper tilt angles.

3.3.2 Refrigeration and Storage Capacities

The sum of the results for steps 2 through 6 gives the average daily air-conditioning requirement (in Btu). Divided by 12,000 Btu/ton-hr, the average daily refrigeration requirement in ton-hours is obtained. If it is assumed that 3 tons and

25 tons are the only sizes in which absorption chillers can be obtained,[10] and if proper allowance in capacity reduction is made to drive these chillers at lower input water temperatures (175°F) (see Section 3.5 and Fig. 3.2), the effective cooling capacities are 2 tons and 15 tons, respectively. Because of the cold-storage facility that should be part of every solar-driven air-conditioning system,[11] it is permissible to figure that a chiller runs anywhere from 5 to 20 hr per day. This makes it easy to select a suitable chiller (combination), even though the selection of available sizes (2 tons and 15 tons, effectively) is very limited. Table 3.3 lists suggested combinations and choices of chiller sizes for various refrigeration requirements.

Table 3.3 also lists the suggested storage volumes (hot and cold) for the various system sizes. The generation of these figures is quite involved and is subject to uncertainties stemming from differences in the assumed air-conditioning use pattern. Two cases are considered: (a) residential systems, where a constant air-conditioning load during the period 1200–2200 hours is assumed, and (b) commercial systems, with an air-conditioning requirement between 900 and 1700 hours that again is assumed constant during that period. Note that, generally, systems with chillers sized for an average of 6–10 hr of running time and used in commercial applications require the least storage-tank capacities. In fact, in some of the cases, the cold storage could be even smaller or could be deleted entirely, if one does not object to on–off cycling of the chiller. However, because chiller cycling dramatically affects the system C.O.P., it is not good practice to delete cold storage. It can also be seen from Table 3.3 that the increased expense for the next larger chiller size (e.g., from one to two 15-ton chillers for a 200-ton-hr refrigeration requirement in a commercial installation) can be more than compensated by a substantial decrease in storage requirements. It can also be seen that the storage requirements for commercial and residential systems can vary substantially. It is generally suggested that the cold-storage tank be sized too large rather than too small.

3.4 Auxiliary Air-Conditioning Supply

Although it is advisable to provide an adequate auxiliary source for air conditioning, this may not be as important as in the case of solar space heating, where a full-size conventional backup system is essential. This is mainly because prolonged periods without sunshine are rare in geographic locations where air-conditioning demands are high, and generally air-conditioning demands are highest on sunny days.

Two methods have been used to supply auxiliary air conditioning. The first takes advantage of the fact that the chiller is already installed, and it uses a gas-heated boiler to supply hot water if the solar tank is low in temperature. The second includes a separate conventional compressor-type air conditioner that is automatically activated if the temperature of the heat-storage tank is below the preset lower limit (170°F–180°F) (see Section 3.5 and Fig. 3.2), at which time the absorption chiller is turned off.

Although the first method seems to be the most natural, it is almost never the most economical. In a very few cases, when the solar collectors are sized to supply virtually 100% of the air-conditioning energy needs, and when natural gas

TABLE 3.3 Suggested Chiller and Storage Size Combinations for Various Refrigeration Requirements[a]

Daily Ton-Hours Refrigeration Requirement	Suggested Combinations of Chillers; Size A 2 tons, Size B 15 tons (Effective Capacity)	Daily Running Time of Chillers (hr)	Approximate Storage Requirements (1000 gal)				Suggested Running Time of Chillers (time of day)
			Residential Systems[b]		Commercial Systems[c]		
			Hot	Chilled[d]	Hot	Chilled[d]	
10	1A	5	0.8–1.2	0.5–0.7	0.8–1.2	0.5–0.7	12–17
12	1A	6	0.8–1.5	0.5–1.0	1.0–1.5	0.7–1.0	12–18
16	1A	8	0.8–1.5	0.5–1.0	1.2–2.0	0.8–1.5	10–18
20	1A	10	0.8–1.5	0.5–1.0	2.0–3.0	1.5–2.0	10–20
25	1A	12.5	1.0–1.5	0.7–1.0	2.0–3.5	1.5–2.5	10–23
30	1A	15	1.5–2.0	1.0–1.5	3.5–4.5	2.5–3.0	10–1
	2A[e]	7.5	1.5–2.0	1.0–1.5	1.5–2.0	1.0–1.5	10–18
35	2A	9	1.5–2.0	1.0–1.5	1.5–3.0	1.0–2.0	10–19
40	2A	10	1.5–2.0	1.0–1.5	2.5–4.0	2.0–3.0	10–20
50	2A	12.5	1.5–2.0	1.0–1.5	4.0–6.0	3.0–4.0	10–23
60	2A	15	2.0–4.0	1.5–2.5	6.0–9.0	4.0–6.0	10–1
	3A[e]	10	1.5–2.5	1.0–1.5	3.5–5.0	2.5–3.5	10–20
80	3A	13	1.5–3.0	1.0–2.0	7–10	5–7	10–23
	1B[e]	5.5	3–6	2.5–4.0	3–5	2–3	10–16
100	1B	7			1.5–3.0	1.0–2.0	10–17
120	1B	8			3.0–4.5	2.0–3.0	10–18
150	1B	10			7–11	5–8	10–20
175	1B	12			14–22	10–15	10–22
200	1B	13			20–30	15–20	10–24
	2B[e]	6.5			3.5–6.0	2.5–4.0	10–17
250	2B	8			4–9	3–6	10–18
300	2B	10			18–30	12–20	10–20

[a] Two chiller sizes available (2 and 15 ton, effective). Yazaki [46] has chillers of different capacities.
[b] Main air-conditioning requirements assumed between noon and 10:00 P.M.
[c] Main air-conditioning requirements assumed between 9:00 A.M. and 5:00 P.M.
[d] A usable temperature of 45°F to 55°F for the cold-storage reservoir is assumed. In some of the cases, cold storage is advised only in order to reduce on–off cycling of the chiller. In most cases, the sizing of the hot-storage capacity exceeds substantially the storage capacities suggested for stand-alone space-heating systems.
[e] Preferable for commercial installations.

is relatively inexpensive as compared with electricity, it may be justifiable to choose the simpler method to supply auxiliary air-conditioning energy in the form of gas-boiler-heated hot water. In all other cases, particularly when a substantial amount of auxiliary air-conditioning demand is anticipated, it pays to install a compressor-type backup air conditioner.

To illustrate this important point, the following example may be helpful. Assume that the daily air-conditioning demand is 300,000 Btu (25 tons); 50% of this amount is supplied by solar heat, and the rest must be supplied by auxiliary sources. If this is done by a compressor-type air conditioner working at a performance coefficient of 3.0, then 150,000/3/3412 = 14.7 kW-hr of electricity are needed, costing $0.87 at 6 cents per kilowatt-hour. To supply the same amount of cooling by gas (from a 60% efficient gas boiler) to the 70% efficient absorption chiller, one needs 150,000/0.6/0.7 = 356,000 Btu, costing $1.88 at a gas rate of 50 cents per therm, i.e., more than twice the equivalent electricity cost. In fact, it would then be even less expensive to supply the entire 300,000-Btu cooling load with a compressor air conditioner, without any solar heat at all.

This example also illustrates that the basis for the calculation of the payback period for a solar air-conditioning installation should be the most economical conventional alternative method. The gas saved by an absorption chiller usually does not qualify, as this will give the false impression of the payback period being too favorable.

3.5 Other Design Principles and Controls

The most important design principle for solar air-conditioning systems is to use the lowest reasonable collector fluid temperature to ensure high-efficiency collector operation. This can be achieved by the following measures:

1. Operate the chiller at the lowest reasonable temperature at which its efficiency is still high. As Fig. 3.2 shows, this temperature is 180°F for the Arkla WF 36 (nominal 3-ton chiller) and 170°F for the Arkla WFB 300 (nominal 25-ton chiller).[12] Note that further lowering of the inlet temperature decreases the efficiency of the large chiller only little (165°F is still quite acceptable), whereas the 3-ton chiller cannot be operated with reasonable efficiency below 175°F.
2. Do not use a heat exchanger between the collector loop and the heat-storage tank.[13] Figure 3.1 shows an example of a design that does not use a heat exchanger. Several options for freezing protection for designs not incorporating a heat exchanger are discussed in Section 2.2. The principle shown in Fig. 2.7 has been used especially successfully. It has the advantage that the collector operates at elevated-pressure conditions; this avoids steaming, which might otherwise occur at the temperatures needed for air-conditioning applications.
3. Operate at a collector-loop flow rate sufficiently high to keep the collector inlet–outlet temperature difference at approximately 10°F. Differentials that are too high can cause steaming at the high collection temperatures needed for operation of the chillers, and this will further decrease the collection efficiency (see Section 2.3.2, particularly footnote 34).

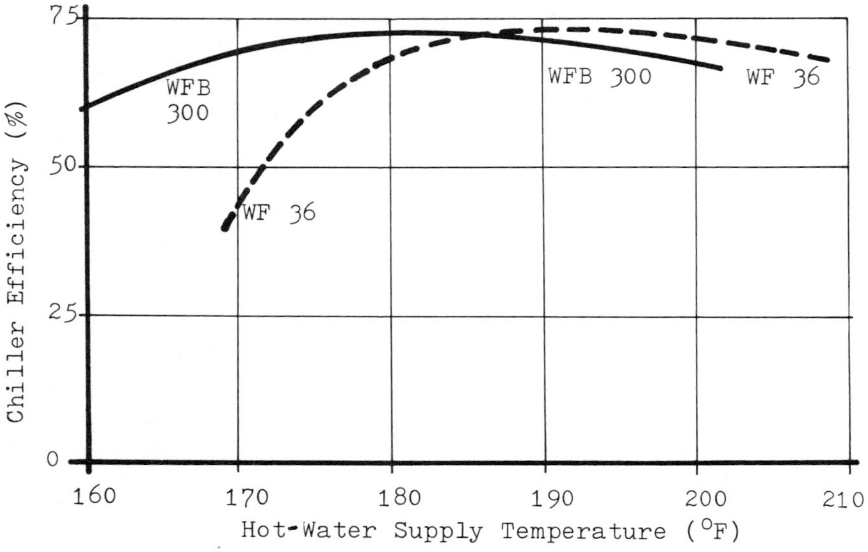

Fig. 3.2. Efficiency (ratio of delivered capacity to hot-water energy input) of two absorption chillers by Arkla (WF 36, with nominal 3-ton capacity; WFB 300, with nominal 25-ton capacity) at 85°F condensing-water inlet temperature and 45°F chilled-water outlet temperature at factory-recommended flow rates (for capacity operation) [5].

It is recommended that provision be made for efficient condensing-water cooling. The factory recommendation is between 80°F and 90°F. However, every 5°F increase (over the 80°F minimum) reduces the chiller efficiency (i.e., overall system efficiency) by approximately 10% (see Tables 3.1B and 3.2B). Therefore, it is strongly recommended that one try to cool the condensing water to 80°F, if possible,[14] and not try to avoid costs by using an inefficient cooling tower.

The insulation requirements for the heat-storage tank and the pipes are higher than in the case of solar space-heating systems. Generally, an insulation R value greater than 38 is recommended for the tank, and pipe insulation equivalent to 1.5 in. of fiberglass should be considered the minimum. The insulation material must be capable of withstanding temperatures greater than 200°F for prolonged periods of time.

For the control system, one should make sure that the number of on–off cycles of the absorption chiller is reduced to the absolute minimum, because the energy loss during each cycle is considerable. The method of obtaining a relatively low number of cycles depends on the design of the chilling capacity and the hot- and cold-storage capacities (see Section 3.4). In most cases it may be desirable to turn on the chiller with a timer at the desired daytime temperature (Table 3.3) and to turn it off whenever a thermostat at the heat-storage tank indicates that the temperature has fallen below the preset lower limit. Whenever the room thermostat calls for cooling, the stored cold water is used to meet the demand. The auxiliary cooling is activated whenever the temperature of the cold-storage tank exceeds a preset maximum temperature (typically 55°F).

3.6 Nocturnal Radiation with Solar Collectors

Under certain circumstances, solar collectors can be used effectively for cooling by black-body radiation [10]. As described in Section 1.1.3, specifically in conjunction with equation 1.16, the nocturnal radiation R_B amounts to

$$R_R = \epsilon_c \sigma (T_c^4 - T_{\text{sky}}^4) \tag{3.1}$$

In this equation, ϵ_c is the effective collector emissivity, T_c is the absorber (emitter) temperature, and T_{sky} is the sky temperature. $\sigma = 1.73 \times 10^{-9}$ [Btu-hr^{-1}-ft^{-2}-°R^{-4}] is the black-body radiation constant, with which R_B results in units of Btu-hr^{-1}-ft^{-2}.

In principle, the radiation occurs day and night and is hardly impeded by the sun, which covers only a very small fraction of the solid angle above the radiator. However, during periods of sunshine, radiation losses usually are overcompensated by solar-radiation gains.

For a solar collector to be an effective radiator, certain conditions must be met. First, the effective collector emissivity must be high. This implies that absorbers with selective coatings[15] are poor radiators. Furthermore, most glazing materials (particularly glass and fiberglass) are poor transmitters of infrared (IR) radiation; i.e., they transmit only a low fraction of the energy in the long-wavelength range in which solar collectors radiate. In practice, this means that collectors used for nocturnal radiation should be unglazed. However, some types of polymer glazing materials, such as FEP (Teflon, Table 1.4), have a low IR absorptivity and are therefore suited principally for solar collectors that are intended for use as nocturnal radiators. However, its low resistance to impact is a severe limitation on the use of Teflon in many applications. Second, the effective sky temperature must be low. For this to occur, the night sky must be clear and the relative humidity must be low (see Section 1.1.3). This condition limits the practical applicability of night-sky radiation for air conditioning to those areas with desertlike climates.

The heat loss from a glazed solar collector, L_g, during night-sky-radiation conditions can be described by the following equation:

$$L_g = R_B - a(T_a - T_c) \tag{3.2}$$

R_B is defined in equation 3.1, and a and $T_a - T_c = \Delta T$ are well known from the collector-efficiency equation 3.1. The term a is now the linear temperature heat-gain coefficient of the collector, and it is identical with the linear temperature heat-loss coefficient that will apply when the collector is operated in the solar-heat-collection mode. Thus the term $a(T_a - T_c)$ describes the reduction in the effective radiative heat loss due to convective heat gain if the temperature of the radiating collector is below the ambient temperature T_a. Equation 3.2 shows that the use of a well-built insulated collector is desirable, as long as the glazing does not impede the IR transmissivity.

If the radiating collector is unglazed, the effective heat loss L_u during night-sky-radiation conditions is

$$L_u = R_B - (a_w + 1)(T_a - T_c) \tag{3.3}$$

where a_w is the wind-velocity-dependent temperature-loss/gain coefficient, as defined in conjunction with equation 1.16 and specified in Fig. 1.16. Equation 3.3 shows that the radiating collector should be in a wind-sheltered location, which is also desirable for unglazed collectors in the solar heating mode.

Figure 3.3 indicates the amount of nocturnal radiation that can be expected of a solar collector/radiator under various conditions. The curves *A*, *B*, and *C* are for $\epsilon_c = 0.9$, $T_c = 60°F$, $a_w + 1 = 2$ Btu-hr^{-1}-ft^{-2}-°F^{-1} (or $a = 2$ Btu-hr^{-1}-ft^{-2}-°F^{-1}), and $T_a = 60°F$ (*A*), 70°F (*B*), and 80°F (*C*). The curves *D*, *E*, and *F* are computed for $\epsilon_c = 0.7$ and correspond otherwise to the curves *A*, *B*, and *C*, respectively.

In principle, nocturnal cooling can be used for air conditioning much as day-time solar heat collection is normally used for space heating (see Fig. 2.18).[16] However, such an air-conditioning system must have a considerably larger storage tank,[17] because the useful temperature differential between the charged ($T_0 \lesssim 55°F$) and discharged ($T_0 \gtrsim 65°F$) conditions is now only $\gtrsim$ 10°F. Furthermore, the heat exchanger in the duct system must be larger[18] in order to assure efficient heat transfer.[19] A natural combination of collector uses for climates that are dry and clear enough to have a low effective night-sky temperature would be pool heating and/or space heating during the daytime (to be used at night) and nocturnal cooling during the night (to be used the next day for air conditioning). In a climate that allows 50 Btu-hr^{-1}-ft^{-2} of radiative cooling, a 400-ft^2 system will pro-

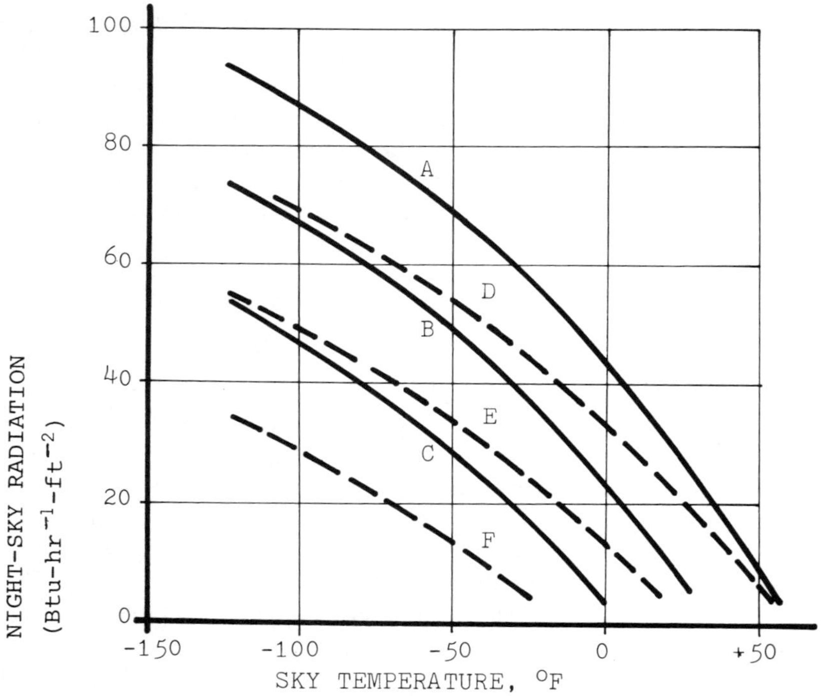

Fig. 3.3. Dependence of night-sky radiation on sky temperature. All curves are for $a = 2$ or $a_w + 1 = 2$ Btu-hr^{-1}-ft^{-2}-°F^{-1} and $T_c = 60°F$. Curves *A*, *B*, and *C* are for $\epsilon_c = 0.9$; curves *D*, *E*, and *F* are for $\epsilon_c = 0.7$; $T_a = 60°F$ for curves *A* and *D*; $T_a = 70°F$ for curves *B* and *E*; $T_a = 80°F$ for curves *C* and *F*.

duce 120,000 Btu (10 ton-hr) of cooling capacity during a 6-hr nocturnal radiation period. This amount may well suffice for air conditioning a well-insulated and well-designed single-family residence in a hot and dry summer climate. The same collector area is also adequate to heat a pool (which, because of the low ambient humidity, probably will require heating even in hot climates), and if it is mounted in a fairly sheltered location, the same (unglazed) collector system may even contribute some energy for winter space heating.

Notes

[1]Particularly, Rankine-cycle engines have been proposed to drive chillers. Their advantage may be lower input-temperature requirements (which is desirable because of increased solar collection efficiency). However, one should carefully study the performance specifications, in particular the overall C.O.P., before deciding to specify a Rankine-cycle engine.

[2]See the work of Anderson [4] for additional information on solar absorption chilling.

[3]Arkla Industries [5].

[4]The Japanese manufacturer Yazaki [46] offers absorption chillers at various capacities (1.5 tons, 2 tons, 7 tons, 10 tons at 170°F inlet-water-temperature operation). These products are quite compatible with the Arkla products. However, the manufacturer indicates effective operation of the low-capacity models at inlet temperatures as low as 160°F for 75°F condensing water temperature and 50°F chilled water outlet temperature [47]. But such performance can be extrapolated, as well, from the Arkla chiller specifications.

[5]For details, see Section 2.1.1. See Appendix J for the HP 67/97 program for heat-loss and window-gain calculations.

[6]See Tables 3.1B and 3.2B (last column); compare also Fig. 3.2.

[7]This is undesirable because of its inefficiency as compared with electric, compressor-type chillers (Section 3.3.2).

[8]Make the appropriate hot-storage-tank temperature assumption; try to get as low as possible. For further details and a more exact calculation with the HP 67/97 program, consult Appendix I.

[9]See Section 5.2.7 for details, in particular determination of the optimum reflector and collector tilt angles. Often a sawtooth arrangement is practical, with the collectors tilted 45 to 60 deg facing south and the reflectors tilted 20 to 30 deg facing north.

[10]See Tables 3.1 and 3.2. Recently, other manufacturers [46, 47] have entered the U.S. market with different sizes.

[11]It would also be desirable to use cold-storage facilities in conventional air-conditioning systems, independent of any solar-energy utilization. This would allow one to control the chillers in such a way that they would be turned off during the peak electric-demand hours, a practice that is extremely desirable from the viewpoint of electric-energy supply and conservation.

[12]Some of the smaller units of the chillers manufactured by Yazaki [46, 47] can be operated efficiently at temperatures as low as 160°F if the condensing-water temperature is below 75°F and if the chilled-water temperature is above 50°F.

[13]The penalty for a heat exchanger is usually an increase of 10°F to 20°F in the collector operating temperature (see Appendix E and Section 5.2.5).

[14]This requirement may be difficult to meet in humid climates, a circumstance that has probably contributed extensively to the disappointing results that have been achieved with some experimental solar absorption-chilling air-conditioning installations. On the other hand, the indications are that solar-driven Rankine-cycle engines are less affected by the condensing-water temperature. Yazaki [46, 47] makes chillers that are quite advantageous for use in low-humidity cooler climates, where a condensing-water temperature as low as 75°F is feasible, and where a chilled-water temperature of $>$ 50°F is permissible (see also Section 3.2).

[15]See the topic Absorptive Coatings under the subsection Materials Aspects in Section 1.1.2.

[16]A conventional air conditioner would replace the furnace, and the thermostat settings and switching functions would have to be adjusted properly.

[17]However, because storage is now at low temperature, a plastic or wooden tank may be used, and no insulation is required.

[18]As is customary in the design of air-conditioning systems.

[19]The difference between room temperature and storage-tank temperature is only about 20°F, which compares with at least 40°F for the case of solar heating.

4

Solar Water-heating Systems

In this chapter we shall deal with those solar-energy-utilization systems that are genuinely used for heating water for domestic or commercial-industrial use and are not part of solar space-heating or air-conditioning systems.[1] The following categories will be discussed: (a) solar domestic water heating, (b) solar commercial/industrial-process water heating, and (c) solar swimming-pool heating. Of these applications, solar domestic water heating (DWH) and solar pool heating have become very common; a number of companies offer packages and install such systems routinely. Solar DWH is simple and most amenable to being generalized. On the other hand, solar pool heating, which is equally common, requires much more tailoring, and its design is quite intricate in the sizing aspect. Therefore, we shall devote more attention to those aspects of solar pool heating that are related to sizing (collection efficiency of unglazed collectors, heat-loss determination for pools). Solar water-heating systems for industrial use are usually of large size and require thorough engineering. We shall discuss those aspects in which they differ from systems discussed earlier.

4.1 Domestic Water-heating Systems

4.1.1 Collector Area and Collector Tilt Angle

Solar domestic water-heating systems are typically sized at 15,000 Btu per person per day. Even if only two persons reside in a regular one-family residence, it is good practice to size the solar DWH system for a family of four, if optimum resale value of the house is of concern. Therefore, a regular solar DWH system should collect approximately 60,000 Btu on a sunny day. Using a collection of 1000 Btu per square foot of collector area per day as rule of thumb,[2] approximately 60 ft^2 of solar-collector area will be required.

Most marketed solar collectors come in two sizes: a nominal 4-ft × 8-ft size with 30 ft^2 of gross collector area and a nominal 3-ft × 7-ft size with about 20 ft^2 of gross collector area. Thus a properly sized solar DWH system should have two collectors of the 4-ft × 8-ft size or three collectors of the smaller size.

There are two important circumstances that might require adjustments in the sizing. First, in some geographic locations the ground-water temperature is significantly different from 55°F, on which the rule of 15,000 Btu per person per day was based. Second, some residences have hot-water recirculation systems for instant hot water at any time. These systems usually have comparatively high heat losses; one or two extra solar collectors, along with a larger storage tank, should be provided to take care of the heat lost in the recirculation loop.[3]

For locations within the continental United States, and for year-round domestic water-heating requirements, a tilt angle of 45 deg is advised, as can be seen from Fig. 4.1. The graph, taken from a brochure describing a 60-ft^2 solar DWH system,[4] indicates the approximate clear-day performance for a 37.5° latitude location in a mild climate (3000 degree-days per year) at three collector tilt angles. An 18.5-deg tilt, corresponding to the common 4/12 roof pitch, clearly has peak collection in the summer and registers low collection values in the winter. Even at a 30-deg tilt (7/12 roof pitch), a pronounced collection peak occurs during the summer months. The clear-day collection is approximately constant throughout the year at a 45-deg collector tilt angle (see also Fig. 5.5). However, overcast during the winter months often reduces the monthly average collection, when compared to collection during the summer months, even at steep collector tilt angles. This circumstance may at times justify the concession to an architecturally more pleasing lower collector tilt angle if it can be shown that the solar heat

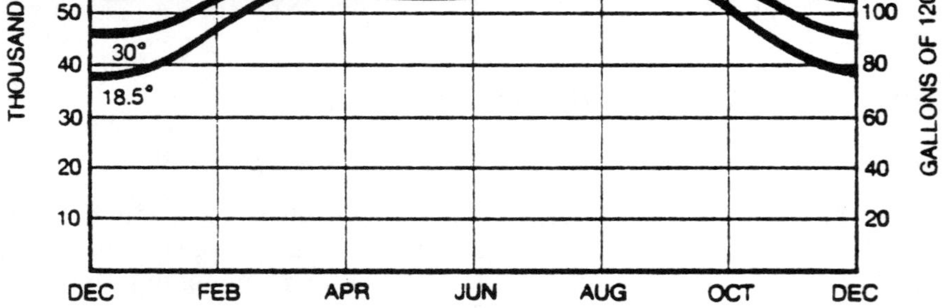

Fig. 4.1. Approximate clear-day solar heat collection with an HW-1 solar domestic water-heating system with 60 ft^2 of gross collector area and an 82-gal storage tank under typical use conditions in a central California location (37.5° N. latitude, 3000 degree-days per year). (Courtesy Alten Corporation [3].)

collected during the summer can all be utilized. In terms of total annual heat collection, any tilt angle between 14° and 45° produces roughly the same result and is, therefore, roughly equally cost-efficient in most climates in the United States.

4.1.2 Storage-Tank Size

Most manufacturers offer an 82-gal storage tank with their standard systems. At a 55°F cold-water inlet temperature, 60,000 Btu will raise 82 gal to 143°F. If some hot water is used during the daytime solar collection hours, the solar heat is effectively distributed over a larger volume, and the water temperature in the storage tank will be correspondingly lower at the end of the solar day. Therefore the volume of the solar domestic water-storage tank should be selected in accordance with (a) the daily use pattern and (b) the average water temperature desired in the evening after a sunny day. As is true for all solar systems, a lower temperature increases the solar system efficiency. Thus, one should choose a storage volume that delivers the lowest reasonable temperature. Table 4.1 can be used for quick reference. It shows the storage-tank temperature that can be expected at the end of a sunny day for various readily available storage volumes and for various amounts of domestic hot-water consumption during the daytime collection hours. Table 4.1 was computed for two constant values of total daily energy collection (60,000 and 40,000 Btu/day). The first approximates the conditions for a solar DWH system consisting of two 30-ft² collectors; the second is for a system using two 20-ft² collectors. Table 4.1 indicates, for example, that an 82-gal tank in connection with a 60-ft² collector system is the correct sizing for an average use pattern (10,000 Btu used during the daytime collection hours, 128°F end temperature at 5:00 P.M.), but a storage volume of 82 gal is definitely too small if little energy is used during the daytime and if the owner wants to maximize solar collection by keeping the collector operating temperature low, being content with 110°F to 120°F water temperature for evening consumption. On the other hand, appreciable use during the daytime hours (20,000–30,000 Btu) justifies a 60-gal tank.

TABLE 4.1 Storage-Tank Temperature[a]

DWH Consumption during Collection Hours (Btu)	Storage-Tank Temperature for 60,000 Btu at Storage-Tank Tank Volume (gal) of				Storage-Tank Temperature for 40,000 Btu at Storage-Tank Volume (gal) of			
	40	60	82	110	40	60	82	110
0	—	175	143	120	175	135	133	99
5,000	—	165	135	115	160	125	106	93
10,000	—	155	128	110	145	115	99	88
15,000	—	145	121	104	130	105	92	82
20,000	175	135	113	99	115	95	84	—
25,000	160	125	106	93	100	85	—	—
30,000	145	115	99	88	85	—	—	—
40,000	115	95	84	—	—	—	—	—

[a] Approximate storage-tank temperatures at 5:00 P.M. for 60-ft² and 40-ft² solar-collector systems (1,000 Btu/ft² per day collection assumed) after a sunny day for various storage volumes and various consumptions during the solar collection hours.

4.1.3 Freezing Protection

Figure 4.2 shows the two most common system layouts for solar domestic water heating. The arrangement on the left is used extensively in mild climates where freezing occurs only seldom and the collectors can be protected by the "hot-shot" method (see Fig. 2.6). Backup freezing protection is provided (in a small solar-collector installation) by a manually operated bypass valve around the check valve. An automatic solenoid valve is offered by some manufacturers as an option.[5] It is characteristic of this type of system that all parts, including the collectors, operate under city water pressure. This increases the boiling point to

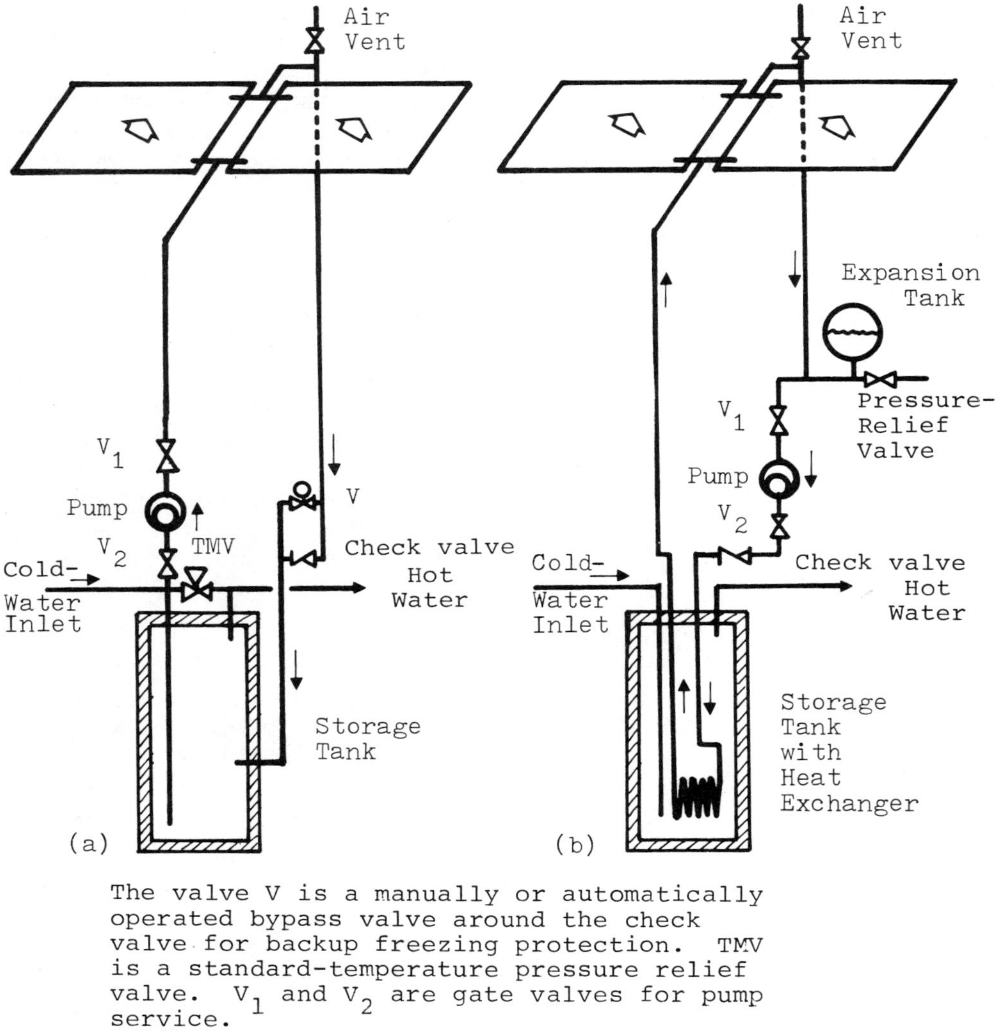

Fig. 4.2. Schematic diagrams of typical solar domestic water-heating installations: (a) for a mild climate, using "hot-shot" freezing protection; (b) for a severe climate, using antifreeze in collector loop and heat exchanger in storage tank.

such a high temperature that steaming in the collectors (which usually are single-glazed in this type of system) seldom occurs, even during prolonged periods of accidental stagnation during hot ambient-temperature conditions.

The arrangement shown in Fig. 4.2b contains a heat exchanger and allows the use of a nonfreezing and nonboiling heat-transport fluid in the collector loop.[6] The storage tank with built-in (or wraparound) double-wall heat exchanger (see Section 1.4.3) is manufactured by various boiler manufacturers.

4.1.4 Other Design Principles and Controls

A cutaway view of a typical solar DWH installation (version with heat exchanger) is shown in Fig. 4.3. Long piping runs between the collectors and the tank (or between the tank and the hot-water outlets) should be avoided (or compensated by additional collector area). Armaflex 0.75-in. insulation often is suggested by

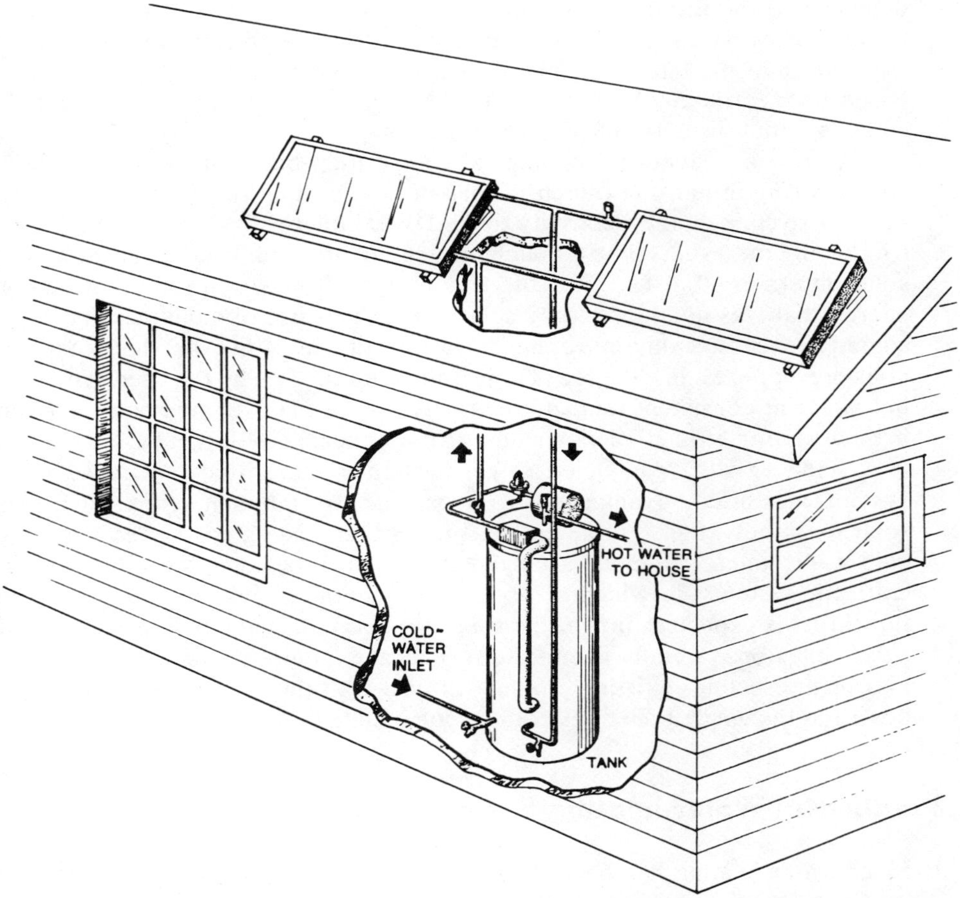

Fig. 4.3. Typical installation of a solar domestic water-heating system (with heat exchanger). (Courtesy Alten Corporation [3].)

the manufacturers and installers. If the pipe runs are longer, a higher insulation value should be considered (see Section 1.4.4).

In most cases the storage tank comes as a complete unit with pump, all valves, and the solar-loop control system. Four piping connections (indicated by arrows in Fig. 4.3) must be made to complete the system plumbing installation. The control system senses the collector and storage temperatures and activates the circulation pump whenever the collector temperature is higher. For the system shown in Fig. 4.2a (''hot-shot'' freezing protection), the pump is also activated if the collector temperature falls below a lower limit (36°F–40°F).

The height at which the solar-heated water is returned into the storage tank after a pass through the collectors is chosen by most manufacturers to be one-third to one-half of the tank height, measured from the bottom of the tank. This is an acceptable compromise for various domestic hot-water use patterns, and it allows some degree of temperature stratification. Those systems that operate at low flow rates (i.e., have higher collector output temperatures) should have the return into the tank at a higher location.

The auxiliary heat to the tank should be applied to (and its need should be detected at) the top half of the tank only. This can easily be done with electric heaters. However, gas boilers are manufactured only with the heating element at the bottom of the tank. For this reason, most marketed solar DWH systems are based on electric auxiliary heat. Their heater elements must be connected to a suitable outlet (usually 208 V, 5000 W).

In order to avoid uneconomical competition between the (relatively slow) solar-heat input and the (much faster) electric-heating recovery time, it is good practice to connect the electricity to the DWH boiler with a timer that allows electricity only for a short time (enough for heating one tank loading) at the end of the solar day (at 5:00 P.M.). Thus the electricity will make up only that amount of energy that was not supplied by solar energy because of cloudiness or was consumed in excessive daytime domestic hot-water use. However, this use of a timer does presuppose some degree of adjustment of life-style by the user, who will find hot water at consistent temperatures only during the evening hours. Section 5.4 presents parametric efficiency studies for a typical solar DWH system.

If an existing gas-fueled DWH system is retrofitted with a solar DWH system, it is often advantageous to install the two tanks in series, such that the solar tank (hot-water outlet, Fig. 4.2) feeds the existing gas boiler, replacing its cold-water inlet. Such an arrangement is advantageous for undersize solar-collector systems, in which a substantial amount of makeup heat with conventional fuel is expected, in which case gas is less expensive than electricity. If the solar-collector system is sized to fully meet the requirements, the additional heat loss of the second (existing) boiler will outweigh this advantage in makeup fuel cost, and the single-boiler system is recommended.

4.2 Industrial Water-heating Systems

Solar water heating has been demonstrated to be particularly cost-effective for such commercial and industrial applications as laundries, hotels, motels, photographic laboratories, fruit/food-processing factories, and chemical-manufacturing plants. Common to most of these applications are the following factors: low-grade heat requirements (often even below 100°F); year-round heating requirements; daytime heating hours, often at constant load, between 9:00 A.M. and 4:00 P.M.; in many cases, holding tanks of sufficient size already exist, and additional heat-storage tanks may not be necessary.

4.2.1 Collector Area and Collector Tilt Angle

One should first determine the average water-heating requirements for the particular industrial process and then determine the percentage of this load to be supplied by solar heat. As in the case of solar space heating, coverage somewhat less than 100% during a sunny day is most economical.[7] The required collector area can then be estimated by the method of Table H.3 or H.4 using an assumed average daytime ambient temperature and the desired collector operating temperature,[8] the difference between which gives the applicable ΔT value.

For many solar commercial/industrial-process water-heating systems, special consideration must be given to heat collection during weekends and holidays. If ample storage capacity can be provided, one can reduce the collector area correspondingly (to about 5/7 of the area determined earlier). If this is not possible, provision should be made for heat dissipation to protect the collectors and the service system from overheating.

The collector tilt angle should be selected in accordance with the principles established in Sections 4.1.1 and 5.4.2. In most applications in which year-round water heating is required, a 45-deg tilt will prove satisfactory.

4.2.2 Other Design Principles

Selection of the appropriate storage capacity will depend on a variety of factors, particularly the disposition of solar heat during the weekends and the pattern of heat consumption during a regular working day. Thus it is not possible to make generalized recommendations. Rather, it is suggested that the design engineer follow the principles for sizing storage capacity described in Sections 2.1.2, 3.3.2, and 4.1.2.

The larger the solar installation, the more attention should be directed to the ease and cost of installing collectors and piping connections to the collectors. A satisfactory solution is illustrated by the solar water-heating installation at the Disneyland Hotel in Anaheim, California (Fig. 4.4). A collector 4 ft × 20 ft was used, thus reducing the entire 8000-ft^2 solar-collector area to an installation of 102 collectors. Most of these collectors were directly connected to one another with 6-in. pieces of silicone hose and hose clamps. As the piping diagram of Fig. 4.5 indicates, only relatively short pipe runs were required for this installation. A total of 600 linear feet of pipe was used on the hotel roof for the entire 8000-ft^2 installation. This computes to a ratio of 1 ft of pipe per 13 ft^2 of collector area and is comparable to the ratio of 1 ft of pipe per 2–4 ft^2 of collector area for standard solar installations.[9]

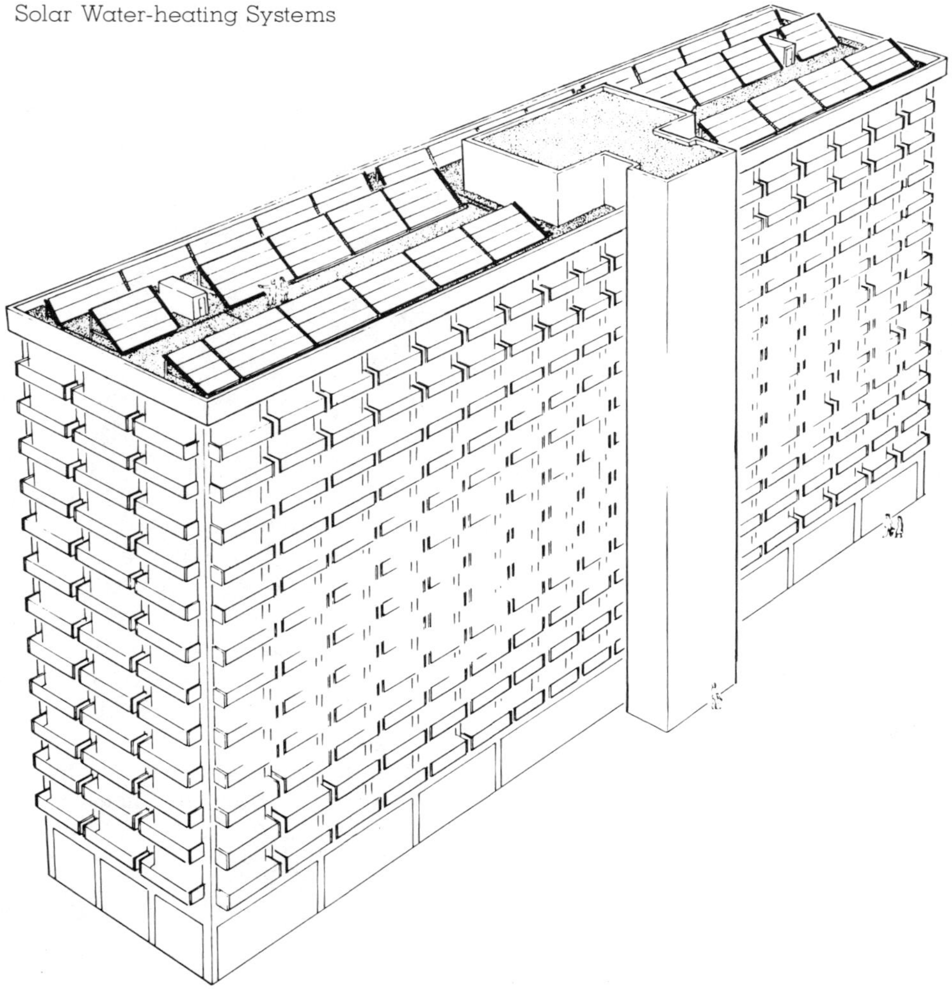

Fig. 4.4. The 8000-ft^2 solar-collector area of the Disneyland Hotel; 102 single-glazed collectors (4 ft × 20 ft) were used to reduce installation cost. (Courtesy Alten Corporation [1].)

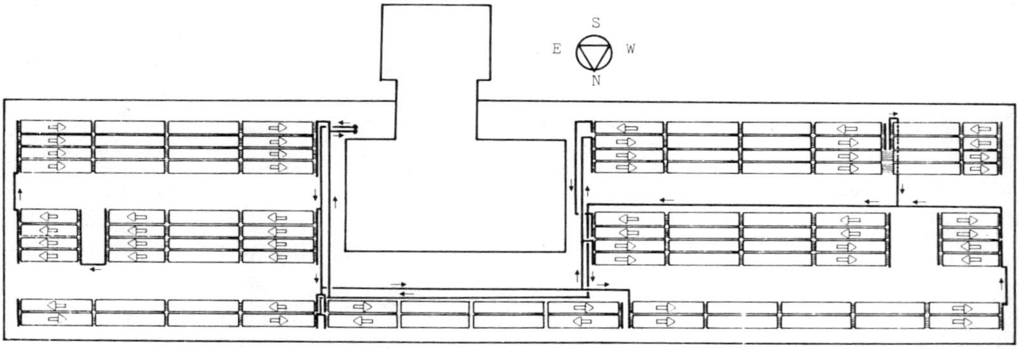

Fig. 4.5. Collector and piping layout for the 8000-ft^2 solar-collector installation at the Disneyland Hotel. Only 600 linear feet of piping were required. Open arrows indicate flow directions in the collectors; solid arrows indicate flow directions in pipes. (Courtesy Alten Corporation [1].)

Fig. 4.6. View of large solar-collector installation; 4-ft × 20-ft collectors, with minimum piping cost.

4.3 Swimming-Pool-heating Systems

4.3.1 Heat Loss from Pools

Like air-conditioning systems, swimming-pool-heating systems have traditionally been sized conservatively because an extra reserve in the Btu rating of the heater is relatively inexpensive in installed cost. For correct sizing of solar pool heaters, which often are installed in place of conventional heaters (without backup), knowledge of the heat loss from the pool is essential. Rules of thumb that relate the collector area to the pool surface area[10] are widely used, but often they are inaccurate because they do not take into account important parameters such as wind velocity, relative humidity, and pool temperature. Thus the use of a more accurate procedure to determine the solar-collector area required for a given pool is strongly advised.

The heating demand of a pool depends mainly on the following parameters: the pool surface area (A_p), the wind velocity over the pool (v), the ambient relative humidity, the ambient temperature (T_a), the pool temperature (T_p), the exposure of the pool to the sun, and the use pattern and quality of a pool cover. In agreement with a report by de Winter [16],[11] these parameters fit the following equation:

$$Q_p = A_p[\rho_p I_p - \epsilon_p \sigma(T_p{}^4 - T_a{}^4) - (a_w + 1)(T_p - T_a) - 200a_w(P_w - P_a)] \tag{4.1}$$

with

$$I_p = I_0(\cos \vartheta_0 + k_1) \qquad (4.2)^{12}$$

The absorptivity of a pool, ρ_p, is typically 0.7 to 0.85, depending on the cleanliness of the pool water and the color of the pool bottom and walls.[13] The infrared emissivity ϵ_p is usually 0.9. The wind-speed-dependent temperature-loss coefficient a_w was explained in Section 1.1.3 (Fig. 1.16). P_w is the water-vapor pressure immediately above the pool surface; P_a is the ambient water-vapor pressure. According to de Winter [16], the value of $P_w - P_a$ can, to a good approximation, be determined from Fig. 4.7. P_w is that value of the vapor pressure that results for 100% relative humidity (i.e., directly above the pool surface) at the pool water temperature (0.52 lb/in.^{2} at 80°F), and P_a equals the water-vapor pressure at the ambient conditions distant from the pool surface. The value found for P_a is in many climates nearly constant day and night; e.g., during the night at 55°F ambient temperature and 100% relative humidity, one finds P_a = 0.22 lb/in.^{2}; about the same value is obtained during typical daytime conditions of 75°F ambient temperature and 50% relative humidity. Consequently, $P_w - P_q \approx 0.30$, and the term $200a_w(P_w - P_a)$ in equation 4.1 simplifies in this example to $60a_w$.[14]

As was pointed out in Section 1.1.3, the black-body radiation into the sky often is approximated with a constant value of ≈ 25 Btu-hr^{-1}-ft^{-2} [9, 16]. It should be noted that as in the case of unglazed collectors, this value can be significantly higher in desertlike climates.

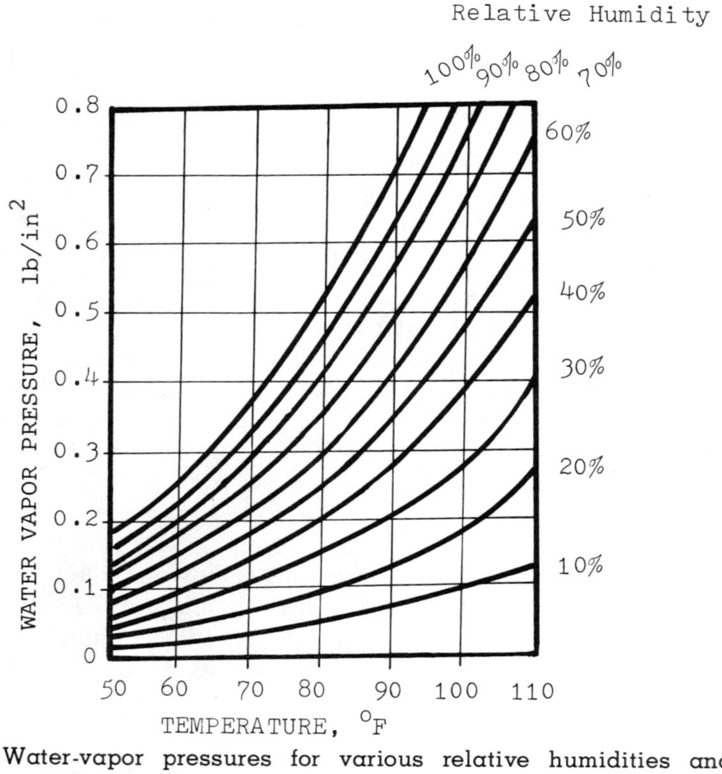

Fig. 4.7. Water-vapor pressures for various relative humidities and ambient temperatures [16].

Using the preceding assumptions for the water-vapor pressures and the radiation losses, equation 4.1 can be simplified to

$$Q_p \approx A_p[\rho_p I_p - 23 - (a_w + 1)(T_p - T_a) - 60a_w] \tag{4.3}$$

The following significant facts can be concluded from the pool heat-balance equation 4.1 or 4.3:

1. Because a_w is practically always quite high ($\gg 1.0$, see Fig. 1.16), the losses due to evaporation usually represent the most substantial heat losses. Consequently, pool blankets provide a highly efficient means of reducing heat losses from pools, as will be described later.
2. The strong increase in a_w with increases in the wind velocity v (see Fig. 1.16) causes a significant increase in the heat loss from the pool if the wind over its surface increases. As a rough approximation, an increase in wind velocity by a factor of two will increase the average losses by approximately 50% [19] (see Table 4.2 for a more detailed distribution of this effect during the year).
3. Because P_w increases strongly with increases in pool temperature, the heat losses of the pool also depend strongly on the water temperature. As a rough approximation, each 1°F increase produces a 5% increase in the heat losses [19] (see Section 4.3.2 for a more detailed assessment of this effect).
4. The direct insolation into the pool is a substantial factor offsetting its heat losses (Fig. 4.8). It is therefore appropriate to consider the pool water to be a solar collector that is 75% to 90% efficient, and it is essential to design the pool so that it will be fully exposed to the sun. It also becomes evident that any significant contribution by solar collectors must involve a collector area comparable to the surface area of the pool.
5. The heat loss from a pool is strongly dependent on the ambient relative humidity. In humid, hot climates the evaporation loss can become insignificant. Concomitantly, the black-body radiation loss into the sky is reduced, because the effective sky temperature increases with increasing relative humidity. Thus

TABLE 4.2 Heat Loss from a Swimming Pool[a]

Wind Speed (MPH)	Jan.	Feb.	Mar.	Apr.	May	June	July	Aug.	Sept.	Oct.	Nov.	Dec.
No cover												
2	2340	1960	1510	1150	860	680	610	760	1040	1520	2050	2370
5	3250	2840	2350	1970	1640	1420	1330	1480	1770	2300	2890	3270
10	4770	4300	3760	3320	2930	2660	2520	2680	2980	3600	4290	4760
Pool covered												
2	1210	840	400	60	0	0	0	0	0	440	940	1240
5	1680	1300	840	470	180	0	0	70	340	840	1370	1710
10	2480	2060	1560	1170	840	610	520	680	960	1500	2100	2490

[a] Heat losses from a pool on a sunny day in a mild climate (San Francisco Bay area) at 2-, 5-, and 10-mph wind velocities over the water surface. Pool cover (R value 0.5) applied from 1700 to 800 hours. All values are Btu per square foot of pool surface per day. Assumptions: Palo Alto, California average ambient temperature conditions (similar to those used in Fig. 5.4a); no clouds; pool exposed to sun fully and year-round; pool water temperature 80°F; wind velocity reduced by 50% during 1700–800 hours; 23 Btu-hr^{-1}-ft^{-2} constant radiation loss from pool to sky; 1 Btu-hr$^{->}$-hr^{-2}-°F^{-1} $(T_p - T_a)$ radiation loss to surroundings; 80% solar absorptivity of pool water; $P_w - P_a = 0.30$ lb/in.^{2}.

the climate has a decisive influence on the proper sizing for a solar pool-heating systems, and rules of thumb that neglect this important aspect are highly questionable.

A pool blanket has several immediate consequences on the heat balance of the pool. Using equation 4.1, the main consequences are the following:

1. Depending on the transmissivity of the cover material in the solar spectrum, the effective absorptivity of the pool, ρ_p, is reduced. It is therefore good practice to select a translucent cover material only *if the cover is also to be used during the sunshine hours*, not only during the night hours. Opaque material will eliminate a valuable source of "passive" solar heat, but it is preferable, because of its better R value, in applications in which the blanket is consistently removed during the daytime (sunshine) hours.
2. The black-body-radiation term may be significantly reduced, depending on the IR emissivity and transmissivity of the cover material.
3. The temperature-loss coefficient a_w will be reduced and less dependent on the wind velocity.
4. The evaporation-loss term will become negligible.

Table 4.2 and Fig. 4.8 include comparative calculations for the case in which a pool blanket is applied during the night hours (between 1700 and 800 hours). The significant improvement due to the use of a pool cover is clearly apparent. Note that even in the (mild) climate assumed for these calculations, it will be impractical to attempt solar heating of a pool year-round[15] unless a pool blanket can be applied in the winter months for almost 24 hrs daily, i.e., whenever the pool is not in use.

4.3.2 Choice of Collectors and General Sizing Principles

A solar pool-heating system is no exception to the general rule that the daily collection depends strongly on the collector tilt angle as well as on the type of collector. As Table H.5 shows, unglazed collectors are severely inferior to glazed collectors under high-ΔT conditions and when the wrong collector tilt angle is chosen. Qualitatively, this feature of solar pool-heating systems is summarized in Fig. 4.9. The curves represent approximate collection values under full-sunshine conditions for a solar pool-heating system located in a mild climate zone (Palo Alto, California). They were computed for a specific unglazed-collector system operated at wind velocities of 5 mph (Fig. 4.9a) and 10 mph (Fig. 4.9b). Figure 4.9c shows the respective curves for a relatively low cost glazed (fiberglass) collector system. Because the absolute values of such curves obviously depend strongly on weather conditions and on a number of other system parameters,[16] they should be used with caution, but the relative trends of the curves with respect to one another are quite representative. Figure 4.9 indicates that during the winter months (a) a steep collector tilt angle is needed to obtain adequate collection, and (b) glazed collectors are highly recommended if year-round pool heating is desired. On the other hand, if only seasonal pool heating is required, and if the collectors can be mounted in a location where they are reasonably well sheltered, unglazed collectors usually are more cost-effective because of their lower initial cost.

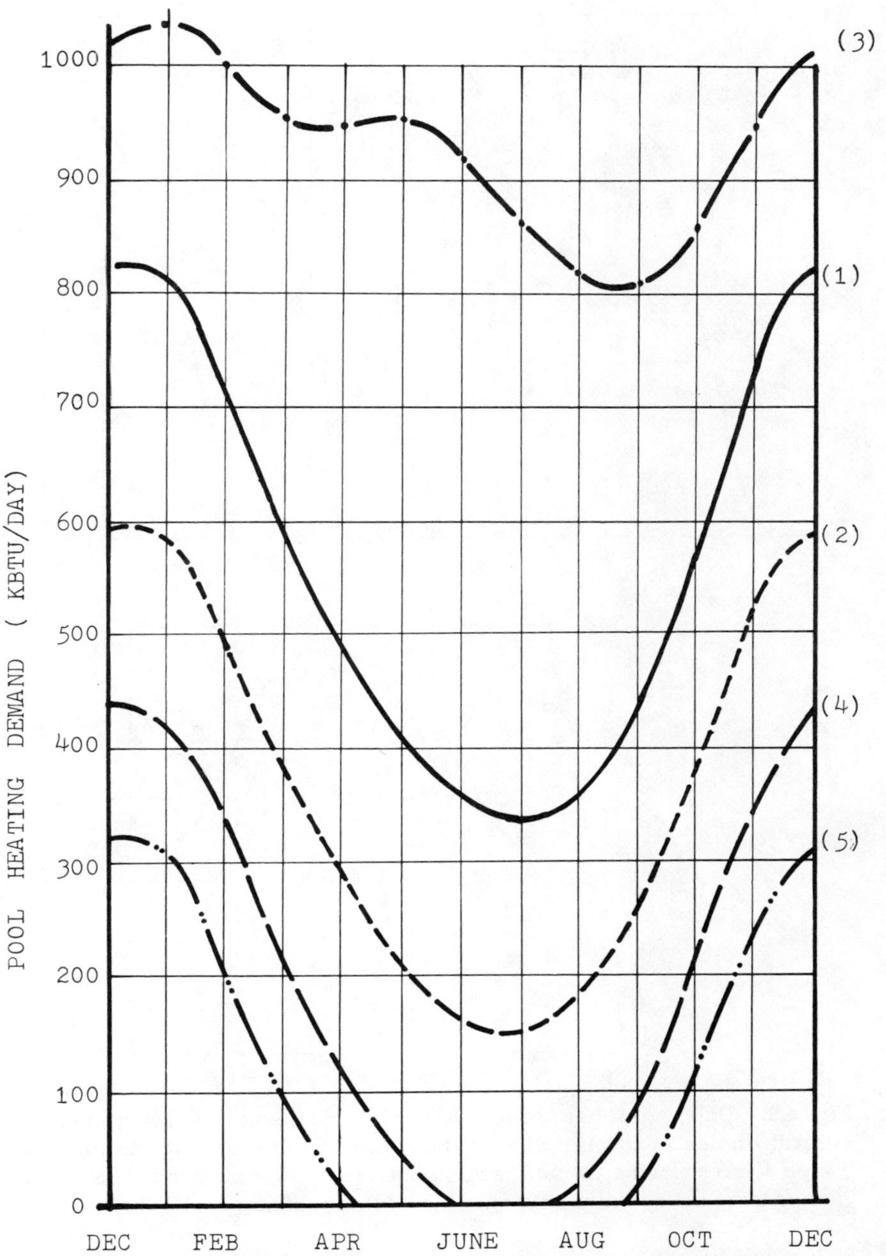

Fig. 4.8. Approximate heat-demand characteristic for a pool with 250 ft^2 of surface area in Palo Alto, California, under various conditions: Curve 1: pool in full sunshine; 5-mph wind over pool surface; sun every day. Curve 2: same as 1, but less than 2-mph wind conditions. Curve 3: pool totally shaded; 5-mph wind. Curve 4: pool covered with good material (R-0.5) during 1700 to 800 hours; 5-mph wind. Curve 5: pool covered as in curve 4, but 2-mph wind.

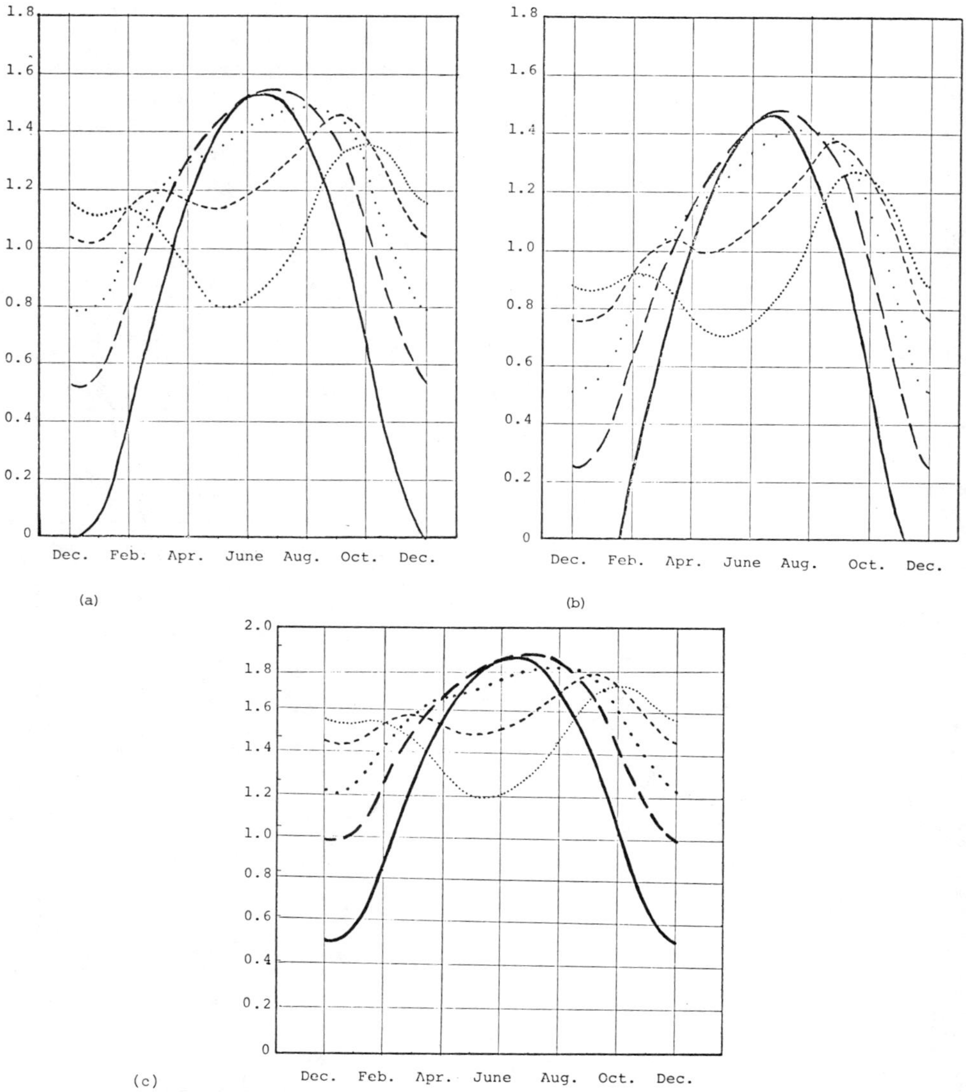

Fig. 4.9. Daily solar heat collection for a typical solar pool-heating system at various tilt angles in a mild climate (Palo Alto, California): (a) unglazed collectors at 5 mph wind velocity; (b) unglazed collectors at 10 mph wind velocity; (c) relatively inexpensive collectors with single (fiberglass) glazing.

The choice of collectors must also involve other considerations. Even if the solar pool-heating system is desired only for the swimming season (late spring through early fall), it might be appropriate to consider a combination with a space-heating system, in which case glazed collectors will be advisable rather than unglazed collectors.

Proper sizing for a solar pool-heating system usually involves the following steps:

1. Determine the season during which the pool is to be heated.
2. Determine how many hours daily a pool cover can be tolerated during the coldest months of the desired swimming season.
3. Estimate the approximate daily heating demand for the pool, per square foot of pool surface, using Table 4.2. If the climate is substantially different than the climate used to compute Table 4.2,[17] make the following approximate adjustments: (a) Add (subtract) 50 Btu/ft^2 per day heat loss for each degree that the ambient temperature is colder (warmer)[17] if the wind velocity is 2 mph or less; use 60 Btu/ft^2 per day if the wind velocity is 5 mph; use 85 Btu/ft^2 per day if the wind velocity is 10 mph (interpolate for other wind velocities). (b) Add (subtract) 100 Btu/ft^2 per day for each degree that the pool is warmer (colder) than 80°F, if the wind velocity is less than 2 mph; use 130 Btu/ft^2 per day for a 5-mph wind velocity and 200 Btu/ft^2 per day for 10 mph (interpolate for other wind velocities). (c) Subtract (add) 10 Btu/ft^2 per day for each percent by which the estimated average relative humidity during the daytime hours is higher (lower) than 50%, if the wind velocity is less than 2 mph; use 20 Btu/ft^2 per day for 5 mph and 30 Btu/ft^2 per day for 10 mph.
4. Multiply the result by the pool surface area and obtain the daily heat requirement.
5. Go through steps 1–4 for one day (usually the 21st) of each month during the heating season.
6. For each month, divide the result of step 4 by an approximate average expected collection per square foot of collector area, using Table H.3 (single-glazed collectors) or Table H.5 (unglazed collectors). Compute ΔT as the difference between the pool temperature and the average daytime ambient temperature, which can be approximated as the average maximum ambient temperature (available in many tables) minus 2°F to 4°F. The result is the recommended collector area.
7. Take into account all pertinent parameters (such as available roof orientation, roof tilt, roof area, desired length of heating season, use of excess solar heat for other purposes, flexibility in the use of the pool blanket) to come up with collector area, tilt, and azimuth specifications that will best meet one's requirements at minimum risk of generating excess (waste) solar heat, which would decrease the cost-effectiveness of the solar system.[18]

4.3.3 System Design

A typical schematic flow diagram for a residential solar pool-heating system is shown in Fig. 4.10. The bypass line with valve, indicated as optional, allows one to regulate the flow through the collectors to the desired rate.[19] This line, which generally should have the same diameter as the regular piping connection between the filter and the heater, aids in relieving the flow restrictions introduced by some types of collectors and/or collector system designs. The bypass line generally should be installed close to the top of the collector system. It will then facilitate expulsion of the air from the collectors during start-up after drain-down.[20] The bypass valve should be of the ball type, and it may be plastic.

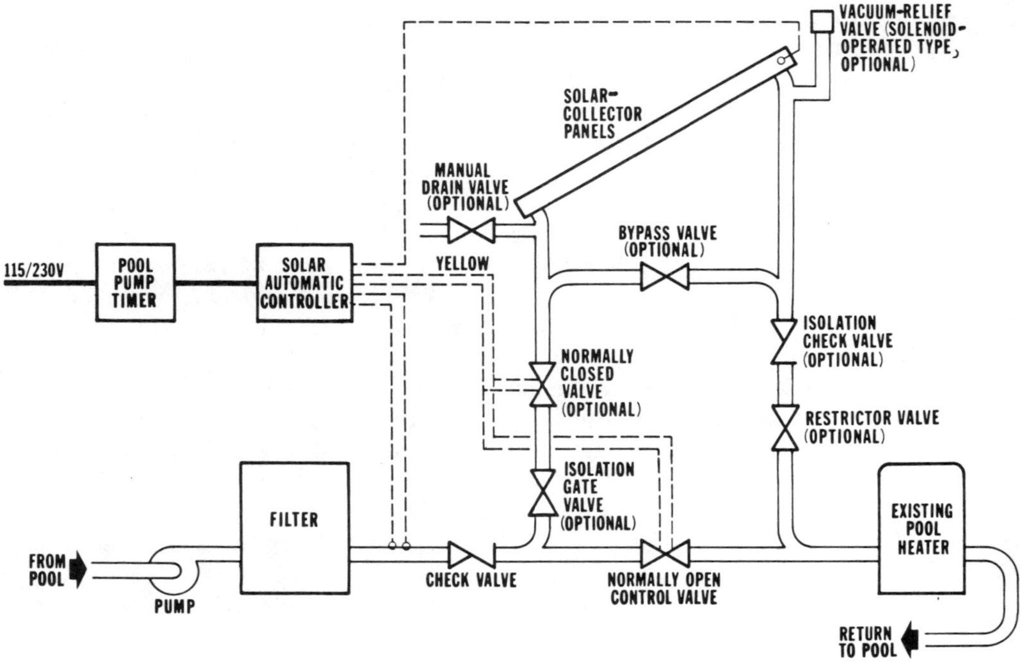

Fig. 4.10. Typical schematic flow diagram for a solar pool-heating system for residential application. "Normal" valve positions denote operation without solar collection system. (Courtesy Alten Corporation [2].)

It is good practice to make sure that the supply and return pipes between the bypass line and the collectors are made of a material that can withstand high temperatures. If the collectors are unglazed, CPVC is adequate. If the collectors are glazed, it is essential to specify copper pipes. In such cases the T connections between the bypass and the collector pipes should be specified as either CPVC or copper.

In some installations the vacuum-relief valve may have to be replaced by an air vent that does not break the vacuum.[21] This should be considered for all installations in which there is a substantial vertical height difference between the collectors and the pool (the return water will consequently pull a vacuum at the top of the collectors). Normal vacuum-breaker valves will admit air during normal operation, a circumstance that may be undesirable for reasons of corrosion and noise.

The restrictor valve in the return pipe may be used as an adjustment to eliminate a vacuum problem at the relief valve at the top of the collectors. However, at excessive heights of the collectors above the pool surface, the restrictor valve will have to be closed so much that a significant back-pressure will be generated. This will reduce the total system flow rate below the tolerable minimum and therefore is not desirable.

The optional normally closed (solenoid-operated) valve and the isolation check valve prevent the collectors from being filled with water because of system back-pressure under conditions in which the pool pump is operated but solar heat is not called for or is not available (i.e., when the normally open control valve is open). The extra expense for this option is usually well advised.

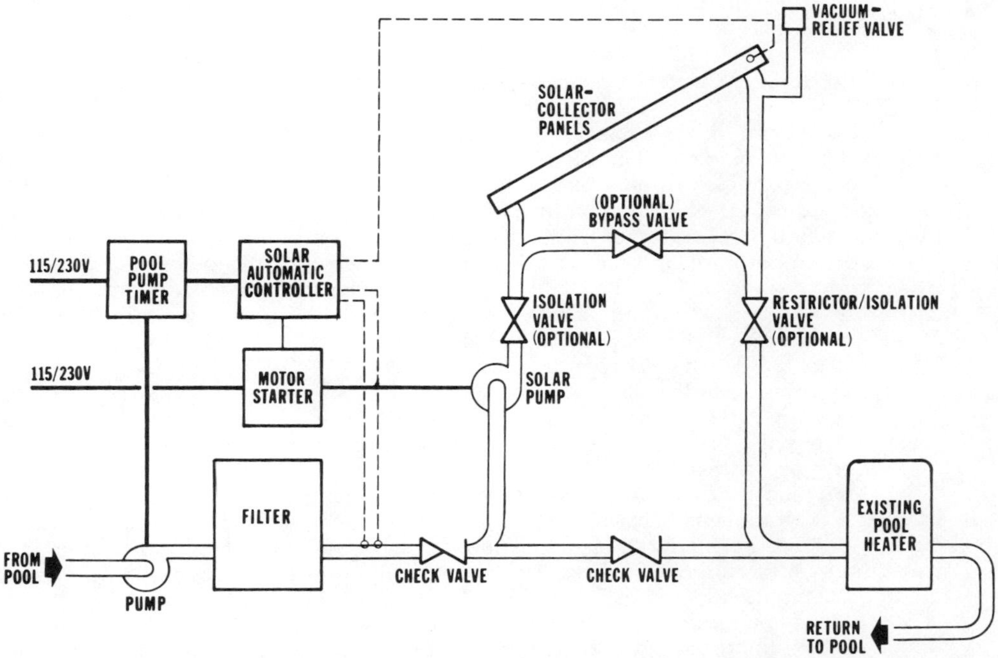

Fig. 4.11. Typical schematic flow diagram for a solar pool-heating system of larger size than is common for residential applications. (Courtesy Alten Corporation [2].)

The schematic flow diagram shown in Fig. 4.11 differs from that in Fig. 4.10 only in the use of a separate solar pump that is activated whenever solar heat is called for and available. The pump replaces the two solenoid-driven solar valves in Fig. 4.10. However, because usually a pump costs more than these valves, and because extra energy is required to activate it, this alternative is advised only for larger pool installations and for those cases in which the existing pump is under-size but otherwise functions satisfactorily.

Fig. 4.12. Typical installation of solar collectors for a community swimming pool. The low collector tilt angle and the use of unglazed metal collectors indicate that the system was designed for seasonal pool heating (May–October). (Courtesy Alten Corporation [2].)

Notes

[1]The domestic water-heating aspects of these systems are discussed in Section 2.5.

[2]See Appendix H (Tables H.3 and H.4) for more accurate average collection values and justification of this rule of thumb.

[3]As an example, the case of a recirculation system of 200 ft of 0.75-in. copper pipe, insulated with 1-in. fiberglass, may be considered (1-in. insulation is already quite thick; usually 0.5-in. insulation is installed, having even higher heat losses). At 130°F hot-water temperature and 60°F ambient temperature near the pipes, the daily heat loss computes to 1410 (Btu/100°F per 100-ft length, see Section 1.4.4) × 24 (hr/day) × (130 − 60)/100 (temperature difference) × 200/100 (length factor) ≈ 47,400 Btu. The additional installation of two solar collectors would be advised in this case.

[4]Model HW-1, Alten Corporation [3].

[5]See valve V_4 in Fig. 2.6.

[6]See also Fig. 2.3. The only difference is that for the DWH system the automatic water makeup (V_4, CV_1, CV_2, PR) and the air separator are usually deleted.

[7]See step 4 in Section 2.1.1.

[8]Many industrial/commercial-process water-heating systems have the characteristic that hot water is used quite constantly throughout the day, which reduces storage-volume requirements and helps to keep the storage-tank temperature at a fairly constant level throughout the day. This, in turn, improves the accuracy of the results obtained using the method of Table H.3 or H.4.

[9]A conventional solar installation would have required 1400 to 3400 additional linear feet of piping costing $21,000 to $51,000 more (at $15 per linear foot of installed and insulated pipe). This additional cost compares with the total expense for the solar collectors of $80,000 ($10 per square foot is assumed) and is therefore a very significant fraction of the total system cost.

[10]A ratio of collector area (A_0) to pool surface area (A_p) anywhere between 0.4 and 1.0 is being used in various locations, frequently with very little rationale.

[11]De Winter [16] included the loss term $1(T_p - T_a)$ to account for radiation losses to the surroundings other than the sky (this is the same term used in the efficiency equation for unglazed collectors, equation 1.16). It could well be argued that this term can be neglected for regular, built-in swimming pools.

[12]Compare equation 1.2 using $\vartheta \equiv \vartheta_0 = 90 - \alpha$ for pools.

[13]A consequence of this high absorptivity of pool water is that the use of a dark surface finish has relatively little influence on the thermal balance of the pool.

[14]See Table K.7 in Appendix K for numerical values for the term $200(P_w - P_a)$ under various pool-temperature and relative-humidity conditions.

[15]The collector area will have to be about four times the pool surface area, even in a sheltered location, if ≡1200 Btu/ft² per day is assumed as the collection value on a sunny day.

[16]See Appendix K for details about factors that influence the performance of solar pool heaters and for programs that can be used for reasonably accurate performance predictions. See also Appendix H, Tables H.3, H.4, and H.5, in which daily performances for glazed and unglazed collectors under various conditions are calculated for various constant ΔT values.

[17]See Fig. 5.4a for reference data concerning the ambient temperatures used for Table 4.2.

[18]A comprehensive minicomputer program for solar pool heating, performing both heat demand and solar heat-collection computations, was developed by the author for use on TRS-80 Level II Basic (16 k) computers [25]. A simplified version of this program, tailored for HP 67/97 hand-held programmable calculators, is presented in Appendix K.

[19]See Section 2.3.2 and Figs. 2.11–2.15 for details concerning determination of the correct flow conditions for various types of collector banks and interconnections.

[20]Another (often quite significant) advantage of a bypass line close to the collectors is that the danger of overheating the plastic hot-water-return pipes (a potential problem during start-up after drain-down during sunshine periods) can be reduced.

[21]Systems that use an air vent without a vacuum breaker at the top heater must have another means to allow drain-down of the collectors when the pump stops. This can be a solenoid valve or a regular vacuum breaker mounted 2–4 ft above the pool water level (see valve V_4 in Fig. 2.4).

5

Computation of Solar System Heat Collection

5.1 Principles

A variety of procedures have been proposed to compute the expected performance of a solar installation. A partial list of the most commonly used programs is presented in Appendix Table H.2. All the reported studies have at least one of the following characteristics that tend to work against realistic and reasonably accurate results:

1. The computation is not done on an hour-by-hour basis, but is averaged over days or months.
2. The computation is done with major emphasis on the collector, without consideration of the thermodynamics of the rest of the solar system and the dwelling to be heated.
3. The actual working parameters of the system (e.g., effective collector insolation, collection heat losses, building heat loss and gain, storage-tank temperature, ambient temperatures) are not all considered as variables during any given collection computation period; some are approximated with average constant values or are omitted.
4. Cloudiness is accounted for in terms of averaged insolation values rather than in terms on sunshine/overcast probabilities, an approach that will lend itself more directly to solar system performance calculations, as will be shown in this chapter.
5. The calculation requires access to a relatively large computer facility, and this restricts the general applicability of the approach.

This chapter is intended to assist those solar designers who do not have access to expensive computer facilities. It is also intended to provide the highest reasonable degree of accuracy in the computation while at the same time reducing the required climatic input to only a few data that usually are readily available for any location and microclimate. The approach involves careful analysis of all parameters that must be considered in such calculations. The results of this analysis are then integrated into a computer programming structure that can be operated conveniently and rapidly with desk-size computers[1] or (somewhat more slowly) even with small portable desk or pocket calculators like the Hewlett-Packard model 67/97[2] or Texas Instruments TI-59.

The solar system performance computation is done with reference to the temperature of the heat-storage tank. This temperature is increased because of solar collection, and it is decreased by demands for space heating, cooling, and domestic water heating, as well as system heat losses. The tank temperature changes continually during the day, depending on a variety of parameters: effective insolation into the collectors, collector area and tilt, augmentation of insolation by reflectors,[3] difference between collector temperature and ambient temperature, wind velocity over the collectors, heat loss and direct solar heat gain of the building, storage-tank volume, heat-exchanger effectiveness characteristic, piping and tank heat losses, and cloudiness. Because most of these parameters vary significantly during the day, successive computations at fixed time intervals are necessary. It has been found to be sufficient[4] to limit the number of computations to one every hour. The improvement in accuracy that would be gained by using 30-min or 15-min intervals can be shown to be insignificant when compared with the uncertainties in the available weather data that must be used for the calculations. Also, shorter time intervals would increase the computing time significantly.[5] This is, of course, of little consequence if larger computer facilities are available. However, if the computation is to be done with a relatively inexpensive and readily available programmable calculator like an HP 67/97 or a TI-59, the computing time does become significant.[6]

Typically, the collection computation is performed for the 21st day of each month[7] during the season the collector is used.[8] As will be explained in detail later, the results for each calculation will then be multiplied by the sunshine probability[9] and the number of days in the month and added to give the amount of solar heat collected annually.

5.2 Variables in Solar Efficiency Computations

5.2.1 Insolation at Normal Incidence

It is customary to use measured insolation values for solar system collection computations [29, 34]. However, here we are recommending an alternative method that does not require detailed insolation data but nevertheless ensures high accuracy, often improved accuracy. The measured insolation values are usually averaged over longer periods (e.g., months) that include periods of cloudiness. Because the insolation during cloudiness periods is almost entirely diffuse, the fraction of diffuse insolation versus direct insolation becomes quite high for such averaged periods [29] and special care must be taken to differentiate between the effects of the diffuse and direct insolation components in the computations. Often, the relationships given by Liu and Jordan [30] either directly [20] or in some modified or improved version [11, 34, 41] may lead to underestimation of the influences of collector tilt angle and azimuth on the collection result.

In typical domestic solar water-heating, space-heating, and air-conditioning applications, the solar collectors operate at moderate ΔT conditions, i.e., conditions for which the collector efficiency is already significantly affected by the collector thermal-loss coefficients (see Section 1.1.2). In this case the insolation exceeds the collector thermal losses only rarely and insignificantly during overcast periods.[10] Thus it is not only possible but also mandatory in most practical cases to consider the collection negligible during overcast periods. This reduces the calculation to a simple on/off choice for the collector fluid circulation during sunshine and overcast periods, respectively. Two major conclusions can be drawn: First, during periods of collection, the insolation is highly directional; i.e., the ratio of diffuse radiation I_1 to direct radiation I_0 is small ($\sim$ 0.08–0.15).[11] Consequently, the collection depends strongly on the azimuth orientation and on the tilt of the collector. Second, the only significant weather data that must be known for a particular area is the sunshine probability during the mean collection hours (1000 to 1400 hours for due-south-oriented collectors in the Northern Hemisphere).[12] On clear sunny days the total insolation (diffuse plus direct) varies surprisingly little from a constant value (I = 295 Btu-hr^{-1}-ft^{-2} or 930 W-m^{-2}) during the entire collection day (i.e., during that priod for any given day when the solar incidence into the collector is direct enough to ensure significant collection). This value, in turn, changes only slightly from season to season.[13] It is therefore justifiable to make solar system performance calculations using specific monthly constant insolation values at normal incidence on a clear day for near-sea-level elevations.[14] It is again emphasized that for this computational approach it is not permissible to use insolation values that are modified to account for cloudiness conditions. Instead, cloudiness is accounted for in the computation by multiplying the results by the sunshine probability (see Section 5.2.8).

A comment should be made concerning the suppression of insolation because of haziness and air pollution. In a recent study, Peterson and Flowers [35] presented data concerning the influence of air pollution on UV and total solar radiation transmission in different selected U.S. urban areas. That study found that suppression of available solar energy by haze and air pollution was not too severe and that it could be considered counterbalanced by (a) increased diffuse radiation values leading to higher actual collection values in the early morning and late afternoon hours, as compared with the calculations presented here, and (b) by some cloudiness periods that actually provide some collection, whereas our calculational approach neglects these collection amounts entirely.

5.2.2 Fraction of Insolation Intercepted by the Collector

The cosine of the angle ϑ between the sun direction and the normal to the panel surface is the single most important variable in solar collection computations. It determines the amount of insolation incident on the collectors, and it is highly dependent on collector tilt angle, azimuth orientation of the collectors with respect to true south, geographic latitude, date, and time of day.[15] This circumstance, as well as others that will be explained later, necessitates an hourly iterated computation procedure in order to establish, for any entire day, reasonable accuracy in the computation of the collected solar heat.

5.2.3 Collector Efficiency Characteristic

Knowledge of the collector efficiency parameters a, b, and ϵ (see Section 1.1.2) is important in order to compute the correct solar collection at any given difference between absorber-plate temperature and ambient temperature, ΔT. Good approximations can be made to estimate these constants if they are not known from measurements or from data sheets (see Section 1.1.2). These approximations are based on considerations involving the absorptivity of the absorber coating in the visible and near IR range (often referred to as emissivity), the transmissivity of the glazings, the fin efficiency of the absorber plate, convection losses, and the insulation quality of the collector housing [43, 44].

5.2.4 Ambient Temperature

The ambient temperature profile of the day, $T_a(t)$ should be known for accurate collection computation. However, this profile can be sufficiently well approximated if the temperatures at 6:00 A.M. and 2:00 P.M. are known. These temperatures are often listed in tables available for different locations. The lowest average temperature may be equated with the temperature at 6:00 A.M., and the highest average temperature may be equated with the temperature at 2:00 P.M. To a sufficient approximation it can then be assumed and programmed that the ambient temperature at 6:00 A.M. remains constant until 8:00 A.M., increases rapidly between 8:00 and noon, peaks at 2 P.M., and then decreases slowly and linearly until it reaches the 6:00 A.M. value at some time before midnight. This approach gives sufficient accuracy for solar heat collection computations,[16] although it requires only a few temperature data and programming steps. Specifically, the following can be assumed: $T_a = T_a$(6:00 A.M.) until 8:00 A.M.; T_a(12:00 noon) = T_a(2:00 P.M.) − 1°F; temperature rises linearly between 8:00 and noon: T_a(4:00 P.M.) = T_a(2:00 P.M.) − 2°F; temperature decreases linearly by 2°F per hour beyond 4:00 P.M. until T_a(6:00 A.M.) is reached. Figure 5.1 shows an example of such an approximated ambient temperature profile.

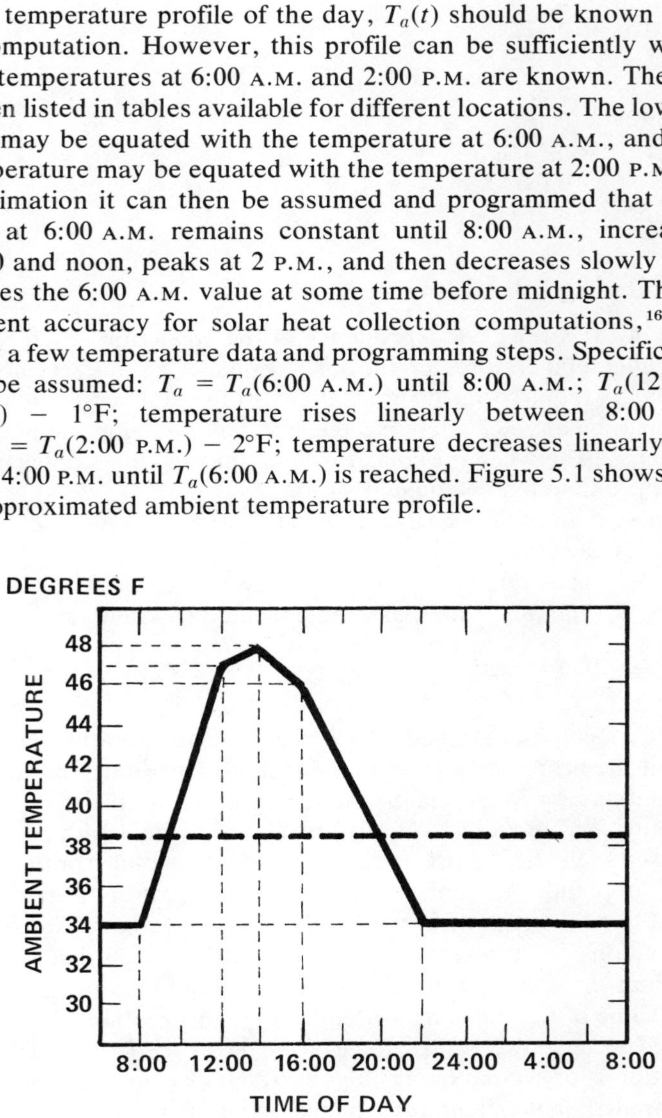

Fig. 5.1. Ambient temperature profile for T_a (6:00 A.M.) = 34°F and T_a (2:00 P.M.) = 48°F for use in solar heat collection calculations. The average ambient temperature T_a is indicated by the bold dash line.

If the average max/min ambient temperatures are not known, it is accurate enough for the purpose of solar system heat collection computation to deduce an average ambient temperature, T_a, from monthly or even annual degree-day tables. If D_m is the monthly degree-day value for a particular month (65°F base temperature), then $T_a = 65 - D_m/30$. The ambient temperature profile can be assumed as a variation around T_a, as indicated in Fig. 5.1. If only the annual degree-day value, D_a, is known, the monthly degree-day values can be derived from D_a by the following rough but acceptable approximation:

D_m(Oct.) $= D_m$(May) $= 0.05D_a$
D_m(Nov.) $= 0.12D_a$
D_m(Dec.) $= 0.18D_a$
D_m(Jan.) $= 0.2D_a$
D_m(Feb.) $= 0.16D_a$
D_m(Mar.) $= 0.13D_a$
D_m(Apr.) $= 0.09D_a$
$D_m = 0$ for summer months

5.2.5 Collector Inlet Temperature

Because of the strong ΔT dependence of the collection efficiency, knowledge of the fluid inlet temperature into the absorber plate, T_c, at every hour during the collection day is required for good accuracy in the daily collection computation. T_c can, in turn, be directly related to the storage-tank temperature T_0. If no heat exchanger is used between the collector-loop fluid and the storage medium, T_c can to a good approximation be equated with T_0.[17]

In a system with a heat exchanger,[18] the collector inlet temperature T_c is raised above the storage-tank temperature by an amount δT_c; i.e., $T_c = T_0 + \delta T_c$, whereby δT_c is a function of the heat-exchanger area A, the heat-exchanger efficiency E_x, the flow rate i, and the collected solar energy q:

$$\delta T_c = \frac{q}{E_x A} - \frac{C_0 q}{i} \tag{5.1}$$

where q/E_xA is the difference between the exit temperature of the collector-loop fluid from the heat exchanger and the mean temperature of the storage tank and C_0q/i is a decrease in ΔT due to the temperature stratification[19] that occurs normally in this type of closed-loop system (i.e., a system with a heat exchanger that is immersed in the heat-storage tank). C_0 is a constant factor that with water as the collector-loop fluid amounts to 1×10^{-3} Btu^{-1}-gal-hr-°F-min^{-1}, if q and i are given in Btu-hr^{-1} and gpm, respectively. Usually $C_0q/i << q/(E_xA)$.[20]

Equation 5.1 represents a certain simplification in that E_x is assumed to be independent of q and i; in fact, this is an acceptable approximation only if one uses the value of E_x that is measured for the most probable operating parameters. However, because the entire effect of q/E_xA in the overall daily heat collection computation is of second order only, extremely accurate values for E_x and A are not required. For the same reason, the value of q, which itself is dependent on q/E_xA and C_0q/i and must be known in equation 5.1, can be taken in an hour-by-hour computational program from the results of the iteration of the respective previous hour. If m and $m - 1$ denote the present and previous hours, respectively, equation 5.1 can be expressed in the following form for use in an hourly iterated computer program:

$$\delta T_c^{(m)} = T_c^{(m)} - T_0^{(m)} = q^{(m-1)}/E_x A - C_0 q^{(m-1)}/i \qquad (5.2)$$

It should be emphasized that because the collection efficiency strongly depends on ΔT, and because any increase in ΔT is generally undesirable for that reason, a heat exchanger should be avoided in the design whenever possible. See Appendix E for a comparison of the collection expectances for typical solar heating systems with and without heat exchangers.

5.2.6 Storage-Tank Temperature

The major function of a typical computer/calculator program designed to compute how much solar heat can be collected with a given system is to calculate the temperature of the storage tank (volume V_0) as it develops during the day. All energy spent for space heating or other uses is expressed as a decrease in $T_0(t)$, and all heat gains due to solar energy collection are temperature gains in the storage medium. The tank temperature at the end of the mth iteration, $T_0^{(m+1)}$, is given by

$$T_0^{(m+1)} = T_0^{(m)} + [q^{(m)} - W_0^{(m)} + W_1^{(m)} - W_2^{(m)}]/C_1 V_0 \qquad (5.3)$$

with $C_1 = 8.34$ Btu-gal^{-1}-°F^{-1} for water.[21] The solar energy collected, $q^{(m)}$, is computed as

$$q^{(m)} = A_0 I^{(m)} \left[\epsilon - a \frac{\Delta T - c}{I^{(m)}} - b \left(\frac{\Delta T - c}{I^{(m)}} \right)^2 \right] \qquad (5.4)$$

with

$$I^{(m)} = I_0[\cos \vartheta^{(m)} + k_1] \qquad (5.5)$$

See Section 1.1.2 for equations 5.4 and 5.5. Note that c is a constant factor that takes into account an attenuation in the ΔT-dependent collection loss in panel arrangements where the temperature directly outside the collectors is different from the general ambient temperature as listed in weather data tables. This is, for example, significant for flat roofs with solar collectors in a sawtooth arrangement, where c can assume values of 5°F to 15°F, depending on the prevailing wind velocities. A_0 is the collector area.

In equation 5.3 the function $W_0^{(m)}$ represents all heat taken from the heat-storage tank for space heating or air conditioning. The value can be computed as the product of the building heat-loss coefficient[22] and the difference between a preset inside temperature and the ambient outside temperature. $W_1^{(m)}$ is the direct solar heat gain of the building (window gain). This function should be computed with a separate subroutine (or program) using the window areas of the building and their respective azimuth orientations and considering appropriate shading factors. Entering the exact daily $W_1(t)$ profile may require an undue amount of storage registers and programming steps. For programming with an HP 67/97 calculator it is necessary,[23] instead, to enter into the solar system collection program an hourly averaged value of W_1 that is the total computed accumulated solar window gain for the entire day divided by the number of computation iterations. Good judgment must be used in entering W_1. The building layout may prevent a uniform distribution of the direct solar window heat gain, in which case appropriate reductions of W_1 must be taken. Additional internally generated heat gain, such as that due to people or appliances, should be added to W_1. The program itself must have a limitation $W_1 < |W_0^{(m)}|$, because excess window gain cannot be stored in the solar storage tank. Note that in air-conditioning applications, W_1 represents an additional heat-removal requirement and therefore must be treated as a negative heat gain.

The function $W_2^{(m)}$ in equation 5.3 represents heat losses from the solar system other than the collector losses that have already been considered, in particular piping and storage-tank heat losses. Because the magnitude of these losses is small when compared with $q^{(m)}$, it is permissible to assume a constant hourly value W_2 for these losses in order to prevent unnecessary complexity in the computation. A constant value for domestic hot-water consumption can also be included as part of W_2.

5.2.7 Reflector Augmentation[24]

The effective insolation into the solar-collector arrays can be increased by reflectors that are located in front of the collectors [22]; a detailed study of this effect was reported elsewhere [24]. If a reflector area equal to the collector area is

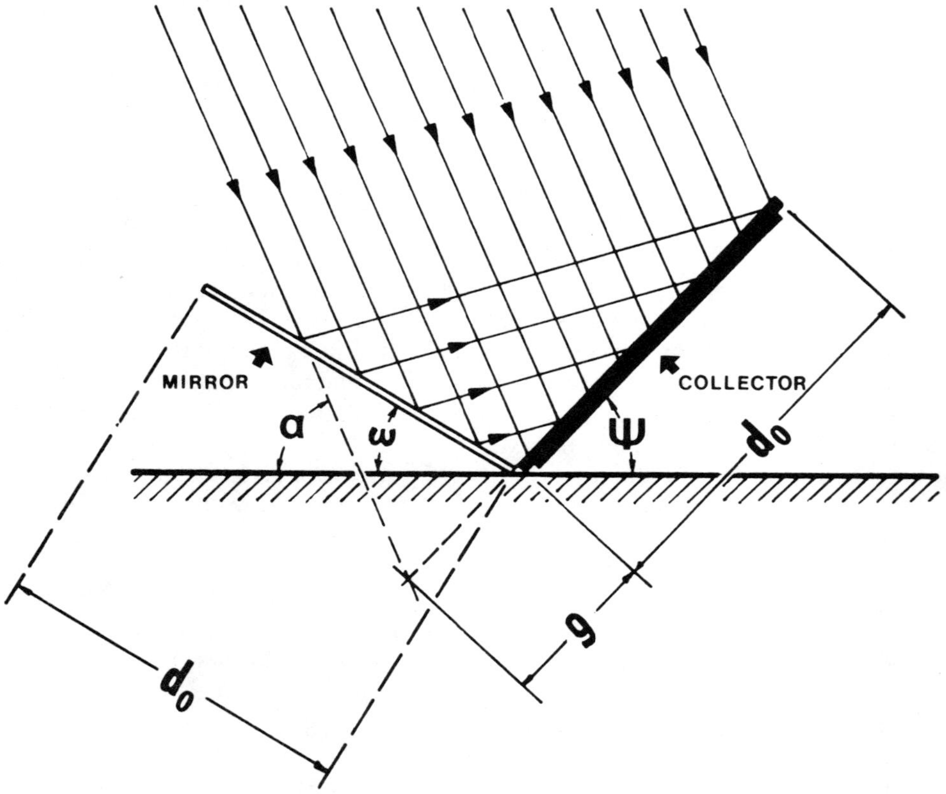

Fig. 5.2a. Explanation of parameters for the computation of solar heat collection in systems aided by reflective surfaces in front of collectors [24].

assumed, the total insolation I_2 into the collectors augmented by reflectors can be calculated as

$$I_2 = I_0(1 + gr_0)\cos\vartheta + I_1 \tag{5.6}$$

In equation 5.6, r_0 is the relativity of the reflector. I_0 and ϑ were defined earlier; g is a function of collector tilt angle ψ, reflector tilt angle ω, and sun altitude α (see Fig. 5.2a):

$$g = \begin{cases} \dfrac{\sin(\psi - \alpha + 2\omega)}{\sin(\psi + \alpha)} & \text{for} \quad \sin(\psi - \alpha + 2\omega) < \sin(\alpha - \omega) \\[2ex] \dfrac{\sin(\alpha - \omega)}{\sin(\psi + \alpha)} & \text{for} \quad \sin(\psi - \alpha + 2\omega) > \sin(\alpha - \omega) \end{cases} \tag{5.7}$$

The effective insolation for the special example of $\psi = 45$ deg and $\omega = -28$ deg for 38° N. latitude (collectors and reflectors equal sizes) is given in Fig. 5.2b for various months and times during the day. It can be seen that the reflector-augmented collection (solid curves) is substantially higher than the collection without reflector augmentation. If vertical reflectors can be positioned at the ends of the collector/reflector banks, the augmentation is virtually independent of the solar azimuth angle.

The reflector tilt angles ω can be optimized for specific applications. Reflector augmentation for solar air-conditioning systems is found to be optimal if ω is equal to the sun elevation α at noon on Dec. 21 and if the collector tilt angle ψ is 45 deg. Optimal collection for winter space-heating applications is found for horizontally placed reflectors in front of collectors that are tilted at an angle numerically equal to the geographic latitude plus approximately 30 deg.[25]

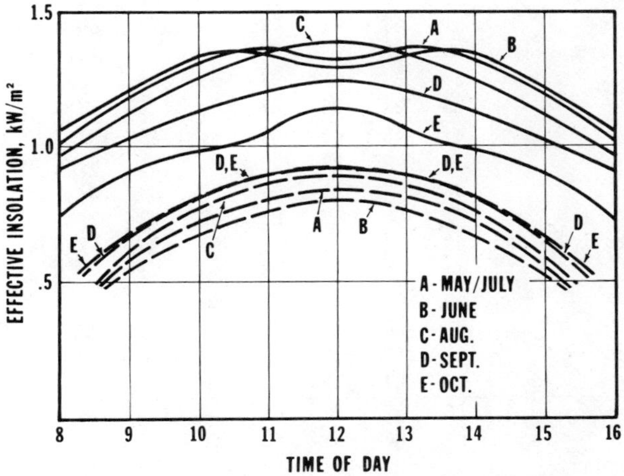

Fig. 5.2b. Effective insolation at various times of the solar day for south-oriented regular flat-plate collectors at 45-deg tilt (dash curves) and for reflector-augmented south-oriented collectors at 45-deg collector tilt and −28-deg reflector tilt (solid curves) for the 21st day of various months; calculated for 38°N. latitude and equal sizes of reflectors and collectors [24].

5.2.8 Sunshine Probability

A significant parameter for solar system collection computations is the probability of sunshine during the particular month for which the calculation is performed. This is the only way in which weather-related changes in the insolation are (and need to be) taken into account. For reasons explained in Section 5.2.1, tabulated values for daily or monthly insolation should not be used directly in the collection computation.[26]

Average sunshine probabilities usually are readily available from weather stations and airports. A convenient feature of such data (as opposed to average insolation data) is that the designer can use personal judgment directly in applying data for the particular microclimate in which the solar system is to be located. One can easily account for particular fog or cloud patterns.[27]

The mean percentages of possible sunshine are summarized in Table 5.1 for various selected locations in the United States. If for a particular location only insolation data rather than sunshine probability data are available, an easy conversion from insolation to probability values is possible. Details on this topic can be found in Appendix C.

TABLE 5.1 Mean Percentages of Possible Sunshine for Various Locations in the United States

State and Station	J	F	M	A	M	J	J	A	S	O	N	D
Ala., Birmingham	43	49	56	63	66	67	62	65	66	67	58	44
Montgomery	51	53	61	69	73	72	66	69	69	71	64	48
Alaska, Anchorage	39	46	56	58	50	51	45	39	35	32	33	29
Fairbanks	34	50	61	68	55	53	45	35	31	28	38	29
Juneau	30	32	39	37	34	35	28	30	25	18	21	18
Nome	44	45	48	53	51	48	32	26	34	35	36	30
Ariz., Phoenix	76	79	83	88	93	94	84	84	89	88	84	77
Yuma	83	87	91	94	97	99	92	91	93	93	90	83
Ark., Little Rock	44	53	57	62	67	72	71	73	71	74	58	47
Calif., Eureka	40	44	50	53	54	56	52	46	52	48	42	39
Fresno	46	63	72	83	89	94	97	97	93	87	73	47
Los Angeles	70	69	70	67	68	69	80	81	80	76	79	72
Red Bluff	50	60	65	75	79	86	95	94	89	77	64	50
Sacramento	44	57	67	76	82	90	96	95	92	82	65	44
San Diego	68	67	68	66	60	60	67	70	70	70	76	71
San Francisco	53	57	63	69	70	75	68	63	70	70	62	54
Colo., Denver	67	67	65	63	61	69	68	68	71	71	67	65
Grand Junction	58	62	64	67	71	79	76	72	77	74	67	58
Conn., Hartford	46	55	56	54	57	60	62	60	57	55	46	46
D.C., Washington	46	53	56	57	61	64	64	62	62	62	54	47
Fla., Apalachicola	59	62	62	71	77	70	64	63	62	74	66	53
Jacksonville	58	59	66	71	71	63	62	63	58	58	61	53
Key West	68	75	78	78	76	70	69	71	65	65	69	66
Miami Beach	66	72	73	73	68	62	65	67	62	62	65	65
Tampa	63	67	71	74	75	66	61	64	64	67	67	61

TABLE 5.1 Mean Percentages of Possible Sunshine for Various Locations in the United States—Continued

State and Station	J	F	M	A	M	J	J	A	S	O	N	D
Ga., Atlanta	48	53	57	65	68	68	62	63	65	67	60	47
Hawaii, Hilo	48	42	41	34	31	41	44	38	42	41	34	36
Honolulu	62	64	60	62	64	66	67	70	70	68	63	60
Lihue	48	48	48	46	51	60	58	59	67	58	51	49
Idaho, Boise	40	48	59	67	68	75	89	86	81	66	46	37
Pocatello	37	47	58	64	66	72	82	81	78	66	48	36
Ill., Cairo	46	53	59	65	71	77	82	79	75	73	56	46
Chicago	44	49	53	56	63	69	73	70	64	61	47	41
Springfield	47	51	54	58	64	69	76	72	73	64	53	45
Ind., Evansville	42	49	55	61	67	73	78	76	73	67	52	42
Fort Wayne	38	44	51	55	62	69	74	69	64	58	41	38
Indianapolis	41	47	49	55	62	68	74	70	68	64	48	39
Iowa, Des Moines	56	56	56	59	62	66	75	70	64	64	53	48
Dubuque	48	52	52	58	60	63	73	67	61	55	44	40
Sioux City	55	58	58	59	63	67	75	72	67	65	53	50
Kans., Concordia	60	60	62	63	65	73	79	76	72	70	64	58
Dodge City	67	66	68	68	68	74	78	78	76	75	70	67
Wichita	61	63	64	64	66	73	80	77	73	69	67	59
Ky., Louisville	41	47	52	57	64	68	72	69	68	64	51	39
La., New Orleans	49	50	57	63	66	64	58	60	64	70	60	46
Shreveport	48	54	58	60	69	78	79	80	79	77	65	60
Maine, Eastport	45	51	52	52	51	53	55	57	53	50	37	40
Mass., Boston	47	56	57	56	59	62	64	63	61	58	48	48
Mich., Alpena	29	43	52	56	59	64	70	64	52	44	24	22
Detroit	34	42	48	52	58	65	69	66	61	54	35	29
Grand Rapids	26	37	48	54	60	66	72	67	58	50	31	22
Marquette	31	49	47	52	53	56	63	57	47	38	24	24
Sau. St. Marie	28	44	50	54	54	59	63	58	45	36	21	22
Minn., Duluth	47	55	60	58	58	60	68	63	53	47	36	40
Minneapolis	49	54	55	57	60	64	72	69	60	54	40	40
Miss., Vicksburg	46	50	57	64	60	73	69	72	74	71	60	45
Mo., Kansas City	55	57	59	60	64	70	76	73	70	67	59	52
St. Louis	48	49	56	59	64	68	72	68	67	65	54	44
Springfield	48	54	57	60	63	69	77	72	71	65	58	48
Mont., Havre	49	58	61	63	63	65	78	75	64	57	48	46
Helena	46	55	58	60	59	63	77	74	63	57	48	43
Kalispell	28	40	49	57	58	60	77	73	61	50	28	20
Nebr., Lincoln	57	59	60	60	63	69	76	71	67	66	59	55
North Platte	63	63	64	62	64	72	78	74	72	70	62	58
Nevada, Ely	61	64	68	65	67	79	79	81	81	73	67	62
Las Vegas	74	77	78	82	85	91	84	85	92	84	83	75
Reno	59	64	69	75	77	82	90	89	86	76	68	56
Winnemucca	52	60	64	70	76	83	90	90	86	75	62	53
N.H., Concord	48	53	55	53	51	56	57	58	55	50	43	43
N.J., Atlantic City	51	57	58	59	62	65	67	66	64	54	58	52
N.M., Albuquerque	70	72	72	76	79	84	76	75	81	80	79	70
Roswell	69	72	75	77	76	80	76	75	74	74	74	69
N.Y., Albany	43	51	53	53	57	62	63	61	58	54	39	38
Binghamton	31	39	41	44	50	56	54	52	47	43	29	26
Buffalo	32	41	49	51	59	67	70	67	60	51	31	28
Canton	37	47	50	48	54	61	63	61	54	45	30	31
New York City	49	56	57	59	62	65	66	64	64	61	53	50
Syracuse	31	38	45	50	58	64	67	63	56	47	29	26
N.C., Asheville	48	53	56	61	64	63	59	59	62	64	59	48

TABLE 5.1 Mean Percentages of Possible Sunshine for Various Locations in the United States—Continued

State and Station	J	F	M	A	M	J	J	A	S	O	N	D
Raleigh	50	56	59	64	67	65	62	62	63	64	62	52
N. Dak., Bismarck	52	58	56	57	58	61	73	69	62	59	49	48
Devils Lake	54	60	59	60	59	62	71	67	59	56	44	45
Fargo	47	55	56	58	62	63	73	69	60	57	39	46
Williston	51	59	60	63	66	66	78	75	65	60	48	48
Ohio, Cincinnati	41	46	52	56	62	69	72	68	68	60	46	39
Cleveland	29	36	45	52	61	67	71	68	62	54	32	25
Columbus	36	44	49	54	63	68	71	68	66	60	44	35
Ok., Oklahoma City	57	69	63	64	65	74	78	78	74	68	64	57
Oregon, Baker	41	49	56	61	63	67	83	81	74	62	46	37
Portland	27	34	41	49	52	55	70	65	55	42	28	23
Roseburg	24	32	40	51	57	59	79	77	68	42	28	18
Pa., Harrisburg	43	52	55	57	61	65	68	63	62	58	47	43
Philadelphia	45	56	57	58	61	62	64	61	62	61	53	49
Pittsburgh	32	39	45	50	57	62	64	61	62	54	39	30
R.I., Block Island	45	54	47	56	58	60	62	62	60	59	50	44
S.C., Charleston	58	60	65	72	73	70	66	66	67	68	68	57
Columbia	53	57	62	68	69	68	63	65	64	68	64	51
S. Dak., Huron	55	62	60	62	65	68	76	72	66	61	52	49
Rapid City	58	62	63	62	61	66	73	73	69	66	58	54
Tenn., Knoxville	42	49	53	59	64	66	64	59	64	64	53	41
Memphis	44	51	57	64	68	74	73	74	70	69	58	45
Nashville	42	47	54	60	65	69	69	68	69	65	55	42
Tx., Abilene	64	68	73	66	73	86	83	85	73	71	72	66
Amarillo	71	71	75	75	75	82	81	81	79	76	76	70
Austin	46	50	57	60	62	72	76	79	70	70	57	49
Brownsville	44	49	51	57	65	73	78	78	67	70	54	44
Del Rio	53	55	61	63	60	66	75	80	69	66	58	52
El Paso	74	77	81	85	87	87	78	78	80	82	80	73
Fort Worth	56	57	65	66	67	73	78	78	74	70	63	58
Galveston	50	50	55	61	69	78	72	71	70	74	62	49
San Antonio	48	51	56	58	60	69	74	75	69	67	55	49
Utah, Salt Lake City	48	53	61	68	73	78	82	82	84	73	56	49
Vt., Burlington	34	43	48	47	53	59	62	59	51	43	25	24
Va., Norfolk	50	57	60	62	67	66	66	66	63	64	60	51
Richmond	49	55	59	63	67	66	65	62	63	64	58	50
Wash., North Head	28	37	42	48	48	48	50	46	48	41	31	27
Seattle	27	34	42	48	53	48	62	56	53	36	28	24
Spokane	26	41	53	63	64	68	82	79	68	53	28	22
Tatoosh Island	26	36	39	45	47	46	48	44	47	38	26	23
Walla Walla	24	35	52	63	67	72	86	84	72	59	33	20
Yakima	34	49	62	70	62	74	86	86	74	61	38	29
W. Va., Elkins	33	37	42	47	55	55	56	53	55	51	41	33
Parkersburg	30	36	42	49	56	69	63	60	60	53	37	29
Wisc., Green Bay	44	51	55	56	58	64	70	65	58	52	40	40
Madison	44	49	52	53	58	64	70	66	60	56	41	38
Milwaukee	44	48	53	56	60	65	63	67	62	56	44	39
Wyo., Cheyenne	65	66	64	61	59	68	70	68	69	69	65	63
Lander	66	70	72	66	65	74	76	75	72	67	61	62
Sheridan	56	61	62	61	61	67	76	74	67	60	53	52
Yellowstone Park	39	51	55	57	56	63	73	71	65	57	45	38
P. Rico, San Juan	64	69	71	66	59	61	65	67	61	63	63	65

Source: U.S. Weather Service.

5.3 Typical Block Diagram for Computation

A block diagram for computation of daily solar heat collection is shown in Fig. 5.3. It contains the system parameters[28] discussed earlier, and it allows computation, in n intervals per hour, of the storage-tank temperature and the total accumulated solar system heat collection. At the end of the computation for one day, the amounts of solar heat collected and used are calculated, as well as solar system coverage (collected solar heat divided by consumed heat). The following data must be entered into a program of this nature:

Geographic latitude (γ);

Date (as declination angle, β, or as the number of the day in the year);

Collector tilt angle (ψ);

Collector azimuth (ϕ);

Collector area (A_0);

Collector efficiency parameters (ϵ, a, b);

Storage-tank temperature at 8:00 A.M. [T_0(8:00)];

Heat-exchanger data (A, E_x, C_0, i);

Ambient temperatures at 6:00 A.M. and 2:00 P.M.;

Building heat-loss/gain coefficient (C_B);[29]

Insolation, direct component (I_0);

Insolation, diffuse component ($I_1 = k_1 I_0$);

Temperature difference between ambient near collectors and ambient distant from collectors (c);

Average hourly heat gain due to direct window insolation (W_1);

Average hourly heat gain due to people and appliances;

Storage-tank volume (V_0);

Storage medium (type);

System heat losses, hourly average estimate (W_2);

Chiller efficiency (if air conditioning is included);

Average hourly heat consumption for DWH;

Sunshine probability.

An example of such a program is given in Appendix I. It is written for a Hewlett-Packard 67/97 programmable calculator. Appendix I contains the program listing, the register loading instructions, and special recommendations for use, as well as examples. Although this program is fairly simple and does not require large expensive computer facilities, it is perfectly suited for collection computations for a wide variety of custom-designed systems [24].[30]

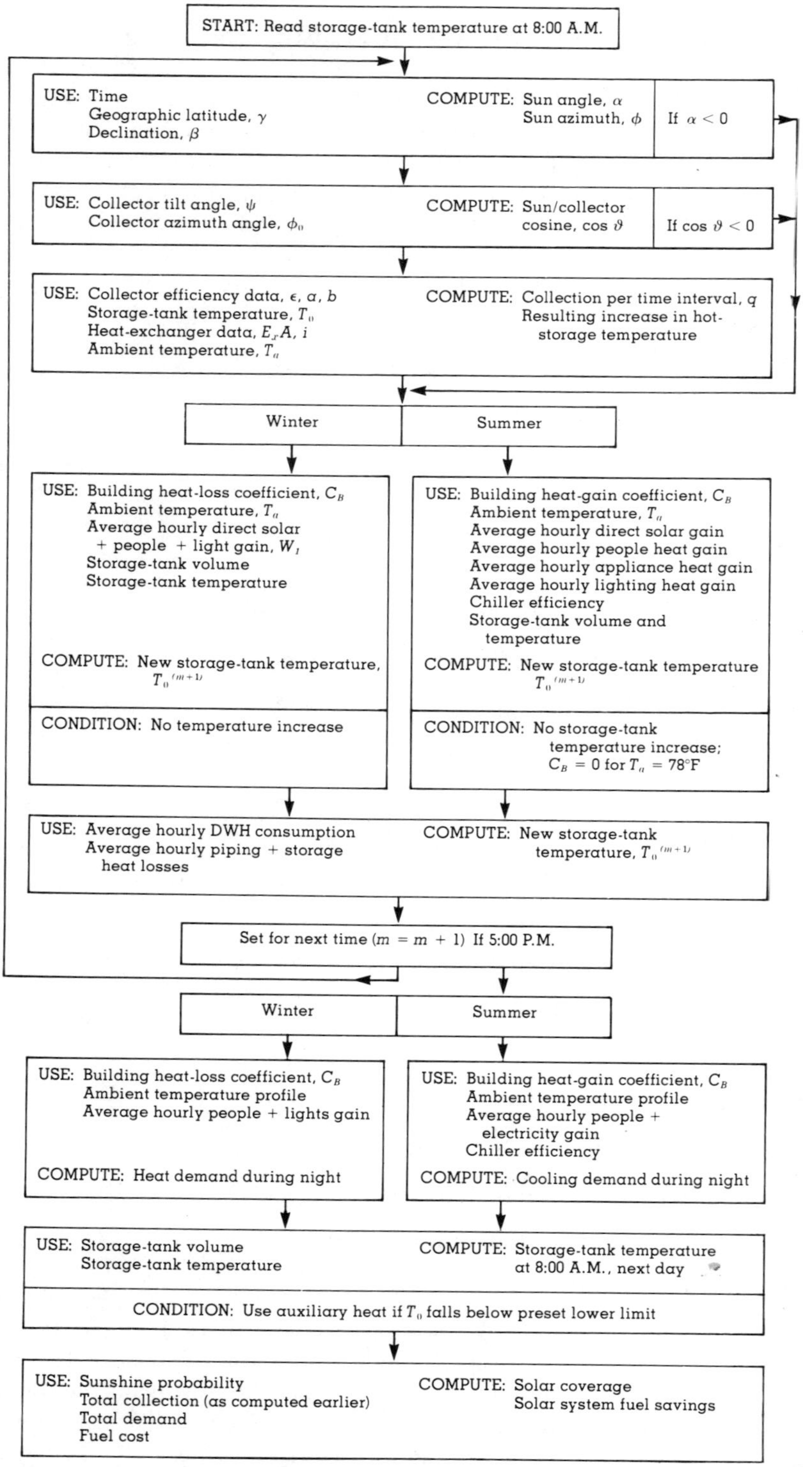

Fig. 5.3. Block diagram for solar system collection computation for any given day. Not included are reflective surfaces, collector shading, and incident-angle modifier.

5.4 Parametric Studies

In designing an efficient solar heat-collection system, it is important to know how variations in specific system parameters, such as collector tilt or azimuth, can influence system collection. The parametric studies presented here were performed for a solar domestic water-heating system. However, this does not significantly restrict the generality of the results. In all cases it was assumed that the system consisted of a 56-ft^2 net collector area[31] and a heat-storage tank of 82 gal volume and that 500-Btu/hr average piping and storage heat losses occurred during the 9-hr interval between 800 and 1700 hours. Furthermore, a constant domestic hot-water consumption of 1500 Btu/hr was assumed between 800 and 1700 hours. It was also assumed that during the evening and night hours enough heat (but no more) was extracted from the solar storage tank to reduce its mean temperature to 80°F by 8:00 A.M. the following morning. In all cases the total collected heat (including the 9 × 500-Btu/day heat losses) was calculated. The Palo Alto, California weather conditions were chosen, except for those cases in which the ambient temperature itself was the parameter. In all cases the clear-day collection was computed.

5.4.1 Variation in Collector Efficiency Parameters

The dependence of the clear-day system collection on the collector efficiency characteristic, i.e., on the absorption/transmission parameter ϵ, and on the thermal-loss constants a and b is shown in Fig. 5.4. Five types of collectors were considered: a good version and a poor version each of single- and double-glazed collectors and an unglazed collector operated under average wind velocity conditions of 10 mph. The calculations were performed for two climate zones: Figure 5.4a shows the results for a mild climate (Palo Alto, California, 3050 degree-days per year); Fig. 5.4b was computed for a more severe climate (6500 degree-days per year), which also necessitated the use of a heat exchanger. The exact temperature data entered for the computations are also listed. It is noteworthy that in both climate zones the high-quality single-glazed collector outperforms the lower-quality double-glazed collector. The unglazed collectors supply, on an annual basis, about only half as much as the single-glazed collectors.

5.4.2 Variation in Collector Tilt Angle and Geographic Latitude

The parameter that most significantly affects the solar system collection is the collector tilt angle. The results are shown in Fig. 5.5. They clearly indicate that steep collector tilt angles of 45 deg are required if solar water heating with year-round even effectiveness is desired.[32] For space heating only, with emphasis on winter effectiveness and little collection in the summer (to avoid overheating), even steeper tilt angles (60 deg) are advised.

It is of interest to further examine this collector tilt influence for systems located at different geographic latitudes. The results are shown in Fig. 5.6. Although the use of the same temperature profile over the entire latitude range between 0° and 70° does not conform to reality, the results clearly indicate (a) that turning to steeper collector tilt angles for more northern geographic latitudes is required for increased average annual collection and (b) that the expected

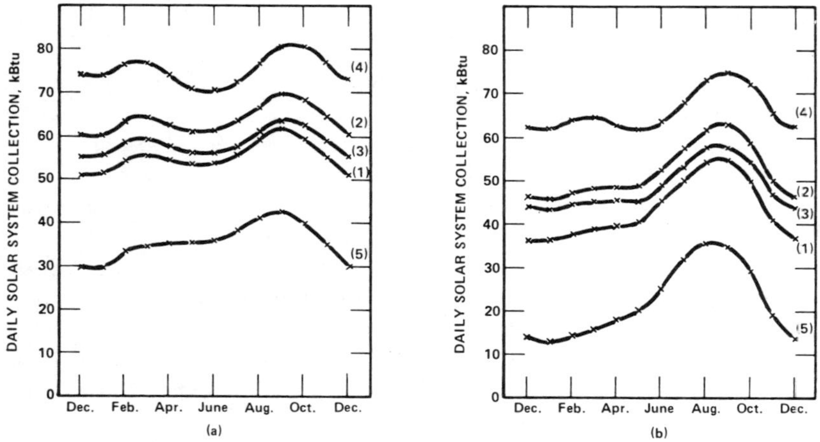

	Ambient Temperatures (°F)	Jan.	Feb.	Mar.	Apr.	May	June	July	Aug.	Sept.	Oct.	Nov.	Dec.
(a)	Ta (8:00)	37	40	43	45	48	51	53	53	51	48	42	39
	Ta (12:00)	55	59	63	67	70	74	76	75	75	70	64	56
	Ta (14:00)	56	60	64	68	72	76	78	77	77	72	65	57
(b)	Ta (8:00)	26	27	29	34	40	49	60	62	54	42	34	27
	Ta (12:00)	33	34	38	46	56	66	75	77	74	63	44	35
	Ta (14:00)	34	35	40	48	58	69	78	80	76	65	46	37

Fig. 5.4. Dependence of clear-day system collection on the collector efficiency characteristic under otherwise identical conditions for a typical solar domestic water-heating system: (a) for a mild climate zone; (b) for a severe climate zone requiring a heat exchanger (400 Btu-hr^{-1}-°F^{-1} heat-exchange capacity); 56 ft^2, net absorber area; 37.5° N. latitude. Curves 1: $\epsilon = 0.79$; $a = 1.4$; $b = 1.2$ (poor single-glazed collector). Curves 2: $\epsilon = 0.84$; $a = 1.15$; $b = 1.0$ (good single-glazed collector). Curves 3: $\epsilon = 0.66$; $a = 0.75$; $b = 1.1$ (poor double-glazed collector). Curves 4: $\epsilon = 0.79$; $a = 0.55$; $b = 0.8$ (good double-glazed collector). Curves 5: $\epsilon = 0.90$; 10 mph average wind velocity (unglazed collector). See Section 1.1.3 for efficiency calculation with unglazed collectors.

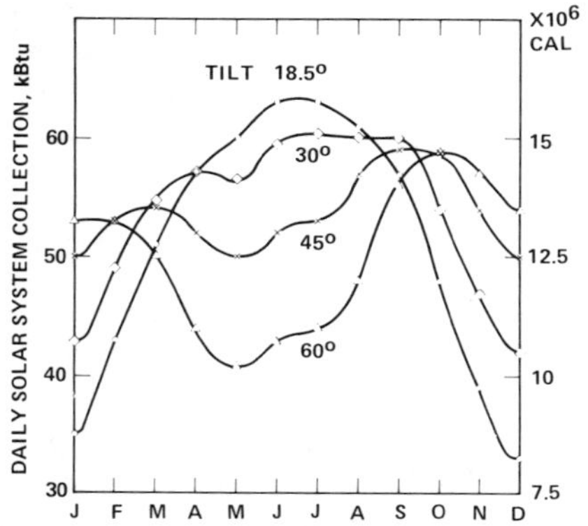

Fig. 5.5. Dependence of clear-day system collection on collector tilt angle under otherwise identical conditions for a typical solar domestic water-heating system: $\epsilon = 0.84$; $a = 1.15$ Btu-hr^{-1}-°F^{-1}-ft^{-2}; $b = 1.0$ Btu^2-hr^{-2}-°F^{-2}-ft^{-4}; 56 ft^2 net absorber area, T_c(8:00 A.M.) = 80°F; 1500 Btu/hr average hot-water energy consumption; 500 Btu/hr average piping and storage heat losses; no heat exchanger; 37.5° N. latitude; ambient temperatures same as in Fig. 5.4a.

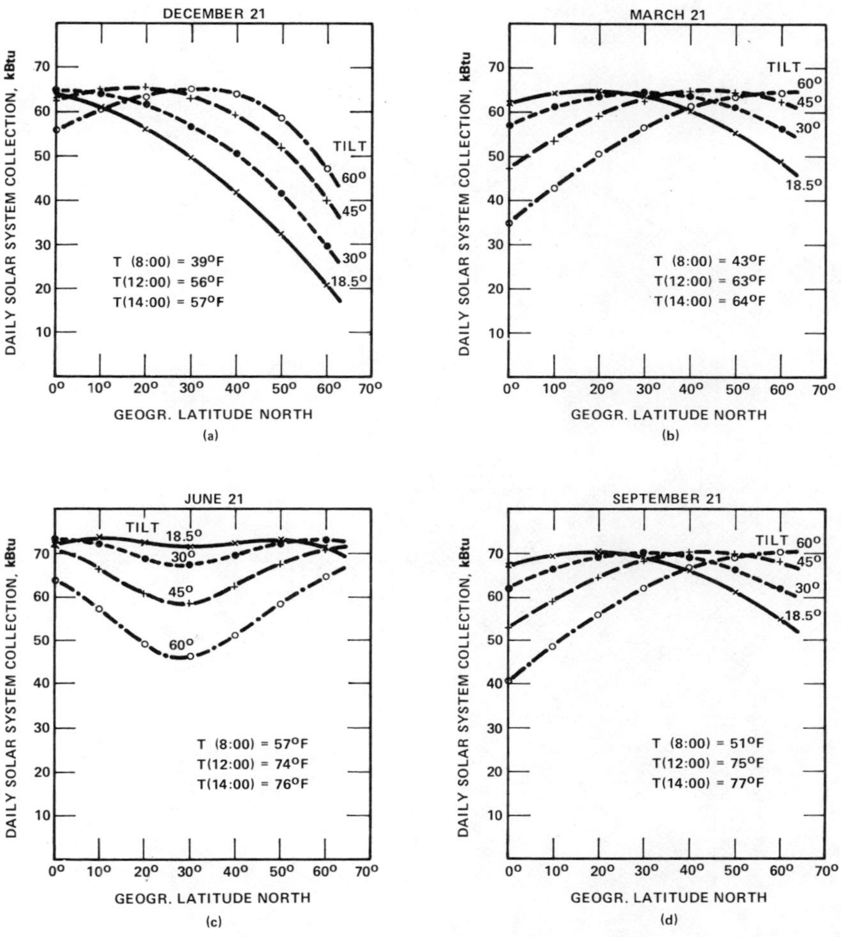

Fig. 5.6. Dependence of clear-day solar system collection on geographic latitude for a typical water heating system: $\epsilon = 0.84$; $a = 1.15$; $b = 1.0$; 56 ft^2 collector area (net); 80°F storage-tank temperature at 8:00 A.M.; 1500 Btu/hr average hot-water consumption; 500 Btu/hr average piping and storage heat losses; no heat exchanger.

decrease in the daily collection during the summer months at increased collector tilt angles is strongest at 25° to 30° latitude and decreases considerably for $\gamma \to 0$ and $\gamma \to 90°$.

Often this question arises: How much reduction in daily system collection will be expected if the collectors are not mounted in due-south (azimuth) orientation but at some deviation east or west of south? The results of such computations are shown in Fig. 5.7. The general trend is that off-orientations toward the west result in less reduction than off-orientations toward the east. This is because of the

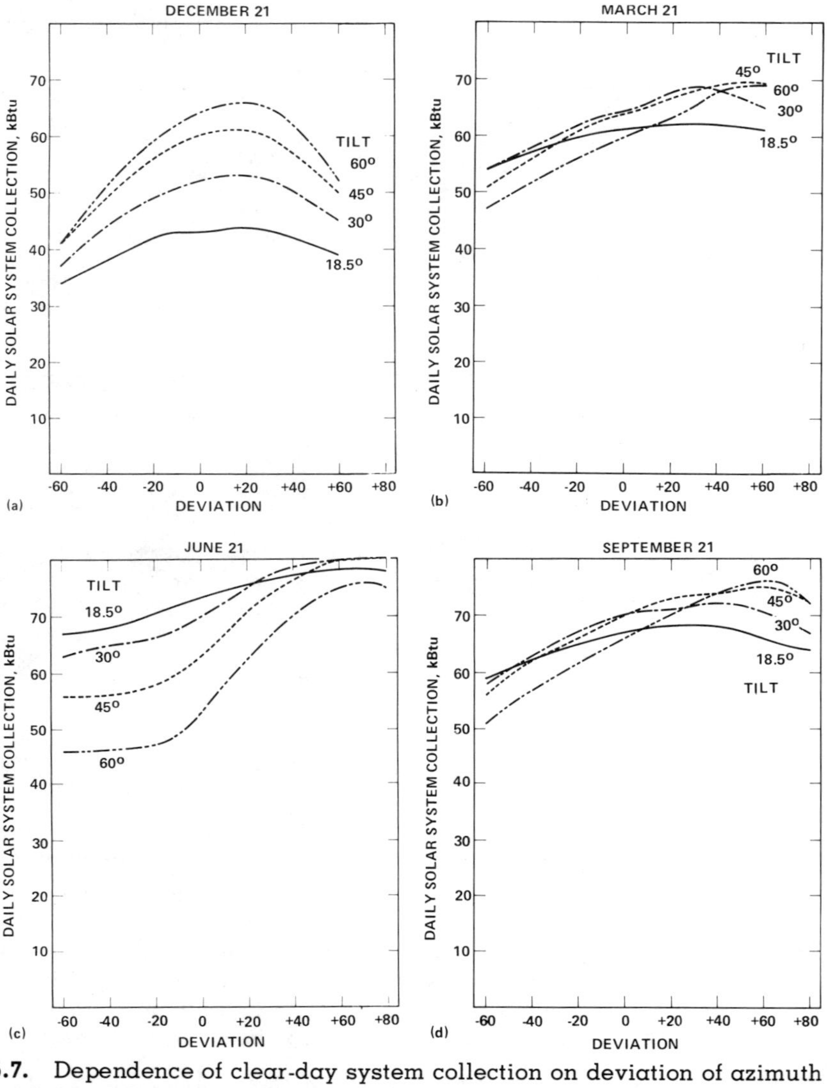

Fig. 5.7. Dependence of clear-day system collection on deviation of azimuth orientation of the collector from due south for a typical solar domestic water-heating system. Parameters same as in Fig. 5.6.

higher afternoon temperatures and consequently higher collection efficiencies at those collection times. However, the magnitude of this effect (sometimes there is even a gain when compared with due-south orientations) cannot be generalized quantitatively for other types of solar installations. It is in part due to the particular conditions used for the calculations, i.e., the assumption of only 1500 Btu/hr domestic hot-water consumption and 500 Btu/hr heat losses starting at 8:00 A.M., whereas collection for west orientations does not start until later in the day. At that time, the tank temperature is already further reduced (to as low as 68°F by noon for due-west collector azimuth orientation), which further lowers ΔT and increases the collection efficiency. However, it is evident that collector orientations considerably west of south may still be more acceptable than is widely believed. This result is in agreement with the results of an earlier report by Felske [20].

5.5 Computation Examples for Various Types of Solar Installations

The results for a series of solar installations are given in Table 5.2. The majority of the data in the table indicate the various parameters that describe the solar system installation used in the calculations. The lower seven lines are the results of the performance calculation obtained with an HP 97 calculator.[33]

In columns 1, 2, and 3 a solar space-heating system is analyzed for a typical single-family home. For column 1 the house is located in a mild climate zone, and the collectors are single-glazed. For columns 2 and 3 a severe climate zone was chosen, requiring double-glazed collectors (for the same type of building). The main difference between columns 2 and 3 is a reflector system that augments the insolation for column 3. The reflectors increase the solar coverage during the coldest winter months by about 50%. Because the financial investment for reflectors usually is considerably less than 50% of the cost of the solar system installation, reflectors can be considered cost-effective and are advised wherever architectural layout and/or snow loading considerations allow their application.

Two typical combined space-heating and air-conditioning systems, one for a private residence and one for a commercial building, are presented in columns 4 and 5. It is apparent that the annual distribution of the solar coverage is quite different for the two types of buildings. It is strongly dependent on the internal heat-generation values. Columns 6 and 7 give the parameters and results for two types of water-heating installations. Both are designed for maximum cost-effectiveness at lowest tolerable operating temperature. The application of reflectors in conjunction with single-glazed collectors for the industrial-process water-heating system (column 6) was economically warranted. The reflector tilt angle of 20 deg was chosen to obtain significant augmentation during the spring, summer, and fall seasons when the sunshine probability is highest. Large monthly variations in solar coverage was assumed to be of no particular concern to the owner, who was primarily interested in highest annual solar heat collection and had ample use for any solar heat that was collected.

TABLE 5.2 Computation Examples

		1	
1	Type of system[a]	H	
2	Type of building[b]	RS	
3	Floor area of building (ft^2)	1500	
4	Building heat-loss coefficient ($Btu\text{-}hr^{-1}\text{-}°F^{-1}$)	800	
5	Average daily window gain (kBtu)	150	145
6	Average hourly people gain (kBtu)	n.a.	
7	Average hourly lighting gain (kBtu)	n.a.	
8	Ambient temperature at 6:00 A.M.	39	40
9	Ambient temperature at 12:00 noon	57	60
10	Ambient temperature at 2:00 P.M.	58	61
11	Inside temperature from 8:00 A.M. to 10:00 P.M.	68	68
12	Inside temperature from 10:00 P.M. to 8:00 A.M.	65	65
13	Geographic latitude (North)	40	
14	Month	12	2
15	Collector area (ft^2)	312	
16	Collector tilt angle (0 deg = horizontal)	60	
17	Collector azimuth angle (deg = due south)	10	
18	Collector absorptivity/transmissivity, ϵ (%/100)	0.84	
19	Collector thermal-loss coefficient (linear equation)	1.6	
20	Collector system wind factor, c (°F)	6	
21	Reflector area (ft^2)	0	
22	Reflector tilt angle (deg)	n.a.	
23	Reflector effectivity factor (%/100)	n.a.	
24	Storage-tank volume (U.S. gal)	900	
25	Storage-tank temperature at 8:00 A.M.	120	
26	Storage + piping heat losses (Btu/hr)	720	
27	Heat-exchanger effectivity ($Btu\text{-}hr^{-1}\text{-}°F^{-1}$)	n.a.	
28	Flow rate in collector loop (gpm)	n.a.	
29	Fluid heat capacity ($Btu\text{-}gal^{-1}\text{-}°F^{-1}$)	8.34	
30	Average normal insolation ($Btu\text{-}hr^{-1}\text{-}ft^{-2}$	295	
31	Average hourly DWH consumption, daytime (kBtu/hr)	1.8	
32	Total daily DWH consumption (kBtu/day)	60	
33	Total daily lighting, people, appliance gains (kBtu)	40	
34	Probability of sunshine (%)	52	63
35	Storage-tank temperature at 5:00 P.M. (°F)	147.6	147.7
36	Total solar system collection (kBtu/day)	238.6	238.8
37	Total collection per square foot of collector area (Btu)	764.8	765.5
38	Heat content of storage tank at 5:00 P.M. (kBtu)	206	208
39	Degree-F-hours between 5:00 P.M. and 8:00 A.M.	614	556
40	Solar coverage, full sunshine (%/24 hr)	77.2	84.6
41	Total solar coverage, with clouds (%/24 hr)	40.2	53.2

[a] H: heating. A/C: air conditioning. W: water heating only. Any full-size solar air-conditioning and/or space-heating installation also includes solar domestic water heating.
[b] RS: residential. C: commercial. I: industrial. RM: multiunit residential.

TABLE 5.2 Computation Examples—Continued

2	3	4
H	H	A/C + H
RS	RS	RS
1500	1500	2500
800	800	1150
150 145	150 145	80 200 190
n.a.	n.a.	1 0 0
n.a.	n.a.	1 0 0
20 22	20 22	70 39 42
32 35	32 35	90 53 57
34 37	34 37	94 54 59
68 68	68 68	78 68 68
65 65	65 65	78 65 65
40	40	36
12 2	12 2	8 12 2
312	312	560
60	60	18.5
0	0	0
0.82	0.82	0.82
0.9	0.9	0.9
6	6	15 6 5
0	312	0
n.a.	0	n.a.
n.a.	0.7	n.a.
900	1500	2000
120	120	180 120 120
1080	1080	1440
5000	n.a.	n.a.
12	n.a.	n.a.
6	8.34	8.34
295	295	295
1.8	1.8	2.52
60	60	90
40	40	60 50 50
52 63	52 63	95 52 63
142.2 143.1	144 146.1	197.1 139.6 147.9
321.2 313.2	452.4 465.4	603 384 513
1030 1004	1450 1492	1077 685 915
166 173	298 325	282 327 463
1188 1123	1188 1123	76(C) 664 560
46.7 48.9	65.9 72.3	115 77 121
24.3 30.7	34.3 45.6	109 40 76

TABLE 5.2 Computation Examples—Continued

			5	
1	Type of system[a]		A/C + H	
2	Type of building[b]		C	
3	Floor area of building (ft^2)		15000	
4	Building heat-loss coefficient ($Btu\text{-}hr^{-1}\text{-}°F^{-1}$)		5800	
5	Average daily window gain (kBtu)	200	600	550
6	Average hourly people gain (kBtu)	22	15	15
7	Average hourly lighting gain (kBtu)	124	154	154
8	Ambient temperature at 6:00 A.M.	70	39	42
9	Ambient temperature at 12:00 noon	90	53	57
10	Ambient temperature at 2:00 P.M.	94	54	59
11	Inside temperature from 8:00 A.M. to 10:00 P.M.	78	68	68
12	Inside temperature from 10:00 P.M. to 8:00 A.M.	78	65	65
13	Geographic latitude (North)		36	
14	Month	8	12	2
15	Collector area (ft^2)		2000	
16	Collector tilt angle (0 deg = horizontal)		45	
17	Collector azimuth angle (deg = due south)		0	
18	Collector absorptivity/transmissivity, ϵ (%/100)		0.82	
19	Collector thermal-loss coefficient (linear equation)		0.9	
20	Collector system wind factor, c (°F)	15	15	15
21	Reflector area (ft^2)		1600	
22	Reflector tilt angle (deg)		28	
23	Reflector effectivity factor (%/100)		0.7	
24	Storage-tank volume (U.S. gal)		6000	
25	Storage-tank temperature at 8:00 A.M.	175	120	120
26	Storage + piping heat losses (Btu/hr)		6480	
27	Heat-exchanger effectivity ($Btu\text{-}hr^{-1}\text{-}°F^{-1}$)		n.a.	
28	Flow rate in collector loop (gpm)		n.a.	
29	Fluid heat capacity ($Btu\text{-}gal^{-1}\text{-}°F^{-1}$)		8.34	
30	Average normal insolation ($Btu\text{-}hr^{-1}\text{-}ft^{-2}$		295	
31	Average hourly DWH consumption, daytime (kBtu/hr)	7.2	9	9
32	Total daily DWH consumption (kBtu/day)	80	100	100
33	Total daily lighting, people, appliance gains (kBtu)	1600	1700	1700
34	Probability of sunshine (%)	95	52	63
35	Storage-tank temperature at 5:00 P.M. (°F)	171.9	162.9	169
36	Total solar system collection (kBtu/day)	2806	2271	2579
37	Total collection per square foot of collector area (Btu)	1494	1136	1317
38	Heat content of storage tank at 5:00 P.M. (kBtu)	×159	2131	2440
39	Degree-F-hours between 5:00 P.M. and 8:00 A.M.	76(C)	664	560
40	Solar coverage, full sunshine (%/24 hr)	75	124	145
41	Total solar coverage, with clouds (%/24 hr)	71	64	91

[a]H: heating. A/C: air conditioning. W: water heating only. Any full-size solar air-conditioning and/or space-heating installation also includes solar domestic water heating.
[b]RS: residential. C: commercial. I: industrial. RM: multiunit residential.

TABLE 5.2 Computation Examples—Continued

	6			7	
	W			W	
	I			RM	
	n.a.			n.a.	
	n.a.			n.a.	
	n.a.			n.a.	
	n.a.			n.a.	
	n.a.			n.a.	
39	45	51	39	45	51
57	68	76	57	68	76
58	70	79	58	70	58
	n.a.			n.a.	
	n.a.			n.a.	
	38			38	
12	4	8	12	4	8
	1600			800	
	45			45	
	0			0	
	0.84			0.84	
	1.6			1.6	
15	15	15	15	15	15
	1400			0	
	20			n.a.	
	0.7			n.a.	
	4000			2000	
	100			80	
	2520			6120	
	n.a.			12500	
	n.a.			30	
	8.34			6	
	295			295	
	300			30	
	2700			800	
	n.a.			n.a.	
52	75	95	52	75	95
84.2	100	103.8	117	120.4	124.3
2202	2716	2854	939	993	1060
1384	1755	1841	1174	1242	1326
−521	−6.6	129	614	669	735
	n.a.			n.a.	
81	100	105	110	116	124
42	75	100	57	87	118

Notes

[1]A comprehensive program for solar domestic water heating, space heating, and/or air conditioning for residential and commercial buildings was developed by the author for use on TRS-80 level II basic (16k) computers [25].

[2]See Appendix I, where such a program is presented in detail.

[3]See Sections 1.1.1 and 5.2.7 and Appendix F.

[4]This is mandatory for computations with an inexpensive programmable calculator such as a Hewlett-Packard 67/97 calculator, which has only 224 programming steps but has proved to be an excellent tool for such computations. Most other investigators who have used an iteration method have also used hourly intervals [29].

[5]The computation time for a year-round analysis with the "Building Solar Thermal Analysis, KH-80" program [25], computing the system thermal balance for one representative day per month with a TRS-80 level II, 16k computer, is about 2.5 min. The same task performed with an HP 97 takes 6 min per month or 72 min per year (and involves some intermediate data entry because of the limited number of programming steps and registers in an HP 97).

[6]Several examples of solar collection computations are presented in this handbook. We chose to perform these with an HP 97 calculator in order to emphasize the relative ease and general applicability of the approach presented. An example of a typical solar heating system collection program is given in Appendix I.

[7]The 21st day rather than the 15th day of each month is used because the extreme declinations of the Earth fall near these days (June 22, Dec. 22). Also, Mar. 21 and Sept. 23 are the days when sunrise and sunset are at exactly 6 A.M. and 6 P.M.

[8]Typically, Oct.–Apr. for space-heating systems, year-round for domestic water-heating systems. Good design practice is to utilize all collected solar heat fully during all seasons.

[9]See Section 5.2.8 and Appendix C.

[10]Typically, the linear thermal-loss coefficient for a good single-glazed flat-plate collector is $a = 1.0$ Btu/hr-ft²-°F. Such a collector, operating at $\Delta T = 60°F$ (a normal condition for space heating or domestic water heating) will have a heat loss of $60 \times 1.0 \cong 60$ Btu/hr-ft² (without even considering the quadratic loss term). If we further consider a transmission/absorption loss coefficient $\epsilon = 0.8$, one will be able to collect solar heat only if the insolation is above $60/0.8 = 75$ Btu/hr-ft². This condition is rarely fulfilled during overcast periods.

[11]At normal incidence on the collector. For most calculations it is sufficient to arbitrarily set $I_1 = 0.1I_0$. The ratio I_1/I_0 depends on environmental factors (e.g., foreign particles in the air, relative humidity) and on the solar altitude. See Appendix B for further details on this subject.

[12]See also Section 5.2.8 and Appendix C.

[13]In most of the continental United States the insolation at normal incidence varies between a low of approximately 275 Btu/hr-ft² in Dec. and a high of 305 Btu/hr-ft² in June. In the sample calculations in this handbook, we have used 275 in Nov., Dec., and Jan., 281 in Feb. and Oct., 286 in Mar. and Sept., 292 in Apr. and Aug., 297 in May and July, and 303 in June. See Appendix B for more details.

[14]At higher elevations, the absorption by the atmosphere is reduced, and somewhat higher insolation values can be expected. At a 6000-ft elevation the effective increase is 5%; at 9000 ft it is 10% above the values prevailing at sea level (see Appendix B).

[15]The equation to compute cos ϑ as a function of solar altitude, solar azimuth, geographic latitude, time, date, collector tilt angle, and collector azimuth orientation is given in Appendix A along with cos ϑ tables. Knowledge of cos ϑ is also important to determine the incident angle modifier (see Section 1.1.2). For reasons of clarity, we are neglecting the influence of the incident-angle modifier, which is of a second-order nature, in our further discussion in this chapter.

[16]This approximation is justified in particular for solar heat collection computations, because little collection occurs after 4:00 P.M., and the temperature profile during

the night hours is insignificant for this purpose. However, even for heating/cooling-demand calculations this approach has been shown to be quite adequate.

[17]T_0 is the mean storage-tank temperature. It should be noted that in open systems (systems that use no heat exchanger) there is normally no temperature stratification. If a system is designed in which stratification is expected (e.g., if a storage tank with a height-to-diameter ratio $\gtrsim 3$ is used), T_c should be decreased by the expected amount of temperature stratification.

[18]The heat exchanger discussed here is a metal coil immersed in the heat-storage tank. This type of exchanger is favored in smaller solar installations (< 600 ft² of collector area). In larger (mostly commercial) systems, a tube-in-shell type of exchanger is usually specified. This type of heat-transfer device requires a slightly modified mathematical treatment. See Section 1.4.3 for more details on heat exchangers.

[19]This corresponds to a temperature stratification that equals half the temperature difference between the inlet and outlet of the heat exchanger. It is a good approximation for average space-heating systems that require little extraction of heat from the solar storage tank during the main collection hours (the heating period is mainly evening and night); the extraction of heat will cause disturbances that might impede the temperature stratification. The term $C_0 q/i$ should be deleted if stratification is not assured.

[20]For example, an A_0 = 400-ft² collector system collecting $q = 200 \times 400 = 80{,}000$ Btu/hr having a flow rate of i = 15 gpm in the collector loop and using $A = 1/8 \times A_0 \approx 50$ ft² as heat-exchanger area (using E_x = 80 Btu-hr^{-1}-ft^{-2}°F^{-1}—for justification of these rules of thumb see Section 1.4.3)

$$\delta T_c = \frac{80{,}000}{80 \times 50} - \frac{10^{-3} \times 80{,}000}{15} \approx 20 - 5.3 \approx 15°\text{F}$$

[21]10^3 kcal-m^{-3}-°C^{-1} in the metric system.

[22]See Appendix J for details.

[23]Without significant sacrifice of accuracy of the overall system collection computation.

[24]This type of stationary concentrating collector array (see Section 1.1.1) is included in this discussion of solar system collection computation because (a) it can easily be treated in conjunction with regular flat-plate collector systems and (b) many flat-plate collector systems do indeed have some sort of reflective surface in front of them (e.g., white gravel, sheet metal coated with reflective paint, water).

[25]Consult Heinemann [24] for more details on this subject. Appendix F gives a numerical example of the effect of reflective surfaces in front of solar collectors.

[26]Refer to Appendix C for conversion of average insolation data into average sunshine probabilities.

[27]It is sometimes advisable to adapt the collector azimuth orientation to extreme fog patterns, e.g., to orient the collectors westward if fog conditions persist typically during the morning hours in a particular microclimate.

[28]For reasons of clarity, reflector augmentation (as discussed in Section 5.2.7) was not included in the block diagram.

[29]See Appendix J.

[30]Note also the availability of a comprehensive desk-computer program written in basic level II for TRS-80 computers for annual performance calculations [25]. This program includes some aspects of solar collector systems that are not discussed here: shading of the collectors by distant objects and by other collector rows and the incident-angle modifier. The program also includes reflector augmentation.

[31]For reasons of generality in the results, the net absorber area was used, rather than the overall collector area (see Section 1.1.2 for elaboration on the significance of the distinction between net and gross collector areas).

[32]Note that these studies were performed for 37.5° N. latitude. The 45-deg guide figure should be adjusted by about the same amount that the geographic latitude differs from 37.5°.

[33]These computations were performed with a program that was slightly different from the one discussed in Appendix I. In order to be able to accommodate reflective sur-

faces in the program, the relatively unimportant steps concerning the quadratic temperature-loss coefficient were eliminated, and the collector efficiency characteristic was described with ϵ and a only ($b = 0$). As described in Section 1.1.2, this is permissible if the values used for ϵ and a are properly adjusted such that the collector loss characteristic best approaches the quadratic function in the ΔT range in which it is predominantly operated. The values used describe average single-glazed ($\epsilon = 0.84$, $a = 1.6$, $b = 0$) and average double-glazed collectors ($\epsilon = 0.82$, $a = 0.9$, $b = 0$) and correspond approximately to $\epsilon = 0.83$, $a = 1.25$, $b = 1.0$ and $\epsilon = 0.78$, $a = 0.55$, $b = 0.90$, respectively, if the quadratic loss term is included.

6

Solar System Cost-Payback Analysis

A solar installation should be considered a standard component of a building. Just as it is not customary to talk about the payback of a furnace used for space heating or of a gas-fueled domestic water heater, one should not (or should not have to) talk about the payback of a solar installation. In fact, one should realize that it is the general advantage of a solar-collector installation that it *does have* a payback potential. A regular furnace does not have that potential.

Any conventional payback analysis of a solar-collector installation is potentially unfair because it is difficult to assess its fringe benefits, such as the reduction in dependence on conventional energy sources and the fact that the use of solar energy helps conserve scarce and nonreplenishable resources. Furthermore, a solar system represents a value addition to a dwelling and is a low-risk capital investment with a potential rate of return that is approximately equal to that for other types of real estate investment.

Nevertheless, there is a demand for a financial payback analysis scheme, and an attempt is made here to meet this demand. Because the private homeowner is typically more inclined than the owner of commercial real estate to take the not specifically tangible aspects of a solar installation into consideration, the financial analysis scheme presented here addresses primarily a commercial/industrial large-scale solar investment.

If an investment is made for the primary reason that it reduces future operating expenses, its worthiness is often judged by the payback period, i.e., by the period of time that it takes to recapture the investment by conserved operating expenses. Many schemes have been proposed to compute the payback period. Rather than describing the general features of such schemes, we have chosen to be specific and describe one particular method, programmed for a hand-held programmable calculator. We have chosen the Hewlett-Packard 67/97 calculator, which has 26 storage registers and 224 programming steps. However, the program can easily be transposed for the more powerful HP 41C calculator, as well as (with more effort) for other brands like the Texas Instruments TI-59.[1] In Section 6.3 we shall discuss an alternative approach to cost-effectiveness analysis of a commercial-size solar installation. This approach is based on cash-flow considerations.

6.1 Parameters in the Cost-Payback Analysis

The following factors are considered in the program listed in Table 6.1:

1. The total system cost (register S5). Good judgment is necessary to include only those expenses that are directly related to the solar installation, and to subtract all expenses that would have been incurred for a conventional service system. One should subtract the cost of any parts of conventional systems that have become obsolete because of the solar installation. Expenses for separate energy-saving devices, such as insulation and special timers, should not be included in the solar system cost.
2. The amount of financed capital (register S4).
3. The conditions of financing the loan (the number of monthly payments and the loan interest rate, registers S1 and S2).
4. The savings ($/year) of conventional fuel[2] (register S8). This important parameter should be estimated carefully and realistically. The method recommended here is as follows: (a) calculate the solar system performance, using the principles explained in Chapter 5 and in Appendix I;[3] (b) do this for every month, and establish the amount of energy saved during an average year; (e) divide this result by the efficiency of a typical conventional furnace that is being replaced by the solar installation;[4] (d) calculate the dollar amount of the fuel saved using present fuel costs and the cheapest available conventional fuel as the basis.[5]
5. The solar investment tax credit plus the regular investment tax credit that might be applicable (register S9).
6. The annual depreciation (register P9). A straight-line depreciation is assumed, given as a constant percentage of the total system cost (e.g., 5% for 20-year straight-line depreciation). A different depreciation schedule can be approximated by a program change.
7. The tax bracket that would be applicable for tax deduction of eligible interest expenses (register P8). All interest expenses incurred in connection with the solar installation are programmed as eligible expenses. However, deviations from this assumption can easily be accommodated by a minor program change.
8. The interest of savings (register S6). The down payment of the solar system cost is considered an expense that might otherwise have generated income if properly invested. This "lost" interest expense is compounded annually on the down payment. On the other hand, all savings generated as a consequence of the solar installation, such as the investment tax credit, the depreciation, the corporate tax savings due to interest expenses, and the savings of conventional fuel, are considered subject to savings interest generation. Because of the limited (constant) register and programming step capacity of the HP 67/97 calculator, a constant savings interest rate is assumed during the entire payback. period.
9. The rate of fuel cost increase during future years (registers P1–P6). It is assumed that this rate changes from year to year. Three fixed scenarios are built into the program, a fourth scenario can be selected by the user of the program.[6] The details are explained in the following program description.

The most important variable not considered explicitly is the rate of general monetary inflation. The effect of inflation is intimately connected with a counteracting increase in real estate value attributable to the solar installation and is

TABLE 6.1 Program Listing for Computation of Payback Period for a Solar Installation for a Hewlett-Packard 67/97 Calculator

001	*LBLA	21 11	035	8	08	069	÷	−24	103	+	−55
002	2	02	036	STO5	35 05	070	X⇌Y	−41	104	RTN	24
003	0	00	037	7	07	071	÷	−24	105	*LBLa	21 16 11
004	STO1	35 01	038	STO6	35 06	072	RCL4	36 04	106	RCLD	36 14
005	1	01	039	GSBD	23 14	073	×	−35	107	1	01
006	8	08	040	RTN	24	074	1	01	108	+	−55
007	STO2	35 02	041	*LBLC	21 13	075	2	02	109	STOD	35 14
008	1	01	042	9	09	076	×	−35	110	RCLC	36 13
009	6	06	043	STO1	35 01	077	STO3	35 03	111	ST+7	35 −55 07
010	STO3	35 03	044	8	08	078	RTN	24	112	P⇌S	16 −51
011	1	01	045	STO2	35 02	079	*LBLe	21 16 15	113	GSBe	23 16 15
012	4	04	046	7	07	080	RCL0	36 00	114	RCL7	36 07
013	STO4	35 04	047	STO3	35 03	081	RCL2	36 02	115	GSBb	23 16 12
014	1	01	048	6	06	082	×	−35	116	RCLC	36 13
015	2	02	049	STO4	35 04	083	P⇌S	16 −51	117	RCLI	36 46
016	STO5	35 05	050	STO5	35 05	084	STO0	35 00	118	%	55
017	1	01	051	STO6	35 06	085	P⇌S	16 −51	119	+	−55
018	0	00	052	GSBD	23 14	086	RCL3	36 03	120	STOC	35 13
019	STO6	35 06	053	RTN	24	087	−	−45	121	+	−55
020	GSBD	23 14	054	*LBLE	21 15	088	ST+0	35 −55 00	122	RCLB	36 12
021	RTN	24	055	1	01	089	RTN	24	123	P⇌S	16 −51
022	*LBLB	21 12	056	RCL2	36 02	090	*LBLb	21 16 12	124	ST+7	35 −55 07
023	1	01	057	1	01	091	RCL6	36 06	125	+	−55
024	6	06	058	2	02	092	×	−35	126	RCL7	36 07
025	STO1	35 01	059	÷	−24	093	ST+7	35 −55 07	127	P⇌S	16 −51
026	1	01	060	1	01	094	CHS	−22	128	RCL6	36 06
027	4	04	061	+	−55	095	RCL3	36 03	129	×	−35
028	STO2	35 02	062	RCL1	36 01	096	−	−45	130	+	−55
029	1	01	063	CHS	−22	097	P⇌S	16 −51	131	ENT↑	−21
030	2	02	064	Y^x	31	098	RCL0	36 00	132	ENT↑	−21
031	STO3	35 03	065	−	−45	099	RCL8	36 08	133	RCLE	36 15
032	1	01	066	RCL2	36 02	100	×	−35	134	+	−55
033	0	00	067	1	01	101	ST+7	35 −55 07	135	STOE	35 15
034	STO4	35 04	068	2	02	102	P⇌S	16 −51	136	RCL0	36 00

TABLE 6.1 Program Listing for Computation of Payback Period—Continued

137	RCL7	36 07	159	STO7	35 07	181	RCLC	36 13	203	GSBα	23 16 11
138	+	−55	160	P⇌S	16 −51	182	+	−55	204	GSBα	23 16 11
139	X≤Y?	16−35	161	GSBE	23 15	183	RCLA	36 11	205	GSBα	23 16 11
140	GTOc	22 16 13	162	GSBe	23 16 15	184	+	−55	206	GSBα	23 16 11
141	P⇌S	16 −51	163	1	01	185	RCLB	36 12	207	GSBα	23 16 11
142	RTN	24	164	STOD	35 14	186	+	−55	208	GSBα	23 16 11
143	*LBLD	21 14	165	RCL8	36 08	187	STOE	35 15	209	GSBα	23 16 11
144	P⇌S	16 −51	166	P⇌S	16 −51	188	P⇌S	16 −51	210	GSBα	23 16 11
145	RCL4	36 04	167	RCL1	36 01	189	RCL2	36 02	211	GSBα	23 16 11
146	STO0	35 00	168	%	55	190	STOI	35 46	212	RTN	24
147	RCL5	36 05	169	+	−55	191	GSBα	23 16 11	213	*LBLc	21 16 13
148	RCL9	36 09	170	STOC	35 13	192	RCL3	36 03	214	−	−45
149	×	−35	171	P⇌S	16 −51	193	STOI	35 46	215	÷	−24
150	STOA	35 11	172	RCL5	36 05	194	GSBα	23 16 11	216	1/X	52
151	RCL5	36 05	173	RCL4	36 04	195	RCL4	36 04	217	RCLD	36 14
152	P⇌S	16 −51	174	−	−45	196	STOI	35 46	218	−	−45
153	RCL9	36 09	175	STO7	35 07	197	GSBα	23 16 11	219	CHS	−22
154	×	−35	176	GSBb	23 16 12	198	RCL5	36 05	220	PRTX	−14
155	RCL8	36 02	177	RCL4	36 04	199	STOI	35 46	221	PREG	16 −13
156	×	−35	178	+	−55	200	GSBα	23 16 11	222	P⇌S	16 −51
157	STOB	35 12	179	RCL5	36 05	201	RCL6	36 06	223	PREG	16 −31
158	RCLA	36 11	180	−	−45	202	STOI	35 46	224	R/S	51

extremely difficult to quantify. It can be argued that this counteraction approximately cancels a measurable impact of inflation. The payback analysis should be evaluated with much caution. Inaccuracies in some of the parameters used, in particular in the estimation of the fuel savings and the fuel cost increases, may considerably alter the results.

6.2 Program for Determination of Payback Period

The program listing is presented in Table 6.1. A detailed explanation of the programming steps is given so that one can easily change the appropriate steps in the program to adapt it to one's own specific requirements. In order to facilitate the distinction between the two storage registers (P and S), the steps are indicated under P or S, depending on the momentarily acting register position, and an arrow indicates switching from one register to the other.

Register		
P	S	Explanation
001–021		Start with LBL "A" for scenario A, which increases the dollar amount of the fuel savings by 20% in the first year, 18% in the second, 16% in the third, 14% in the fourth, 12% in the fifth, and 10% in the sixth and any subsequent year of system operation.
022–040		Start with LBL "B" for a more conservative scenario B that increases the dollar amount of the fuel savings by 16%, 14%, 12%, 10%, and 8% in the first, second, third, fourth, and fifth years and by 7% in any subsequent year of system operation.
041–053		Start with LBL "C" for the most conservative scenario C starting with 9% fuel-cost increase in the first year, 8% in the second, 7% in the third, and 6% in the fourth and any subsequent year of system operation.
143		Start with LBL "D" if a scenario of custom choice is selected. In this case, enter in registers P1 through P6 the percentage fuel-cost increase of your own choice for the years 1 through 6, respectively, of system operation. The value entered in register P6 will then also represent the fuel-cost increase for any subsequent year.
→	145–146	Store the loan amount in register S0.
	147–150	Calculate the one-time tax credit and store it in register A.
	151	Calculate the annual, constant depreciation (in actual $) and
153–157	←	store it in register B.
158–159		Start a savings account by storing the one-time tax credit in register P7.
→	161 654–078	Calculate the annual loan payment and store it in register S3.
	162	Start subroutine e, steps 079–089.
	080–082	Calculate the annual interest on the loan (balance).
084	←	Store that result in register P0.
→	086–088	Calculate the dollar amount of the principal paid with the first year's loan payment and subtract this amount from register S0 to obtain the loan balance for this year.

Register P	Register S	Explanation
	163, 164	First year of system operation.
	165	Recall first year's fuel savings.
167–170	←	Add first year's fuel-cost increase, and store results in register C.
	172–175	Compute the down payment and store it in register S7.
	176	Start subroutine *b*, steps 090-104.
	091-093	Calculate lost interest on down payment and add to lost capital in register S7.
	094	Consider lost interest as a negative amount and compute the cumulative balance (CB).
	095, 096	Subtract loan payment from CB.
098–100	←	Calculate the annual tax benefit of the interest paid with the loan payment.
101		Add this tax benefit to the savings account.
→	103	Add this tax benefit to the above cumulative balance (CB).
	177, 178	Add loan amount received to CB.
	179, 180	Pay for system (subtract from CB).
	181, 182	Add fuel savings to CB.
	183, 184	Add one-time tax credit to CB.
	185, 186	Add depreciation to CB.
	187	Store CB in register E.
189, 190	←	Store fuel-cost increase for second year in register I.
191		Start subroutine *a*, steps 105-142.
106–109		Second year of system operation.
110, 111		Add last year's fuel savings to savings account.
→	113	Go through subroutine 3 and calculate the new interest and principal parts of the loan payment, and reload the registers S0 and P0 with the respective new values.
	114	Recall from previous year the sum of the down payment and compounded lost interest.
	115	Go through subroutine *b*, calculating this year's tax savings.
	116–120	Calculate this year's fuel savings and store in register C.
	121	Add as gain to CB.
	122	Add this year's depreciation to savings account.
124	←	
125		Add as gain to CB.
126		
→	128–130	Compute savings interest and add to CB.
	131, 132	Needed to retain CB in stack for further use.
	133, 135	Add this year's CB to last year's CB and store in register E.
	136–138	Add loan balance and accumulated losses due to down payment.
	139–142	If losses still >CB, repeat calculation for another year, or else go to final subroutine *c*.
192–194		Subroutine for third year.
195–197		Subroutine for fourth year.
198–200		Subroutine for fifth year.

Register P	Register S	Explanation
201–203		Subroutine for sixth year.
204		Subroutine for seventh year.
204–211		Subroutines for eighth through fourteenth year. For payback periods longer than 14 years, the calculator will stop at the end of each following year. Depress "fa" and "R/S" for each year until result-printout signals end of computation.
213–224		Subroutine to express payback period in decimals, and to print the parameters and results.

The registers must be loaded as follows:

Register P	Register S	Loading Instructions
	1	Number of monthly loan payments
	2	Loan interest rate per year (%/100)
	4	Amount of financed capital ($)
	5	Total solar system cost ($)
	6	Interest rate on savings (%/100)
	8	Fuel savings, annual, at present fuel rates ($)
	9	One-time tax credit (solar + investment, %/100 of total system cost)
8		Tax bracket for deduction of interest expenses (%/100)
9		Straight-line depreciation (1/number of years)

Before starting the calculation, make sure that the P register is in front. Press "A" for fuel-cost increase scenario A, press "B" for scenario B, and press "C" for scenario C. If none of the fuel-cost increase scenarios A, B, and C are adequate, start the program by depressing "D," after loading the following additional registers:

Register P	Register S	Loading Instructions
1		First year's fuel-cost increase (%)
2		Second year's fuel-cost increase (%)
3		Third year's fuel cost increase (%)
4		Fourth year's fuel cost increase (%)
5		Fifth year's fuel-cost increase (%)
6		Fuel-cost increase for any subsequent year (%)

A typical printout of the results of a computation with the program of Table 6.1 is presented in Table 6.2.

TABLE 6.2 Data Printout for the Program of Table 6.1[a]

	(1)	Detailed payback period (years)	5.55	***
S0	(2)	Loan balance at end of gross payback period ($)	123589.27	0
S1	(3)	Number of monthly loan payments[b]	360.00	1
S2	(4)	Annual interest rate on loan (%/100)[b]	0.09	2
S3	(5)	Annual loan payments ($)	12552.11	3
S4	(6)	Financed (loaned) capital ($)[b]	130000.00	4
S5	(7)	Total cost of solar installation ($)[b]	150000.00	5
S6	(8)	Annual interest rate that down payment would have generated (%/100)[b]	2.07	6
S7	(9)	Down payment + accumulated interest after gross payback period ($)	30014.61	7
S8	(10)	Fuel savings in first year at present fuel rates ($)[b]	12000.00	8
S9	(11)	One-time total tax credit (solar + investment, in %/100)[b]	0.30	9
A	(12)	Amount of one-time tax credit ($)	45000.00	A
B	(13)	Amount of annual depreciation ($)	3750.00	B
C	(14)	Fuel savings during last payback year ($)	27934.96	C
D	(15)	Gross payback period (years)	6.00	D
E	(16)	Total accumulated system cash balance after gross payback period ($)	169917.39	E
I	(17)	Fuel-cost increase during last payback year (%)	11.00	I
P0	(18)	Amount of interest on loan payment during last payback year ($)	11241.03	0
P1	(19)	Fuel-cost increase during first year of system operation (%)	20.00	1
P2	(20)	Fuel-cost increase during second year of system operation (%)	18.00	2
P3	(21)	Fuel-cost increase during third year of system operation (%)	16.00	3
P4	(22)	Fuel-cost increase during fourth year of system operation (%)	14.23	4
P5	(23)	Fuel-cost increase during fifth year of system operation (%)	12.00	5
P6	(24)	Fuel-cost increase during sixth year of system operation (%)	11.00	6
P7	(25)	Balance of savings account after gross payback period ($)	196940.56	7
P8	(26)	Income tax bracker (%/100)[b]	0.50	8
P9	(27)	Annual rate of depreciation (%/100)[b]	0.05	9

[a]Typical printout of results of a calculation of the payback period with the program listed in Table 6.1 (if the program steps 221 and 223 are deleted, only the first result, i.e., the payback period, is printed.
[b]Input data. Note further that results 12–17 are repeated at the end and should be discarded.

6.3 Program for Determination of Cost-effectiveness Based on Cash-Flow Considerations

The program (Table 6.3) is identical in all major elements with the payback-period program discussed earlier (Table 6.1), except that because of the nature of a cash-flow analysis on an annual basis no compounded interest gains are computed on the savings,[7] and obviously the printout format is changed. A further minor difference is that the payback scenarios A, B, and C now extend over seven rather than six independent annual fuel-cost increases. Because of the similarity of the programs, a detailed explanation of the program steps for Table 6.3 is not given here; the reader is referred to Section 6.2. The registers of the program must be loaded exactly as in Section 6.2. It must be remembered that if scenario D is chosen, register P7 must now also be loaded (with the fuel-cost increase projected for the seventh year and any subsequent years of system operation). To continue the printout beyond the seventh year, "R/S" must be depressed for every subsequent year. The resulting printout is presented and discussed in Table 6.4 (the same input data as for Table 6.2 were used for the sample calculation).

TABLE 6.3 Program Listing for Cash-Flow Analysis of a Solar Installation Using a Hewlett-Packard 67/97 Calculator

001	*LBLA	21 11	035	1	01	069	CHS	-22	103	P⇌S	16-51
002	2	02	036	0	00	070	Y^x	31	104	RCL0	36 00
003	0	00	037	STO4	35 04	071	−	-45	105	RCL8	36 08
004	STO1	35 01	038	8	08	072	RCL2	36 02	106	×	-35
005	1	01	039	STO5	35 05	073	1	01	107	P⇌S	16-51
006	6	08	040	7	07	074	2	02	108	+	-55
007	STO2	35 02	041	STO6	35 06	075	÷	-24	109	RTN	24
008	1	01	042	6	06	076	X⇌Y	-41	110	*LBLa	21 16 11
003	6	06	043	STO7	35 07	077	÷	-24	111	SPC	16-11
010	STO3	35 03	044	GSBD	23 14	078	RCL4	36 04	112	RCLD	36 14
011	1	01	045	RTN	24	079	×	-35	113	1	01
012	4	04	046	*LBLC	21 13	080	1	01	114	÷	-55
013	STO4	35 04	047	9	09	081	2	02	115	PRTX	-14
014	1	01	048	STO1	35 01	082	×	-35	116	STOD	35 14
015	2	02	049	8	08	063	STO3	35 03	117	P⇌S	16-51
016	STO5	35 05	050	STO2	35 02	084	RTN	24	118	GSBe	23 16 15
017	1	01	051	7	07	085	*LBLe	21 16 15	119	ROL7	36 07
018	1	01	052	STO3	35 03	086	RCL0	36 00	120	GSBb	23 16 12
019	STO6	35 06	053	6	06	087	RCL2	36 02	121	RCLC	36 13
020	1	01	054	STO4	35 04	088	×	-35	122	RCLI	36 46
021	0	00	055	STO5	35 05	089	P⇌S	16-51	123	%	55
022	STO7	35 07	056	STO6	35 06	090	STO0	35 00	124	+	-55
023	GSBD	23 14	057	STO7	35 07	091	P⇌S	16-51	125	STOC	35 13
024	RTN	24	056	GSBD	23 14	092	RCL3	36 03	126	+	-55
025	*LBLB	21 12	059	RTN	24	093	−	-45	127	RCLB	36 12
026	1	01	060	*LBLE	21 15	094	ST+0	35-55 00	128	+	-55
027	6	06	061	1	01	095	RTN	24	129	PRTX	-14
028	STO1	35 01	062	RCL2	36 02	096	*LBLb	21 16 12	130	RCLE	36 15
029	1	01	063	1	01	097	RCL6	36 06	131	+	-55
030	4	04	064	2	02	098	×	-35	132	PRTX	-14
031	STO2	35 02	065	÷	-24	099	ST+7	35-55 07	133	STOE	35 15
032	1	01	066	1	01	100	CHS	-22	134	P⇌S	16-51
033	2	02	067	+	-55	101	RCL3	36 03	135	RTN	24
034	STO3	35 03	068	RCL1	36 01	102	−	-45	136	*LBLD	21 14

TABLE 6.3 Program Listing for Cash-Flow Analysis of a Solar Installation Using a Hewlett-Packard 67/97 **Calculator**—Continued

137	P⇌S	16-51	159	×	-35	181	+	-55	203	ROL2	36 02
138	RCL4	36 04	160	P⇌S	16-51	182	STOC	35 13	204	STOI	35 46
139	STO0	35 00	161	STOB	35 12	183	P⇌S	16-51	205	GSBa	23 16 11
140	RCL5	36 05	162	PRTX	-14	184	RCL5	36 05	206	RCL3	36 03
141	PRTX	-14	163	RCLE	36 06	185	RCL4	36 04	207	STOI	35 46
142	RCL4	36 04	164	EEX	-23	186	−	-45	208	GSBa	23 16 11
143	PRTX	-14	165	2	02	187	STO7	35 07	209	RCL4	36 04
144	RCL2	36 02	166	×	-35	188	GSBb	23 16 12	210	STOI	35 46
145	EEX	-23	167	PRTX	-14	189	RCL4	36 04	211	GSBa	23 16 11
146	2	02	168	GSBE	23 15	190	+	-55	212	RCL5	36 05
147	×	-35	169	PRTX	-14	191	RCL5	26 05	213	STOI	35 46
148	PRTX	-14	170	GSBe	23 16 15	192	−	-45	214	GSBa	23 16 11
149	RCL5	36 05	171	ROL8	36 08	193	RCLC	36 13	215	RCL6	36 06
150	RCL9	36 09	172	PRTX	-14	194	+	-55	216	STOI	35 46
151	×	-35	173	SPC	16-11	195	RCLA	36 11	217	GSBa	23 16 11
152	PRTX	-14	174	1	01	196	+	-55	219	*LBLd	21 16 14
153	STOA	35 11	175	STOD	35 14	197	RCLB	36 12	219	RCL7	36 07
154	RCL5	36 05	176	PRTW	-14	198	+	-55	220	STOI	35 46
155	P⇌S	16-51	177	RCL6	36 08	199	PRTX	-14	221	GSBa	23 16 11
156	RCL9	36 09	178	P⇌S	16-51	200	PRTX	-14	222	R/S	51
157	×	-35	179	RCL1	36 01	201	STOE	35 15	223	GTOd	22 16 14
158	RCL8	36 08	180	%	55	202	P⇌S	16-51	224	R/S	51

TABLE 6.4 Data Printout for the Program of Table 6.3[a]

(1)	Total cost of solar installation ($)	150000.00	***
(2)	Financed (loaned) capital ($)	130000.00	***
(3)	Annual interest rate on loan (%)	9.00	***
(4)	Amount of one-time tax credit (solar + investment, $)	45000.00	***
(5)	Amount of annual depreciation ($)	3750.00	***
(6)	Annual interest rate that down payment would have generated (%)	7.00	***
(7)	Computed annual loan payments ($)	12552.11	***
(8)	Fuel savings in first year at present fuel rates ($)	12000.00	***
(9)	Year of operation (first year)	1.00	***
(10)	Cash balance at end of this year of operation ($)	35047.89	***
(11)	Accumulated cash balance until end of this year ($)	35047.89	***
(12)	(repeat for 2nd year)	2.00	***
(13)		12503.54	***
(14)		47551.43	***
(15)	(repeat for 3rd year)	3.00	***
(16)		15075.61	***
(17)		62627.04	***
(18)	(repeat for 4th year)	4.00	***
(19)		17677.35	***
(20)		80304.38	***
(21)	(repeat for 5th year)	5.00	***
(22)		20204.06	***
(23)		100508.45	***
(24)	(repeat for 6th year)	6.00	***
(25)		22789.81	***
(26)		123298.26	***
(27)	(repeat for 7th year)	7.00	***
(28)		25386.86	***
(29)		148685.11	***

[a] Typical printout of results for a calculation with the program of Table 6.3 (financial analysis on a cash-flow basis).

Notes

[1]A program written in Basic Level II for use on a TRS-80 computer with 16k memory (but easily adaptable other desk-top computers) following the same calculational approach is available [26].

[2]Subtract expenses for servicing and repair per year.

[3]In particular, see the programs shown in Figs. I.1 and I.2 and Tables I.1 and I.2.

[4]In some cases (e.g., if air conditioning and space heating are involved) both electricity and gas may have to be considered. Count the replaced electricity at the appropriate efficiency (C.O.P) rating of the replaced chiller or heat pump.

[5]For example, solar absorption-chilling air-conditioning systems replace gas as heat source for the hot-water input to the chiller; yet the non-solar-air-conditioned building might be equipped more economically with electric compressor-type chillers, not with absorption chillers.

[6]It is, for example, possible to use higher fuel cost increases in the near future and to taper down the rate of increase in the following years. Thus the deregulation of gas and crude oil prices can be accounted for if the annual rates of increase can be reasonably well estimated.

[7]Steps 128–130 in Table 6.1.

References

1. Alten Corporation. Diagram of a commercial installation. Alten Corporation, 2594 Leghorn Street, Mountain View, California 94043.
2. Alten Corporation, 2594 Leghorn Street, Mountain View, California 94043.
3. Alten Corporation. Model HW-1 product brochure. Alten Corporation, 2594 Leghorn Street, Mountain View, California 94043.
4. Anderson, P. 1978. Solar absorption cooling. In *Energy from the sun, a survey of technologies for collection, storage, conversion and utilization*, ed. J. W. White et al., p. 209. Institute of Gas Technology, 3424 South State Street, Chicago, Illinois 60616.
5. Arkla Industries. Product brochure. Arkla Industries, Evansville, Indiana 47704.
6. *ASHRAE standard 93–77*. 1977. American Society of Heating, Refrigeration, and Air Conditioning Engineers, 345 East 47th Street, New York, New York 10017.
7. *ASHRAE handbook of fundamentals*. 1972. Chapter 26. American Society of Heating, Refrigeration, and Air Conditioning Engineers, 345 East 47th Street, New York, New York 10017.
8. Beckmann, W. A., Klein, S. A., and Duffie, J. A. 1977. *Solar heating design by the F-chart method*. New York: John Wiley & Sons.
9. Bliss, R. W. 1961. Atmospheric radiation near the surface of the ground: a summary for engineers. *Solar Energy* 5:103.
10. Boldrin, B., Scalabrin, G., Lazzarin, R., and Sovrano, M. 1978. Cooling of a fluid by a nocturnal radiator. In *Energy conservation in heating, cooling, and ventilating buildings, Vol. 2*, ed. C. J. Hoogendoorn and N. H. Afgan, p. 839. Washington, D.C.: Hemisphere Publishing.
11. Bruno, R. 1978. A correction procedure for separating direct and diffuse insolation on a horizontal surface. *Solar Energy* 20:97.
12. California Energy Commission. 1979. Report in compliance with California Assembly Bill 1512. California Energy Commission, 1111 Howe Avenue, Sacramento, California 95825.
13. *Climatic atlas of the United States*. 1977. U.S. Department of Commerce, National Climatic Center, Asheville, North Carolina.
14. Cole, R., Nield, K., Rohde, R., and Wolosewicz, R. 1979. Design and installation manual for thermal energy storage. Report ANL-79–15, Argonne National Laboratory, 9700 South Cass Avenue, Argonne, Illinois 60439. Second Edition.

15. de Winter, F. 1978. Solar energy and the flat plate collector. In *Energy from the sun, a survey of technologies for collection, storage, conversion, and utilization*, ed. J. W. White et al., p. 19. Institute of Gas Technology. 3424 South State Street Chicago, Illinois 60616.
16. de Winter, F. 1974. How to design and build a solar swimming pool heater. Copper Development Association, 405 Lexington Avenue, New York, New York 10017.
17. DSET Laboratories. 1978. Specification DSET 75-SE2 for outdoor thermal performance testing of hydronic solar collectors. DSET Laboratories, P.O. Box 8529, Phoenix, Arizona 85029.
18. Duncan, R. T., Doering, E. 1975. The Atlanta solar heating and cooling experiment. *Baltimore Engineer*, May 1975.
19. Farber, J. 1978. In *Proceedings of the 1978 Annual Meeting of the American Section of the International Solar Energy Society, Vol. 2.1*, p. 235.
20. Felske, J. D. 1978. The effect of off-south orientation on the performance of flat plate collectors. *Solar Energy* 20:29.
21. Granryd, E. 1978. Theory of solar assisted heat pumps. In *Energy from the sun, a survey of technologies for collection, storage, conversion, and utilization*, ed. J.W. White, et al., p. 241. Institute of Gas Technology, 3424 South State Street, Chicago, Ill. 60616.
22. Grassie, L. S., and Sheridan, N. R. 1977. The use of planar reflectors for increasing the energy yield of flat plate collectors. *Solar Energy* 19:663.
23. Grundfos Pumps Corporation, 2555 Clovis Avenue, Clovis, California 93612; telephone: (209) 299–9741.
24. Heinemann, K. 1977. Design and construction of a pilot solar space heating and air conditioning system using flat plate collectors and an absorption chiller, In *Solar technology for building, Vol 2*, Proceedings of the 1st International Conference on Solar Building Technology, London 1977, ed. C. Stambolis, p. 476. London: Riba Publications.
25. Heinemann, K. Building solar thermal analysis, KH-80. ELORET, 1178 Maraschino Drive, Sunnyvale, California 94087.
26. Heinemann, K. Solar system financial analysis, KH-81. ELORET, 1178 Maraschino Drive, Sunnyvale, California 94087.
27. Heliotrope General. Product brochure. Heliotrope General, 3733 Kenora Drive, Spring Valley, California 92077.
28. Howell, Y., and Bereny, J. 1979. *Engineer's guide to solar energy*. Solar Energy Information Services, P.O. Box 204, San Mateo, California.
29. Hunn, B. D., Willcutt, G. E., Jr., and McSweeney, T. B. 1977. Simulation and cost optimization of solar heating systems of buildings in adverse solar regions. *Solar Energy* 19:33.
30. Liu, B. Y. H., Jordan, R. C. 1960. The interrelationship and characteristic distribution of direct, diffuse, and total solar radiation. *Solar Energy* 4:1.
31. Lorsch, G., and Kauffman, K. W. 1978. Residential and commercial thermal storage. In *Energy from the sun, a survey of technologies for collection, storage, conversion, and utilization*, ed. J. W. White et al., p. 309. Institute of Gas Technology, 3424 South State Street, Chicago, Illinois 60616.
32. Magic Aire Division. Product brochure. Magic Aire Division, United Electric Company, 501 Kell Boulevard, Wichita Falls, Texas 76307.
33. Mattox, D. M., Kominiak, G. J., Sowell, R. R., and Pettit, R. B. Report SAND75–0361, Sandia Laboratories, Albuquerque, New Mexico.
34. Ong, K. S. 1976. An improved computer program for thermal performance of a solar water heater. *Solar Energy* 18:183.
35. Peterson, J. T., and Flowers, E. C. 1977. Interactions between air pollution and solar radiation. *Solar Energy* 19:23.

36. Pettit, R. B., and Sowell, R. R. 1976. Solar absorptance and emittance properties of several solar coatings. *J. Vac. Sci. Technol.* 13:596.
37. Popplewell, J. M. 1975. Corrosion considerations in the use of aluminum, copper, and steel flat plate collectors. In *Proceedings of the International Solar Energy Society Conference, U.C.L.A., July 1975.*
38. Solaron Corporation. Product brochure. Solaron Corporation, Denver, Colorado 80222.
39. Taco Corporation, Cranston, Rhode Island 02910.
40. Telkes, M. 1978. Latent heat storage techniques. In *Energy from the sun, a survey of technologies for collection, storage, conversion, and utilization*, ed. J. W. White et al., p. 325. Institute of Gas Technology, 3424 South State Street, Chicago, Illinois 60616.
41. Tuller, S. E. 1976. The relationship between diffuse, total and extraterrestrial solar radiation. *Solar Energy* 18:259.
42. Turbonics Corporation. Product brochure. Turbonics Corporation, 3160 West 106th Street, Cleveland, Ohio 44111.
43. Whillier, A. 1967. Design factors influencing solar collector performance. In *Low temperature applications of solar energy*, ed. R. C. Jordan, American Society of Heating, Refrigeration, and Air Conditioning Engineers, 345 East 47th Street, New York, New York 10017.
44. Woodman, T. P. 1977. The effect of design and operating parameters on the performance of flat plate solar collectors—calculation method and detailed appraisal. *Solar Energy* 19:263.
45. Wurm, J., and Kono, M. T. 1978. A contribution to the technical review of solar heat pumps. In *Energy from the sun, a survey of technologies for collection, storage, conversion, and utilization*, ed. J. W. White, p. 267. Institute of Gas Technology, 3424 South State Street, Chicago, Illinois 60616.
46. Yazaki Corporation, 17th floor, Mita-Kokusai Bldg. 4–28, 1-Chome, Minato-Ku, Tokyo 108, Japan.
47. Yazaki Corporation. Product information. Yazaki Corporation, Los Angeles Liaison Office, Suite 1112, 5959 West Century Boulevard, Los Angeles, California 90045.

Appendixes

A

Sun Altitude, Azimuth, Sun-Panel Cosine

The relationship among sun/collector angle ϑ, solar altitude α, solar azimuth ϕ, collector tilt angle ψ, and azimuth deviation of the collectors from due south ϕ_0 is (Fig. A.1)

$$\cos \vartheta = \cos \alpha \cos(\phi + \phi_0) \sin \psi + \sin \alpha \cos \psi \tag{A.1}$$

with

$$\sin \alpha = -\cos \beta \cos \gamma \cos(15h) - \sin \beta \sin \gamma \tag{A.2}$$

where γ is the geographic latitude and β the declination:

$$\beta = -23.45 \sin \left[\frac{360}{365} (284 + d) \right] \tag{A.3}$$

where d is the number of the day in the year ($d = 1$ on Jan. 1). The solar azimuth ϕ can be computed as

$$\sin \phi = \cos \beta \sin(15h)/\cos \alpha \tag{A.4}$$

In equations A.2 and A.4, h represents the time of the day in 0 to 24 (hr) nomenclature.[1]

Tables A.1 and A.2 give the results for cos ϑ computations at 37.5° N. latitude for various collector tilt angles and day times. In Table A.2 the collector azimuth ϕ_0 is changed toward west (read left time column) and toward east (read right time column).

[1]The equations are presented in the form that gives $\phi = 0$ at $h = $ 12:00 noon (ϕ is negative during the morning and positive during the afternoon hours). When computing ϕ for summer-month conditions, where ϕ can be >90° or <−90°, the ambiguity sin $(90 + x) = \sin(90 - x)$ may lead to erroneous results for ϕ, unless appropriate precautions are taken.

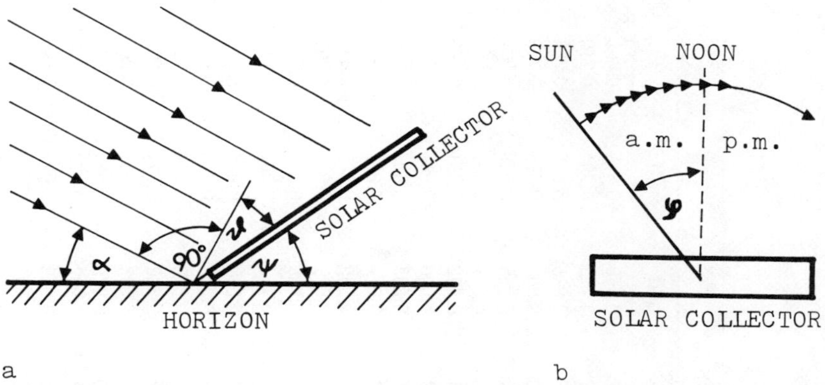

Fig. A.1. North–south side view (a) and top view (b) of south-oriented solar collector indicating the angles relevant to the determination of cos ϑ.

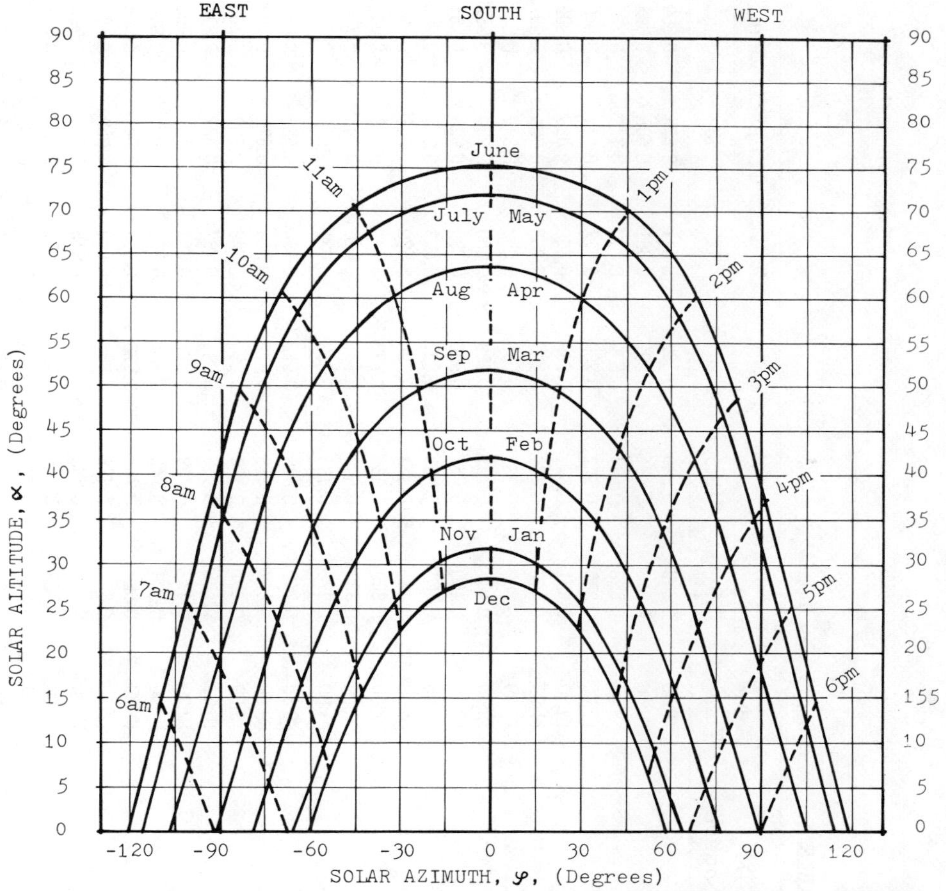

Fig. A.2. Solar altitude versus solar azimuth for various dates (for the 21st of the months indicated). The dash lines indicate the altitude/azimuth conditions for various times of the day. The data were calculated using equations A.1 through A.4 for 38° N. latitude.

TABLE A.1 Values for cos ϑ^a

Declination	Solar Time	Collector Tilt Angle (deg.), Due South									
		0	18.5	30	35	40	45	50	60	70	90
	8:30	0.202	0.399	0.502	0.541	0.576	0.606	0.632	0.668	0.685	0.655
	9:00	0.273	0.485	0.592	0.631	0.666	0.695	0.720	0.752	0.761	0.710
	9:30	0.336	0.559	0.670	0.710	0.745	0.774	0.797	0.825	0.828	0.758
	10:00	0.389	0.622	0.736	0.777	0.812	0.840	0.862	0.886	0.884	0.799
	10:30	0.431	0.673	0.789	0.830	0.865	0.893	0.914	0.935	0.929	0.831
$\beta =$	11:00	0.462	0.709	0.827	0.868	0.903	0.931	0.952	0.971	0.961	0.855
23.45	11:30	0.480	0.731	0.850	0.892	0.926	0.954	0.974	0.993	0.981	0.869
(Dec.)	12:00	0.487	0.739	0.858	0.900	0.934	0.962	0.982	1.000	0.987	0.874
	12:30	0.480	0.731	0.850	0.892	0.926	0.954	0.974	0.993	0.981	0.869
	1:00	0.462	0.709	0.827	0.868	0.903	0.931	0.952	0.971	0.961	0.855
	1:30	0.431	0.673	0.789	0.830	0.865	0.893	0.914	0.935	0.929	0.831
	2:00	0.389	0.622	0.736	0.777	0.812	0.840	0.862	0.886	0.884	0.799
	2:30	0.336	0.559	0.670	0.710	0.745	0.774	0.797	0.825	0.828	0.758
	3:00	0.273	0.485	0.592	0.631	0.666	0.695	0.720	0.752	0.761	0.710
	8:30	0.243	0.428	0.522	0.556	0.586	0.612	0.633	0.660	0.668	0.622
	9:00	0.317	0.515	0.613	0.648	0.678	0.703	0.723	0.745	0.745	0.678
	9:30	0.381	0.592	0.693	0.729	0.759	0.783	0.802	0.820	0.814	0.727
	10:00	0.435	0.656	0.761	0.797	0.827	0.851	0.868	0.883	0.871	0.769
$\beta =$	10:30	0.478	0.708	0.815	0.851	0.881	0.905	0.921	0.933	0.917	0.802
20.24	11:00	0.509	0.745	0.854	0.891	0.921	0.944	0.960	0.970	0.950	0.826
(Jan.,	11:30	0.528	0.768	0.878	0.915	0.945	0.968	0.983	0.992	0.970	0.840
Nov.)	12:00	0.535	0.775	0.886	0.923	0.953	0.976	0.991	0.999	0.977	0.845
	12:30	0.528	0.768	0.878	0.915	0.945	0.968	0.983	0.992	0.970	0.840
	1:00	0.509	0.745	0.854	0.891	0.921	0.944	0.960	0.970	0.950	0.826
	1:30	0.478	0.708	0.815	0.851	0.881	0.905	0.921	0.933	0.917	0.802
	2:00	0.435	0.656	0.761	0.797	0.827	0.851	0.868	0.883	0.871	0.769
	2:30	0.381	0.592	0.693	0.729	0.759	0.783	0.802	0.820	0.814	0.727
	3:00	0.317	0.515	0.613	0.648	0.678	0.703	0.723	0.745	0.745	0.678

	8:30	0.350	0.498	0.565	0.587	0.605	0.618	0.626	0.629	0.612	0.524
	9:00	0.426	0.589	0.660	0.683	0.701	0.713	0.720	0.717	0.693	0.582
	9:30	0.493	0.669	0.744	0.767	0.785	0.797	0.802	0.795	0.764	0.634
	10:00	0.550	0.736	0.814	0.838	0.856	0.867	0.872	0.861	0.824	0.677
$\beta =$	10:30	0.595	0.790	0.871	0.895	0.913	0.923	0.927	0.913	0.872	0.711
11.75	11:00	0.627	0.828	0.911	0.936	0.954	0.964	0.967	0.951	0.907	0.736
(Feb.,	11:30	0.647	0.852	0.936	0.961	0.979	0.989	0.992	0.974	0.928	0.752
Oct.)	12:00	0.654	0.860	0.944	0.970	0.987	0.997	1.000	0.982	0.935	0.757
	12:30	0.647	0.852	0.936	0.961	0.979	0.989	0.992	0.974	0.928	0.752
	1:00	0.627	0.828	0.911	0.936	0.954	0.964	0.967	0.951	0.907	0.736
	1:30	0.595	0.790	0.871	0.895	0.913	0.923	0.927	0.913	0.872	0.711
	2:00	0.550	0.736	0.814	0.838	0.856	0.867	0.872	0.861	0.824	0.677
	2:30	0.493	0.669	0.744	0.767	0.785	0.797	0.802	0.795	0.764	0.634
	3:00	0.426	0.589	0.660	0.683	0.701	0.713	0.720	0.717	0.693	0.582
	8:30	0.483	0.576	0.604	0.608	0.608	0.603	0.594	0.562	0.513	0.370
	9:00	0.562	0.669	0.701	0.706	0.706	0.701	0.690	0.653	0.596	0.430
	9:30	0.630	0.750	0.787	0.793	0.793	0.786	0.774	0.733	0.669	0.482
	10:00	0.688	0.819	0.859	0.865	0.865	0.858	0.845	0.800	0.730	0.526
$\beta =$	10:30	0.734	0.874	0.916	0.923	0.923	0.916	0.902	0.853	0.779	0.562
0	11:00	0.767	0.914	0.958	0.965	0.965	0.958	0.943	0.892	0.814	0.587
(Mar.,	11:30	0.787	0.938	0.983	0.991	0.990	0.983	0.968	0.916	0.836	0.603
Sept.)	12:00	0.794	0.946	0.992	0.999	0.999	0.991	0.976	0.923	0.843	0.608
	12:30	0.787	0.938	0.983	0.991	0.990	0.983	0.968	0.916	0.836	0.603
	1:00	0.767	0.914	0.958	0.965	0.965	0.958	0.943	0.892	0.814	0.587
	1:30	0.734	0.874	0.916	0.923	0.923	0.916	0.902	0.853	0.779	0.562
	2:00	0.688	0.819	0.859	0.865	0.865	0.858	0.845	0.800	0.730	0.526
	2:30	0.630	0.750	0.787	0.793	0.793	0.786	0.774	0.733	0.669	0.482
	3:00	0.562	0.669	0.701	0.706	0.706	0.701	0.690	0.653	0.596	0.430

TABLE A.1 Values for cos ϑ^a—Continued

Declination	Solar Time	Collector Tilt Angle (deg.), Due South									
		0	18.5	30	35	40	45	50	60	70	90
	8:30	0.597	0.630	0.617	0.604	0.586	0.564	0.537	0.472	0.393	0.201
	9:00	0.674	0.721	0.713	0.700	0.682	0.659	0.631	0.561	0.474	0.259
	9:30	0.741	0.801	0.797	0.785	0.767	0.743	0.714	0.639	0.545	0.310
	10:00	0.797	0.868	0.867	0.856	0.838	0.814	0.783	0.705	0.605	0.354
β =	10:30	0.842	0.922	0.923	0.912	0.894	0.870	0.839	0.757	0.653	0.388
−11.75	11:00	0.875	0.961	0.964	0.953	0.936	0.911	0.879	0.795	0.687	0.413
(Apr.,	11:30	0.895	0.984	0.989	0.978	0.961	0.935	0.903	0.818	0.708	0.428
Aug.)	12:00	0.901	0.992	0.997	0.987	0.969	0.944	0.911	0.826	0.715	0.433
	12:30	0.895	0.984	0.989	0.978	0.961	0.935	0.903	0.818	0.708	0.428
	1:00	0.875	0.961	0.964	0.953	0.936	0.911	0.879	0.795	0.687	0.413
	1:30	0.842	0.922	0.923	0.912	0.894	0.870	0.839	0.757	0.653	0.388
	2:00	0.797	0.868	0.867	0.856	0.838	0.814	0.783	0.705	0.605	0.354
	2:30	0.741	0.801	0.797	0.785	0.767	0.743	0.714	0.639	0.545	0.310
	3:00	0.674	0.721	0.713	0.700	0.682	0.659	0.631	0.561	0.474	0.259
	8:30	0.664	0.653	0.611	0.585	0.555	0.521	0.482	0.395	0.295	0.072
	9:00	0.737	0.740	0.703	0.678	0.647	0.612	0.572	0.480	0.373	0.129
	9:30	0.801	0.816	0.783	0.758	0.723	0.692	0.651	0.555	0.441	0.178
	10:00	0.856	0.881	0.850	0.826	0.796	0.760	0.718	0.618	0.499	0.219
β =	10:30	0.899	0.932	0.904	0.881	0.850	0.814	0.771	0.668	0.544	0.252
−20.24	11:00	0.938	0.969	0.943	0.920	0.890	0.853	0.809	0.704	0.578	0.276
(May,	11:30	0.949	0.992	0.967	0.944	0.914	0.877	0.833	0.726	0.598	0.291
July)	12:00	0.955	1.000	0.975	0.952	0.922	0.884	0.840	0.734	0.604	0.296
	12:30	0.949	0.992	0.967	0.944	0.914	0.877	0.833	0.726	0.598	0.291
	1:00	0.930	0.969	0.943	0.920	0.890	0.853	0.809	0.704	0.578	0.276
	1:30	0.899	0.932	0.904	0.881	0.850	0.814	0.771	0.668	0.544	0.252
	2:00	0.856	0.881	0.850	0.826	0.796	0.760	0.718	0.618	0.499	0.219
	2:30	0.801	0.816	0.783	0.758	0.728	0.692	0.651	0.555	0.441	0.178
	3:00	0.737	0.740	0.703	0.678	0.647	0.612	0.572	0.480	0.373	0.129

	8:30	0.685	0.657	0.605	0.575	0.540	0.501	0.458	0.363	0.256	0.023
	9:00	0.757	0.743	0.635	0.665	0.630	0.591	0.547	0.446	0.332	0.078
	9:30	0.820	0.818	0.773	0.744	0.709	0.669	0.624	0.519	0.399	0.126
	10:00	0.873	0.881	0.839	0.811	0.776	0.735	0.689	0.581	0.455	0.167
	10:30	0.915	0.931	0.692	0.864	0.829	0.788	0.741	0.630	0.500	0.199
$\beta =$	11:00	0.946	0.967	0.930	0.902	0.867	0.826	0.778	0.666	0.533	0.223
-23.45	11:30	0.964	0.989	0.953	0.926	0.891	0.849	0.801	0.687	0.552	0.237
(June)	12:00	0.970	0.997	0.961	0.933	0.899	0.857	0.809	0.694	0.559	0.242
	12:30	0.964	0.989	0.953	0.926	0.891	0.849	0.801	0.687	0.552	0.237
	1:00	0.946	0.967	0.930	0.902	0.867	0.826	0.778	0.666	0.533	0.223
	1:30	0.915	0.931	0.892	0.864	0.829	0.788	0.741	0.630	0.500	0.199
	2:00	0.873	0.881	0.839	0.811	0.776	0.735	0.689	0.581	0.455	0.167
	2:30	0.820	0.818	0.773	0.744	0.709	0.669	0.624	0.519	0.399	0.126
	3:00	0.757	0.743	0.695	0.665	0.630	0.591	0.547	0.446	0.332	0.078

[a] Values of cos ϑ for various collector angles and months calculated for 37.5° N. latitude. Collectors are mounted due south. The declination values correspond approximately to the 21st day of each month.

TABLE A.2 Values for cos ϑ^a

Declination	(West) Solar Time	Collector Tilt Angle (30° off Due South) 0	18.5	30	35	40	45	50	60	70	90	(East) Solar Time
	8:30	0.202	0.256	0.276	0.282	0.285	0.287	0.286	0.277	0.260	0.204	3:30
	9:00	0.273	0.351	0.382	0.391	0.396	0.399	0.398	0.388	0.367	0.291	3:00
	9:30	0.336	0.439	0.480	0.492	0.500	0.505	0.505	0.495	0.470	0.378	2:30
	10:00	0.389	0.516	0.568	0.584	0.595	0.602	0.604	0.595	0.568	0.463	2:00
	10:30	0.431	0.582	0.646	0.665	0.680	0.690	0.694	0.687	0.659	0.544	1:30
β=	11:00	0.462	0.635	0.711	0.735	0.753	0.766	0.773	0.769	0.742	0.621	1:00
23.45	11:30	0.480	0.675	0.762	0.791	0.813	0.829	0.839	0.840	0.815	0.693	12:30
(Dec.)	12:00	0.487	0.702	0.800	0.833	0.859	0.879	0.892	0.899	0.877	0.757	12:00
	12:30	0.480	0.713	0.822	0.859	0.890	0.914	0.931	0.944	0.928	0.812	11:30
	1:00	0.462	0.710	0.829	0.871	0.906	0.934	0.955	0.975	0.965	0.859	11:00
	1:30	0.431	0.693	0.821	0.867	0.906	0.938	0.963	0.991	0.989	0.895	10:30
	2:00	0.389	0.661	0.798	0.847	0.890	0.926	0.956	0.992	0.999	0.921	10:00
	2:30	0.336	0.616	0.759	0.812	0.859	0.900	0.933	0.979	0.995	0.936	9:30
	3:00	0.273	0.557	0.706	0.763	0.813	0.858	0.895	0.950	0.976	0.939	9:00
	8:30	0.243	0.284	0.294	0.295	0.293	0.290	0.284	0.266	0.240	0.166	3:30
	9:00	0.317	0.381	0.402	0.406	0.407	0.404	0.399	0.379	0.348	0.255	3:00
	9:30	0.381	0.470	0.502	0.509	0.513	0.513	0.508	0.488	0.454	0.344	2:30
	10:00	0.435	0.549	0.592	0.604	0.610	0.612	0.610	0.591	0.554	0.431	2:00
β =	10:30	0.478	0.617	0.671	0.687	0.697	0.702	0.702	0.685	0.647	0.515	1:30
20.24	11:00	0.509	0.671	0.738	0.758	0.772	0.780	0.782	0.769	0.732	0.594	1:00
(Jan.,	11:30	0.528	0.713	0.791	0.815	0.833	0.845	0.850	0.841	0.807	0.666	12:30
Nov.)	12:00	0.535	0.739	0.829	0.858	0.880	0.896	0.904	0.901	0.871	0.732	12:00
	12:30	0.528	0.751	0.852	0.885	0.912	0.931	0.944	0.947	0.922	0.789	11:30
	1:00	0.509	0.748	0.859	0.897	0.928	0.952	0.968	0.979	0.960	0.836	11:00
	1:30	0.478	0.731	0.851	0.893	0.928	0.956	0.977	0.996	0.985	0.874	10:30
	2:00	0.435	0.698	0.827	0.873	0.912	0.944	0.969	0.997	0.995	0.900	10:00
	2:30	0.381	0.652	0.787	0.837	0.880	0.917	0.946	0.983	0.990	0.915	9:30
	3:00	0.317	0.592	0.734	0.786	0.833	0.874	0.907	0.954	0.972	0.919	9:00

β = 11.75 (Feb., Oct.)	8:30	0.350	0.352	0.335	0.324	0.310	0.293	0.275	0.231	0.181	0.065	3:30
	9:00	0.426	0.454	0.448	0.440	0.428	0.413	0.395	0.350	0.294	0.158	3:00
	9:30	0.493	0.547	0.552	0.548	0.539	0.526	0.509	0.464	0.404	0.251	2:30
	10:00	0.550	0.630	0.647	0.646	0.641	0.630	0.615	0.571	0.509	0.342	2:00
	10:30	0.595	0.700	0.729	0.733	0.731	0.724	0.711	0.669	0.606	0.429	1:30
	11:00	0.627	0.757	0.799	0.807	0.809	0.805	0.795	0.756	0.695	0.511	1:00
	11:30	0.647	0.800	0.854	0.867	0.873	0.873	0.866	0.832	0.773	0.587	12:30
	12:00	0.654	0.828	0.894	0.911	0.922	0.926	0.922	0.894	0.839	0.655	12:00
	12:30	0.647	0.840	0.918	0.940	0.955	0.963	0.964	0.943	0.893	0.715	11:30
	1:00	0.627	0.837	0.925	0.952	0.972	0.984	0.989	0.976	0.933	0.765	11:00
	1:30	0.595	0.819	0.917	0.948	0.972	0.989	0.998	0.993	0.958	0.803	10:30
	2:00	0.550	0.785	0.891	0.927	0.955	0.976	0.990	0.995	0.969	0.831	10:00
	2:30	0.493	0.736	0.850	0.890	0.922	0.947	0.966	0.980	0.964	0.847	9:30
	3:00	0.426	0.674	0.794	0.837	0.873	0.903	0.925	0.950	0.945	0.851	9:00
β = 0 (Mar., Sept.)	8:30	0.482	0.433	0.380	0.352	0.321	0.288	0.253	0.177	0.095		3:30
	9:00	0.559	0.537	0.495	0.470	0.442	0.410	0.376	0.298	0.211	0.021	3:00
	9:30	0.628	0.632	0.602	0.581	0.555	0.526	0.492	0.414	0.323	0.116	2:30
	10:00	0.685	0.716	0.698	0.681	0.659	0.632	0.600	0.523	0.430	0.209	2:00
	10:30	0.731	0.788	0.782	0.770	0.751	0.728	0.698	0.624	0.530	0.298	1:30
	11:00	0.764	0.846	0.853	0.845	0.831	0.811	0.784	0.713	0.620	0.382	1:00
	11:30	0.784	0.890	0.909	0.906	0.896	0.880	0.856	0.790	0.700	0.460	12:30
	12:00	0.791	0.918	0.950	0.952	0.947	0.934	0.914	0.854	0.768	0.530	12:00
	12:30	0.784	0.931	0.975	0.981	0.980	0.972	0.956	0.903	0.823	0.590	11:30
	1:00	0.764	0.928	0.982	0.994	0.997	0.994	0.982	0.937	0.864	0.641	11:00
	1:30	0.731	0.909	0.973	0.989	0.997	0.998	0.991	0.955	0.890	0.681	10:30
	2:00	0.685	0.875	0.948	0.968	0.980	0.986	0.983	0.956	0.900	0.709	10:00
	2:30	0.628	0.825	0.906	0.930	0.947	0.956	0.959	0.941	0.896	0.725	9:30

TABLE A.2 Values for cos ϑ^{a}—Continued

Declination	(West) Solar Time	Collector Tilt Angle (30° off Due South)										(East) Solar Time
		0	18.5	30	35	40	45	50	60	70	90	
	8:30	0.597	0.498	0.410	0.366	0.319	0.270	0.219	0.113	0.002		3:30
	9:00	0.674	0.600	0.522	0.482	0.438	0.390	0.340	0.231	0.116		3:00
	9:30	0.741	0.693	0.627	0.590	0.549	0.503	0.454	0.345	0.226		2:30
	10:00	0.797	0.775	0.721	0.688	0.650	0.607	0.559	0.452	0.330	0.061	2:00
$\beta =$	10:30	0.842	0.846	0.804	0.775	0.741	0.701	0.655	0.550	0.428	0.149	1:30
-11.75	11:00	0.875	0.903	0.873	0.849	0.819	0.782	0.739	0.637	0.516	0.231	1:00
(Apr.,	11:30	0.895	0.946	0.928	0.909	0.883	0.850	0.810	0.713	0.594	0.307	12:30
Aug.)	12:00	0.901	0.974	0.968	0.953	0.932	0.903	0.867	0.776	0.661	0.375	12:00
	12:30	0.895	0.986	0.992	0.982	0.965	0.940	0.908	0.824	0.715	0.435	11:30
	1:00	0.875	0.983	1.000	0.994	0.981	0.961	0.933	0.857	0.754	0.484	11:00
	1:30	0.842	0.965	0.991	0.990	0.981	0.966	0.942	0.874	0.780	0.523	10:30
	2:00	0.797	0.931	0.966	0.969	0.965	0.953	0.934	0.876	0.790	0.551	10:00
	2:30	0.741	0.882	0.925	0.932	0.932	0.924	0.910	0.861	0.786	0.567	9:30
	3:00	0.674	0.820	0.869	0.879	0.883	0.880	0.870	0.831	0.766	0.570	9:00
	8:30	0.664	0.531	0.420	0.366	0.310	0.251	0.190	0.064			3:30
	9:00	0.737	0.629	0.528	0.477	0.423	0.365	0.305	0.178	0.045		3:00
	9:30	0.801	0.718	0.628	0.581	0.529	0.474	0.414	0.287	0.150		2:30
	10:00	0.856	0.797	0.719	0.675	0.627	0.573	0.516	0.389	0.251		2:00
$\beta =$	10:30	0.899	0.864	0.798	0.758	0.713	0.663	0.607	0.483	0.344	0.039	1:30
-20.24	11:00	0.930	0.919	0.864	0.829	0.788	0.741	0.688	0.567	0.429	0.118	1:00
(May,	11:30	0.949	0.960	0.917	0.887	0.849	0.806	0.756	0.639	0.504	0.190	12:30
July)	12:00	0.955	0.987	0.955	0.929	0.896	0.857	0.810	0.699	0.567	0.256	12:00
	12:30	0.949	0.999	0.978	0.957	0.928	0.892	0.850	0.746	0.619	0.313	11:30
	1:00	0.930	0.996	0.986	0.969	0.944	0.913	0.874	0.777	0.657	0.361	11:00
	1:30	0.899	0.978	0.977	0.964	0.944	0.917	0.882	0.794	0.681	0.398	10:30
	2:00	0.856	0.946	0.953	0.944	0.928	0.905	0.875	0.795	0.691	0.424	10:00
	2:30	0.801	0.899	0.914	0.909	0.896	0.877	0.852	0.781	0.687	0.439	9:30
	3:00	0.737	0.840	0.860	0.858	0.849	0.834	0.813	0.752	0.668	0.443	9:00

	8:30	0.685	0.541	0.422	0.364	0.304	0.242	0.177	0.045			3:30
	9:00	0.757	0.636	0.527	0.473	0.415	0.354	0.290	0.156	0.018		3:00
	9:30	0.820	0.724	0.625	0.574	0.519	0.460	0.397	0.263	0.121		2:30
	10:00	0.873	0.801	0.713	0.666	0.614	0.557	0.496	0.363	0.219		2:00
	10:30	0.915	0.867	0.791	0.748	0.699	0.645	0.586	0.455	0.310		1:30
$\beta =$	11:00	0.946	0.920	0.856	0.817	0.772	0.721	0.665	0.537	0.393	0.074	1:00
-23.45	11:30	0.964	0.960	0.906	0.873	0.832	0.784	0.731	0.608	0.466	0.145	12:30
(June)	12:00	0.970	0.987	0.945	0.915	0.878	0.834	0.784	0.666	0.528	0.209	12:00
	12:30	0.964	0.998	0.967	0.942	0.909	0.869	0.823	0.712	0.579	0.265	11:30
	1:00	0.946	0.996	0.975	0.953	0.925	0.889	0.846	0.743	0.616	0.311	11:00
	1:30	0.915	0.978	0.966	0.949	0.925	0.893	0.855	0.759	0.640	0.348	10:30
	2:00	0.873	0.946	0.943	0.929	0.909	0.882	0.847	0.760	0.650	0.374	10:00
	2:30	0.820	0.901	0.904	0.895	0.878	0.855	0.825	0.747	0.646	0.389	9:30
	3:00	0.757	0.842	0.852	0.845	0.832	0.813	0.787	0.718	0.627	0.392	9:00

[a] Values of cos ϑ for various collector angles and months calculated for 37.5° N. latitude. Collectors are mounted 30° off due south. Read left time column for west orientation, right time column for east orientation of collectors with respect to due south. The declination values correspond to the 21st day of each month.

B

Daily Insolation on an Inclined Surface Under Full-Sunshine Conditions

The solar radiation received at normal incidence outside the earth's atmosphere is approximately 420 Btu-hr^{-1}-ft^{-2}.[1] During passage through the atmosphere, part of this radiation is absorbed, part is reflected back, and part is received on the earth as diffuse radiation. Only about 60% of the solar constant is received on the earth as direct radiation. The diffuse radiation is usually between 8% and 15% of the direct component.

Two factors influence the amount of radiation received under clear-sky conditions on the earth: the elevation above sea level and the solar altitude. At increasing elevations the solar radiation must travel through less ''air mass,'' and thus less absorption and backward scattering can be expected.

As Fig. B.1 indicates, an increase in solar radiation by 5% can be expected at 7000 ft above sea level. For lower sun altitudes, in turn, the effective air mass is increased, and a corresponding decrease in received solar radiation can be expected. At noon on December 21 at 40° N. latitude, the sun altitude is only 26.6°, and an insolation of only 220 Btu-hr^{-1}-ft^{-2} is to be expected in this case. However, this is in disagreement with actual insolation measurements, which indicate only slight seasonal variations of the insolation received on the earth.[2] We propose to use the insolation values listed in Table B.1 as sufficiently accurate approximations for any location within the continental United States.[3]

[1]The "solar constant" is actually subject to minor fluctuations, mainly caused by slight seasonal variations in the distance between the sun and the earth.

[2]The reasons for this fact include parameters such as the sun-to-earth distance, absolute humidity in the air, and the amount of particulate matter suspended in the atmosphere.

[3]Use these values if accurately measured data for the particular location are not available. It is certainly advisable to use reliable existing data if they are available. However, doubtful data, or data that are not based on reasonable statistics, should be avoided, and the values given in Table B.1 should be used instead.

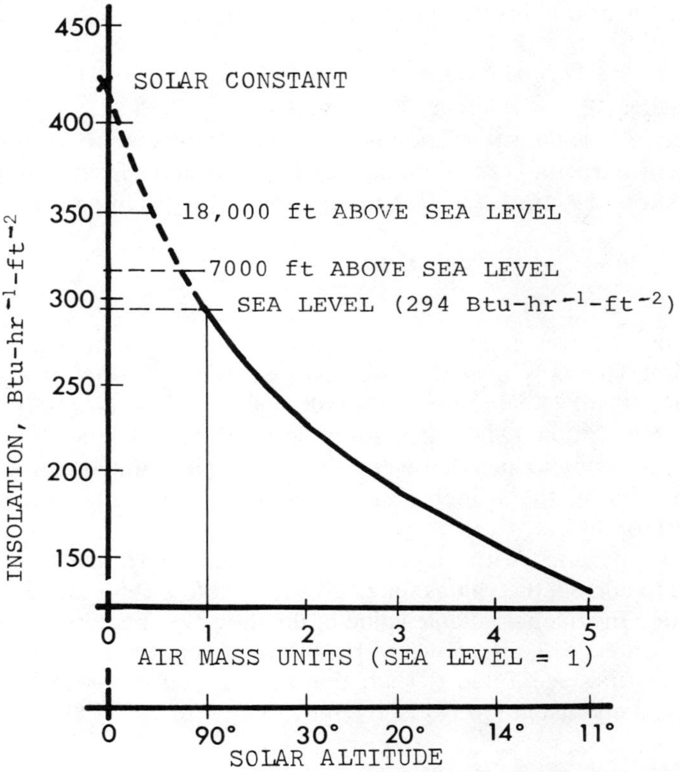

Fig. B.1. Dependence of insolation received at normal incidence on air mass or solar altitude, all other parameters being the same.

TABLE B.1 Direct, Diffuse, and Total Insolation[a]

	Oct. Feb.	Sept. Mar.	Aug. Apr.	July May	June	Nov. Dec. Jan.
I_0 (direct insolation)	255	260	265	270	275	250
I_1 (diffuse insolation)	26	26	27	27	28	25
I (total insolation)	281	286	292	297	303	275

[a] Direct, diffuse, and total insolation (Btu-hr^{-1}-ft^{-2}) received at normal incidence. These values are suggested to be used for sufficiently accurate solar collection computations at locations (anywhere within the continental United States) for which tabulated, accurately measured values are not available.

The clear-day insolation Q can be computed as

$$Q = \frac{1}{n} \sum_{t} [I_0 \cos \vartheta (t) + I_1] \tag{B.1}$$

In equation B.1, the summation is performed from sunrise to sunset and occurs n times per hour. If h is the time between sunrise and sunset, and if I_1 is expressed as a fraction of I_0 (i.e., $I_1 = k_1 I_0$), equation B.1 turns into:

$$Q = I_0(hk_1 + \frac{1}{n} \sum_{t} \cos \vartheta) \tag{B.2}$$

Table B.2 was computed on this basis,[4] using the insolation values listed in Table B.1. The daily insolation values are listed for various geographic latitudes between 32° and 44° and for receiver tilt angles of 0° (horizontal), 18.5° (4/12 roof), 30°, 45°, 60°, 75°, and 90°; calculations were for the 21st day of each month. Note the low insolation values that were obtained for horizontal or slightly tilted (18.5°) receivers during the winter months (or for steeply tilted receivers during the summer months).

In accordance with equation B.2, Q is proportional to I_0. It is therefore possible to correct the values for Q given in Table B.2 for any geographic location or situation in which a reliable value of the (hourly) direct insolation component at normal incidence, I_0, is known to be different from the I_0 value that was used for our calculation (see Table B.1). If that known hourly insolation value is I_0^*, the corrected daily insolation Q^* is given by

$$Q^* = \frac{I_0^*}{I_0} Q \tag{B.3}$$

[4]Two ($n = 2$) iterations were made per hour, and I_0 was assumed to be constant between sunrise and sunset. The program, performed with an HP 97 calculator, contained a subroutine to determine h for each day. The value of $\cos \vartheta$ was computed with the method presented in Appendix A.

TABLE B.2 Computation of Q with Equation B.2

Geographic Latitude (°)	Receiver Tilt (°)	Q [Btu Incident per Cloudless Day (/ft²)] Jan. Nov.	Feb. Oct.	Mar. Sept.	Apr. Aug.	May July	June	Dec.
	00	1276.	1582.	1970.	2363.	2641.	2788.	1145.
	18.5	1762.	1970.	2223.	2429.	2623.	2695.	1631.
	30	1989.	2124.	2280.	2361.	2493.	2516.	1864.
32	45	2180.	2212.	2236.	2150.	2196.	2156.	2068.
	60	2240.	2168.	2060.	1815.	1774.	1677.	2147.
	75	2165.	1995.	1764.	1379.	1256.	1110.	2096.
	90	1961.	1704.	1368.	871.	678.	495.	1919.
	0	1216.	1531.	1932.	2373.	2642.	2800.	1085.
	18.5	1716.	1936.	2205.	2481.	2644.	2724.	1585.
	30	1955.	2102.	2276.	2438.	2525.	2554.	1828.
34	45	2162.	2206.	2250.	2256.	2241.	2204.	2048.
	60	2240.	2182.	2091.	1945.	1829.	1731.	2144.
	75	2183.	2025.	1810.	1526.	1318.	1167.	2111.
	90	1995.	1749.	1426.	1027.	741.	531.	1949.
	0	1131.	1479.	1891.	2355.	2668.	2809.	1025.
	18.5	1612.	1900.	2184.	2485.	2654.	2750.	1537.
	30	1841.	2078.	2269.	2455.	2526.	2590.	1790.
36	45	2043.	2202.	2261.	2290.	2232.	2250.	2025.
	60	2122.	2193.	2119.	1993.	1812.	1784.	2139.
	75	2072.	2053.	1854.	1585.	1295.	1223.	2123.
	90	1697.	1792.	1482.	1093.	717.	606.	1978.
	0	1073.	1425.	1849.	2334.	2667.	2815.	964.
	18.5	1566.	1862.	2162.	2486.	2670.	2776.	1487.
	30	1806.	2052.	2260.	2469.	2552.	2627.	1750.
38	45	2023.	2193.	2270.	2321.	2268.	2300.	2001.
	60	2119.	2202.	2146.	2039.	1857.	1843.	2131.
	75	2086.	2079.	1896.	1642.	1346.	1287.	2132.
	90	1927.	1833.	1537.	1158.	770.	671.	2004.
	0	1013.	1369.	1805.	2310.	2664.	2818.	902.
	18.5	1518.	1822.	2137.	2482.	2689.	2800.	1436.
	30	1768.	2024.	2249.	2478.	2584.	2662.	1708.
40	45	2001.	2181.	2277.	2346.	2316.	2348.	1974.
	60	2113.	2208.	2170.	2078.	1917.	1900.	2121.
	75	2098.	2103.	1936.	1693.	1414.	1351.	2140.
	90	1956.	1873.	1590.	1217.	841.	736.	2029.
	0	953.	1313.	1759.	2285.	2657.	2847.	815.
	18.5	1469.	1786.	2109.	2476.	2702.	2893.	1317.
	30	1729.	1994.	2235.	2485.	2609.	2793.	1575.
42	45	1977.	2168.	2281.	2368.	2355.	2521.	1827.
	60	2106.	2212.	2192.	2115.	1967.	2105.	1969.
	75	2107.	2125.	1974.	1741.	1471.	1576.	1992.
	90	1982.	1910.	1642.	1274.	902.	968.	1893.

C

Determination of Sunshine Probability from Average Insolation Data

The probability of available sunshine, p, often is not known for specific locations. The weather data readily available may instead include average daily insolation values. Most often, such data are available for horizontal receiver surfaces, sometimes also for tilted surfaces facing south. It is possible to estimate the sunshine probability from such averaged insolation data.

In order to obtain reasonable accuracy in this conversion of averaged daily insolation values Q into sunshine probabilities p, it is necessary to consider that in this case the insolation during overcast periods, I_3, must not be neglected.[1] We can, however, express I_3 as a constant fraction of I_0:

$$I_3 = k_2 I_0 \tag{C.1}$$

A good approximation is $k_2 = 0.2$. Tabulated values of Q must then correspond to the following equation:

$$Q = \underbrace{p\, I_0\left(hk_1 + \frac{1}{n}\sum \cos\vartheta\right)}_{\text{clear-sunshine term}} + \underbrace{(1 - p)hk_2 I_0}_{\text{overcast term}} \tag{C.2}$$

The clear-sunshine term is equation B.2 (see Appendix B), which is a summation of the available insolation over the entire day multiplied with the sunshine probability. The period between sunrise and sunset is h, and n is the number of iterations per hour. The diffuse radiation is accounted for as a fraction (k_1) of the direct insolation. During overcast conditions, it is assumed that all radiation is received as diffuse radiation; $k_2 = 0.2 = 2k_1$ is suggested as a reasonable assumption for most locations within the continental United States.

[1]Compare this statement with Section 5.2.1, which is concerned with the collection efficiency. Here the insolation is negligible during overcast periods.

Equation C.2 can then be transposed to express p:

$$p = \frac{Q - hk_2 I_0}{I_0(hk_1 + \sum \cos \vartheta / n) - hk_2 I_0} \quad (C.3)$$

Table C.1 lists the probability of sunshine p, for averaged insolation values Q between 400 and 2200 Btu/day. The calculations were performed using equation C.3 with a programmable calculator (HP 97). The following assumptions were made:

$$\left.\begin{aligned} k_1 &= 0.1 \\ k_2 &= 0.2 \end{aligned}\right\} \text{ for all calculations}$$

I_0 = 250 Btu-hr^{-1}-ft^{-2} for Nov., Dec., Jan.
255 Btu-hr^{-1}-ft^{-2} for Oct., Feb.
260 Btu-hr^{-1}-ft^{-2} for Sept., March
265 Btu-hr^{-1}-ft^{-2} for Aug., April
270 Btu-hr^{-1}-ft^{-2} for July, May
275 Btu-hr^{-1}-ft^{-2} for June

$n = 2$

Dashes indicate Q ranges that are trivial or not applicable. The calculations were performed for geographic latitudes between 34° and 44°. The receiver tilt was 0° (horizontal) and 45° (winter only).

If, for example, at a specific location at 38° N. latitude an average insolation of 1600 Btu/day (horizontal receiver) is listed for the month of March, the sunshine probability is 80.5%, as can be seen from Table C.1.[2] Linear interpolations are permissible for latitudes and insolation values not listed.

[2]For the purpose of performing solar system collection calculations, this determination of p is quite accurate, as long as the correct value for I_0 listed earlier is then used in the actual system performance calculations.

TABLE C.1 Sunshine Probabilities[a]

Geographic Latitude	Q (Btu/day)	p for Horizontal Receiver (%) Jan. Nov.	Feb. Oct.	Mar. Sept.	Apr. Aug.	May July	June	Dec.	p for 45° Receiver (%) Jan. Nov.	Feb. Oct.	Mar. Sept.	Dec.
34°	600	—	—	—	—	—	—	23.6	—	5.3	—	9.3
	700	27.9	18.6	9.4	—	—	—	39.3	12.0	11.1	7.6	15.6
	800	41.9	28.4	16.7	—	—	—	55.0	18.0	17.0	13.5	21.9
	900	55.8	38.2	24.1	12.5	10.2	—	70.8	24.0	22.9	19.5	28.1
	1000	69.8	47.9	31.4	18.4	15.4	11.3	86.5	30.0	28.9	25.5	34.4
	1100	83.7	57.7	38.8	24.4	20.5	16.3	—	36.0	34.7	31.4	40.6
	1200	97.7	67.5	46.1	30.3	25.7	21.2	—	42.1	40.6	37.4	46.9
	1300	—	77.3	53.5	36.2	30.8	26.1	—	48.1	46.5	43.3	53.2
	1400	—	87.1	60.9	42.2	36.0	31.0	—	54.1	52.4	49.3	59.4
	1500	—	96.9	68.2	48.1	41.1	36.0	—	60.1	58.3	55.3	65.7
	1600	—	—	75.6	54.1	46.3	40.9	—	66.1	64.1	61.2	71.9
	1700	—	—	82.9	60.0	51.4	45.8	—	72.1	70.0	67.2	79.2
	1800	—	—	90.3	65.9	56.6	50.7	—	78.1	75.9	73.1	94.5
	1900	—	—	97.6	71.9	61.8	55.7	—	84.2	81.8	79.1	92.7
	2000	—	—	—	77.8	65.5	60.6	—	90.2	87.7	85.1	97.0
	2100	—	—	—	83.7	72.1	65.5	—	96.2	93.6	91.0	—
	2200	—	—	—	89.7	77.2	70.4	—	—	99.5	97.0	—
36°	600	22.0	—	—	—	—	—	26.0	9.4	5.3	—	9.5
	700	36.6	19.6	9.7	—	—	—	43.3	15.6	11.2	7.5	15.8
	800	51.3	29.9	17.2	—	—	—	60.8	21.9	17.1	13.5	22.2
	900	66.0	40.2	24.8	12.6	—	—	78.2	28.2	23.0	19.4	28.5
	1000	80.7	50.5	32.4	18.6	12.8	11.3	95.6	34.5	28.9	25.3	34.9
	1100	95.3	60.9	40.0	24.6	18.0	16.2	—	40.7	34.8	31.2	41.2
	1200	—	71.2	47.6	30.6	23.2	21.1	—	47.0	40.7	37.1	47.6
	1300	—	81.5	55.1	36.6	28.4	26.0	—	53.3	46.7	43.1	53.9
	1400	—	91.8	62.7	42.6	33.7	30.9	—	59.6	52.6	49.0	60.3
	1500	—	—	70.3	48.6	38.9	35.8	—	65.8	58.5	54.9	66.6
	1600	—	—	77.9	54.6	44.1	40.7	—	72.1	64.4	60.8	73.0
	1700	—	—	85.5	60.7	49.4	45.6	—	78.4	70.3	66.7	79.3
	1800	—	—	93.0	66.7	54.6	50.5	—	84.7	76.2	72.7	85.7
	1900	—	—	—	72.7	59.8	55.4	—	90.9	82.1	78.6	92.0
	2000	—	—	—	78.7	65.1	60.3	—	97.2	88.0	84.5	98.4
	2100	—	—	—	84.7	70.3	65.2	—	—	93.9	90.4	—
	2200	—	—	—	90.7	75.5	70.1	—	—	99.9	96.3	—

	600	24.0	—	—	—	—	—	29.1	9.5	5.3	7.5	9.6
	700	40.1	20.7	10.0	—	—	—	48.6	15.8	11.2	13.4	16.1
	800	56.1	31.7	17.8	—	—	—	68.1	22.2	17.2	19.3	22.5
	900	72.2	42.8	25.6	12.8	—	—	87.5	28.6	23.1	22.5	29.0
	1000	88.2	53.5	33.5	18.9	12.8	11.2	—	34.9	29.1	31.0	35.4
	1100	—	64.5	41.3	24.9	18.0	16.1	—	41.3	35.0	36.9	41.9
	1200	—	75.4	49.1	31.0	23.2	21.0	—	47.6	41.0	42.8	48.3
	1300	—	86.3	57.0	37.1	28.5	25.9	—	54.0	46.9	48.7	54.8
38°	1400	—	97.3	64.8	43.2	33.7	30.8	—	60.3	52.8	54.6	61.2
	1500	—	—	72.6	49.3	38.9	35.7	—	66.7	58.8	60.5	67.7
	1600	—	—	80.5	55.3	44.2	40.6	—	73.0	64.7	66.4	74.1
	1700	—	—	88.3	61.4	49.4	45.5	—	79.4	70.7	72.3	80.6
	1800	—	—	96.1	67.5	54.6	50.4	—	85.7	76.6	78.2	87.0
	1900	—	—	—	73.6	59.9	55.3	—	92.1	82.6	84.0	93.5
	2000	—	—	—	79.7	65.1	60.2	—	98.5	88.5	89.9	99.9
	2100	—	—	—	85.7	70.3	65.0	—	—	94.4	95.9	—
	2200	—	—	—	91.8	75.5	69.9	—	—	—	—	—
	600	26.6	10.4	—	—	—	—	33.2	9.6	—	—	—
	700	44.3	22.1	10.3	—	—	—	55.3	16.1	11.3	—	16.4
	800	62.1	33.7	18.4	—	—	—	77.4	22.5	17.3	13.3	22.9
	900	79.8	45.3	26.6	13.0	—	—	99.5	29.0	23.3	19.2	29.5
	1000	97.6	57.0	34.7	19.1	12.8	11.2	—	35.4	29.3	25.1	36.0
	1100	—	68.6	42.2	25.3	18.0	16.1	—	41.9	35.3	30.9	42.6
	1200	—	80.3	50.5	31.5	23.3	21.0	—	48.3	41.2	36.8	49.2
	1300	—	91.5	59.0	37.6	28.5	25.9	—	54.7	47.2	42.7	55.7
40°	1400	—	—	67.1	43.0	33.6	30.8	—	61.2	53.2	48.5	62.3
	1500	—	—	75.2	50.0	39.0	35.6	—	67.6	59.2	54.4	68.9
	1600	—	—	83.3	56.1	44.2	40.5	—	74.1	65.2	60.3	75.4
	1700	—	—	91.5	62.3	49.5	45.4	—	80.5	71.2	66.1	82.0
	1800	—	—	99.6	66.5	54.7	50.3	—	87.0	77.1	72.0	88.5
	1900	—	—	—	74.6	60.0	55.2	—	93.4	83.1	77.9	95.1
	2000	—	—	—	80.8	65.2	60.1	—	99.9	89.1	83.7	—
	2100	—	—	—	87.0	70.5	64.9	—	—	95.1	89.6	—
	2200	—	—	—	93.1	75.7	69.8	—	—	—	95.5	—

TABLE C.1 Sunshine Probabilities[a]—Continued

Geographic Latitude	Q (Btu/day)	p for Horizontal Receiver (%) Jan. Nov.	Feb. Oct.	Mar. Sept.	Apr. Aug.	May July	June	Dec.	p for 45° Receiver (%) Jan. Nov.	Feb. Oct.	Mar. Sept.	Dec.
42°	600	29.8	11.2	—	—	—	—	48.1	9.8	—	—	14.0
	700	49.6	23.6	10.7	—	—	—	72.2	16.3	11.4	—	21.0
	800	69.5	36.1	19.2	—	—	—	96.3	22.5	17.4	13.3	28.0
	900	89.4	48.5	27.6	13.2	—	—	—	29.4	23.5	19.2	35.0
	1000	—	61.0	36.0	19.4	12.6	8.7	—	36.0	29.5	25.0	42.0
	1100	—	73.5	44.4	25.7	18.1	13.6	—	42.5	35.5	30.9	49.0
	1200	—	85.9	52.9	32.0	23.4	16.5	—	49.1	41.6	36.7	56.0
	1300	—	98.4	61.3	38.2	28.6	23.5	—	55.6	47.6	42.6	63.0
	1400	—	—	69.7	44.5	33.9	28.4	—	62.2	53.6	48.4	70.0
	1500	—	—	78.1	50.8	39.1	33.4	—	68.7	59.7	54.3	77.0
	1600	—	—	86.6	57.0	44.4	38.3	—	75.3	65.7	60.1	84.0
	1700	—	—	95.0	63.3	49.7	43.3	—	81.8	71.7	66.0	91.0
	1800	—	—	—	69.6	54.9	48.2	—	88.4	77.8	71.8	98.1
	1900	—	—	—	75.9	60.2	53.2	—	94.9	83.8	77.7	—
	2000	—	—	—	82.1	65.4	58.1	—	—	89.8	83.5	—
	2100	—	—	—	88.4	70.7	63.1	—	—	95.9	89.4	—
	2200	—	—	—	94.6	76.0	68.0	—	—	—	95.2	—
44°	600	33.9	12.0	—	—	—	—	56.1	10.0	—	—	14.2
	700	56.5	25.5	11.2	—	—	—	84.2	16.6	11.5	—	21.4
	800	79.2	38.9	20.0	—	—	—	—	23.3	17.6	13.3	28.5
	900	—	52.3	28.8	13.4	—	—	—	29.9	23.7	19.1	35.7
	1000	—	65.8	37.5	18.8	12.9	8.6	—	36.6	29.8	25.0	42.8
	1100	—	79.2	46.3	26.2	18.2	13.6	—	43.3	35.9	30.8	50.0
	1200	—	92.6	55.1	32.5	23.5	18.5	—	49.9	42.0	36.7	57.1
	1300	—	—	63.9	38.9	28.8	23.5	—	56.6	48.1	42.5	64.3
	1400	—	—	72.6	45.3	34.0	28.4	—	63.3	54.2	48.4	71.4
	1500	—	—	81.4	51.7	39.3	33.4	—	69.9	60.2	54.2	78.6
	1600	—	—	90.2	58.1	44.6	38.3	—	76.6	66.3	60.1	85.7
	1700	—	—	99.0	64.4	49.9	43.2	—	83.3	72.4	65.9	92.8
	1800	—	—	—	70.8	55.2	48.2	—	89.9	78.5	71.7	—
	1900	—	—	—	77.2	60.5	53.1	—	96.6	84.6	77.6	—
	2000	—	—	—	83.6	65.8	58.1	—	—	90.7	83.4	—
	2100	—	—	—	90.0	71.0	63.0	—	—	96.8	89.3	—
	2200	—	—	—	96.3	76.3	67.9	—	—	—	95.1	—

[a] Sunshine probabilities for various average insolation values, geographic latitudes, seasons, and tilts (facing south).

D

Performance Comparison for Various Solar Collectors

As is apparent from the discussion of the multitude of factors that affect the efficiency of solar collectors (see Section 1.1.2), it is very difficult and often misleading to compare collectors on the basis of only a few selected indicators. In particular, the assumption about the ratio of the gross collector area to the net absorber area may lead to false interpretations. A particular collector (e.g., the one shown in Fig. 1.9) will, when compared in accordance with the common methods of comparison, which use only the gross collector area, turn out seemingly inferior to a collector that is identical except for an increased size of the rim (i.e., the window aperture of the collector of Fig. 1.9 would remain 45 × 91 in., but the gross area would be enlarged).

Other obvious factors that are important, but could not be assessed in the following comparison, which is based on performance parameters, are the quality and the cost of the products.

Table D.1 compares the performances and some other characteristic factors of various glazed flat-plate collectors. The information was compiled by the California Energy Commission[1] from certified test data submitted by the collector manufacturers. The tests were performed by one of seven accredited independent testing laboratories. Included in Table D.1 are all results reported in the California Energy Commission study [12], but "grandfathered" collectors were omitted.[2]

[1]1111 Howe Avenue, Sacramento, California; June 1979; the comparison was performed in accordance with California Assembly bill 1512.

[2]"Grandfathering" was used in the California Energy Commission report [12] to indicate testing before CEC inspection and certification of the laboratory equipment had taken place. A more complete list (TIPSE) became available in October 1980 and can be obtained from the California Energy Commission, 1111 Howe Avenue, Sacramento, California 95825.

Table D.1 contains the following details:

Column 1. Manufacturer, in alphabetical order.
Column 2. Model number of tested collector.
Column 3. Number of glazings.
Column 4. Type of absorber plate:
- Cu: all copper [pipe and absorber, compare columns (i) and (iii) in Table 1.3].
- Cu/Al: copper pipe/aluminum fin [see column (iv) in Table 1.3].
- Al: all aluminum [pipe and absorber, see column (ii) in Table 1.3].
- Cu/St: copper tube/steel absorber.
- P: polypropylene [see column (vi) in Table 1.3].

Column 5. Type of glazing:
- G: glass
- FC: fluorocarbon
- A: acrylic
- RP: reinforced plastic.

Column 6. Type of absorber coating:
- S: selective coating
- FB: flat-black coating.

Column 7. Thermal performance under typical operating conditions[3] for (a) swimming-pool-heating application[4] and for (b) domestic hot-water/space-heating application.[5]
Column 8. Linear thermal-loss coefficient[6] (Btu-hr^{-1}-ft^{-2}°F^{-1}).
Column 9. Gross collector area (ft^2)
Column 10. Collector weight, dry (lb).
Column 11. Maximum recommended collector flow rate (gpm).
Column 12. Maximum collector operating pressure (psi).
Column 13. Use restrictions. In addition to the restrictions mentioned below, and referring to letter references in the table (Table D.1),[7] it should be noted that collectors with maximum operating pressures of less than 80–100 psi should not be exposed to city water pressure as a regular operating condition. This restriction might represent a shortcoming in some designs for solar domestic water-heating systems in that it would required a heat exchanger and thus operate under a thermal system efficiency that is noticeably lower than the value indicated in column 7b.

[3]Note that because of variations in climate conditions the *absolute* values of the performance are of little significance. More significant are, instead, the *relative* values. High results in column 7 of Table D.1 may *not always* indicate a superior collector. The values will have significantly more relevance if the ratio of gross area to net absorber area is normalized (see Section 1.12).

[4]Low-ΔT operation

[5]Medium-ΔT operation

[6]The linear fit for the low-to-medium-ΔT range was chosen (see equation 1.6).

[7]These restrictions were copied verbatim from the California Energy Commission report [12].

TABLE D.1 Performance Data for Various Solar Collectors [12]

1	2	3	4	5	6	7a	7b	8	9	10	11	12	13
Alten Corporation 2594 Leghorn St., Mountain View, Ca. 94043, (415) 969-6474	100 G	1	Cu/Al	G	FB	74	45	1.024	30.1	180	2.5	100	
American HCP, Inc., 145 Mitchell Ave., S. San Francisco, Ca. 94080, (415) 952-2811	S-240 -G	1	P	RP	N/A	66	10	1.96	39.7	72	8	80	
American Home Solar Energy Systems, 868 Catalina, Laguna Beach, Ca. 92651	9120-2	1	Cu	FC	FB	66	35	1.085	32.7	135	2	80	
American National Energy Corporation, 9637 Distribution Ave., San Diego, Ca. 92121, (714) 578-2632	401 GP	1	Cu/Al	RP	FB	60	30	1.07	41	147	4	100	
American Solar King Corporation, 6801 New McGregor Hwy., Waco, Tx. 76710	DG 16	2	Cu	G	S	59	38	0.707	23.6	190	0.5	45	
Ametek, Inc., 1025 Polinski Rd., Ivyland, Pa. 18974, (215) 441-8770	D-189 -C	1	Cu	G	S	74	56	0.636	25.5	155	2.5	40	
Amsol, 415 East Branch, Arroyo Grande, Ca. 934200, (805) 481-1010	AMS-S -IC	1	Cu/Al	RP	FB	66	32	1.138	63.4	235	6	150	
Archer Industries, 350 Engle St., Escondido, Ca. 92025, (714) 489-0104	JH 101	1	Cu	G	FB	77	44	1.169	24	116	1	90	a
B&H Refrigeration, 1410 2nd Ave., Yuma, Az. 85364, (602) 726-5564	CSSPV30	1	Cu	G	FB	70	38	1.12	31.5	200	5	100	
	CSSPH30	1	Cu	G	FB	70	38	1.12	31.5	200	5	100	b
	CSSPV15	1	Cu	G	FB	70	38	1.12	15.9	106	5	100	b
	CSSPH15	1	Cu	G	FB	70	38	1.12	15.9	106	5	100	b
Colt, Inc., 71-950 San Jacinto, Rancho Mirage, Ca., (714) 489-0104	A-151	1	Al	G	FB	76	43	1.176	23.3	93	1.5	50	c
Elcam, Inc., 5330 Debbie Lane, Santa Barbara, Ca. 92111, (805) 964-9904	3200	1	Cu/Al	A	FB	63	27	1.266	32.8	100	d	120	
Energy Systems, Inc., 4570 Alvarado Can., San Diego, Ca. 92120, (714) 280-6660	1111-S	1	Cu/Al	G	FB	65	36	1.032	19.4	99	3	150	
Insolarator, 7926 Connoy Ct., San Diego, Ca., (714) 292-1857	18-75 DSS	2	Cu	G	S	76	44	1.13	18.4	116	3.5	120	e
International Solar Leasing Company, 7010 Convoy Ct., San Diego, Ca. 92111, (714) 560-9173	C-9121 T	1	Cu	G	FB	78	40	1.33	22.9	120	3	120	
Jet Air, Inc., 1071 Industrial Pl., El Cajon, Ca. 92020, (714) 442-9786	SGC 4	1	Cu	G	FB	74	42	1.126	18.3	88	3	75	

TABLE D.1 Performance Data for Various Solar Collectors [12]—Continued

1	2	3	4	5	6	7a	7b	8	9	10	11	12	13
Northrup, Inc., 302 Nichols Drive, Hutchings, Tx. 75141, (214) 225-4291	FP2G	2	Cu/Al	FC + RP	FB	62	37	0.897	26.8	180	0.4	125	
RA Energy Systems, 11459 A Woodside Ave.,	C 45	1	Cu	G	FB	70	37	1.16	21.6	150	2	120	
Lakeside, Ca. 92040, (714) 448-6216	C 48	1	Cu	G	FB	70	37	1.16	31.3	168	2	120	f
Revere Solar & Architectural Products,	232	2	Cu	G	S	69	46	0.82	18.8	120	2.5	100	
2107 S. Garfield Ave., Los Angeles, Ca. 90040, (213) 724-6982	432	2	Cu	G	S	69	46	0.82	22.6	149	2.5	100	g
Reynolds Metal Co., 5th and Cary St., Richmond,	1402	2	Al	FC	FB	66	41	0.899	31.8	70	2.5	96	h
Va. 23261, (804) 788-7709	1425	2	Al	+ RP	FB	66	41	0.899	47.8	104	2.5	96	h,i
Scantron Solar Ind., 8925 Carroll Way, A, San Diego, Ca. 92121, (714) 566-6830	7524-01	1	Cu/Al	RP	S	65	25	1.4	27.9	100	2.5	80	
Solar Captivator Systems Inc., 7192 Clairemont	AM 80 CP	1	Cu/St	G	FB	60	33	0.945	16.7	79	4.5	88	
Mesa Blvd., San Diego, Ca. 92111	AM120 CP	1	Cu/St	G	FB	60	33	0.945	25.8	123	4.5	88	j
Solar Energies of Caifornia, 10143 Prospect Ave., Santee, Ca. 92071, (714) 448-4300	C 36 SGI	1	Cu/Al	G	FB	65	36	1.003	18.4	120	5	120	
Solar Unlimited, Inc., 4310 Governors Dr, W,	SH 21	1	Cu/Al	G	FB	67	37	1.04	38.5	147	2	250	
Huntsville, Al. 35805, (205) 837-1320	SH 11	1	Cu/Al	G	FB	67	37	1.04	20.2	85	1	250	k
Sun Energy of Riverside, 25031 Alassandro Blvd., Sunnymead, Ca. 92388, (714) 653-9454	SE 40 T	1	Cu/Al	FC	FB	62	28	1.179	39.1	86	5	80	
Sunray of Solar Heaters, 4898 Ronson Ct., San Diego, Ca. 92111, (714) 278-7720	SR 18-S	1	Cu/Al	G	FB	71	43	0.993	18.4	80	1	75	
Western Energy Inc., 954 Forest Ave., Palo Alto, Ca. 94302, (415) 858-0221	AVIM109	1	Cu	G	FB	64	35	1.01	22.4	112	1.2	80	

[a] Do not exceed recommended flow rate.
[b] Thermal performance based on model CSSPV30.
[c] Use with parafinic oil-base fluid. Thermal performance based on water as heat-transport fluid.
[d] Given as 4 ft/sec.
[e] Install with vent hole down. Proper freeze protection must be used.
[f] Thermal performance based on model C 45.
[g] Thermal performance based on model 232.
[h] Water or water/glycol recommended as heat-transfer fluid. All laboratory tests conducted using water only.
[i] Thermal performance based on model 1402.
[j] Thermal performance based on model AM 80CP.
[k] Thermal performance based on model SH 21.

E

Heat Exchanger in Collector Loop: Effect on System Performance

A direct comparison between typical solar heating systems with and without heat exchangers in their collector loops is shown in Fig. E.1. The effectiveness of the heat exchanger, $E_x A/A_0$, is varied. E_x is the heat-exchange efficiency per unit heat-exchange area (Btu-hr^{-1}-ft^{-2}-°F^{-1}), and A_0 is the solar-collector area.

A recommended practical design goal is $E_x A/A_0 \approx 7$ to 10. At this heat-exchanger size, the daily solar collection with a system using a heat exchanger is only about 7% to 8% less than the collection that can be expected with a comparable system that uses no heat exchanger.[1] A less efficient heat exchanger drastically decreases the solar system efficiency and is not recommended.

Figure E.1 is applicable for heat exchangers that are immersed in the solar heat-storage tank. It is also approximately correct for tube-in-shell exchangers that operate at a high flow rate in the storage loop.

Although it is recommended that one consult the appropriate engineering reference data for the specification of E_x, which is subject to fowling and depends on the heat-exchanger geometry (for exchangers operating with convective heat transfer), the following approximate specifications have been used with satisfactory result:

$E_x \sim 80$ Btu-hr^{-1}-°F^{-1}-ft^{-2} for immersed pipes[2]
$E_x \sim 50$ Btu-hr^{-1}-°F^{-1}-ft^{-2} for immersed vertically arrayed plates
$E_x \sim 350$ Btu-hr^{-1}-°F^{-1}-ft^{-2} for tube-in-shell exchangers operating under turbulent-flow conditions; water is assumed as the heat-transfer fluid in all cases

[1]The graph in Fig. E.1 includes the temperature stratification effects in the heat-storage tank as described in the footnotes to Section 5.2.5.

[2]Cylindrical tank; pipes in the form of coils suspended near bottom of tank.

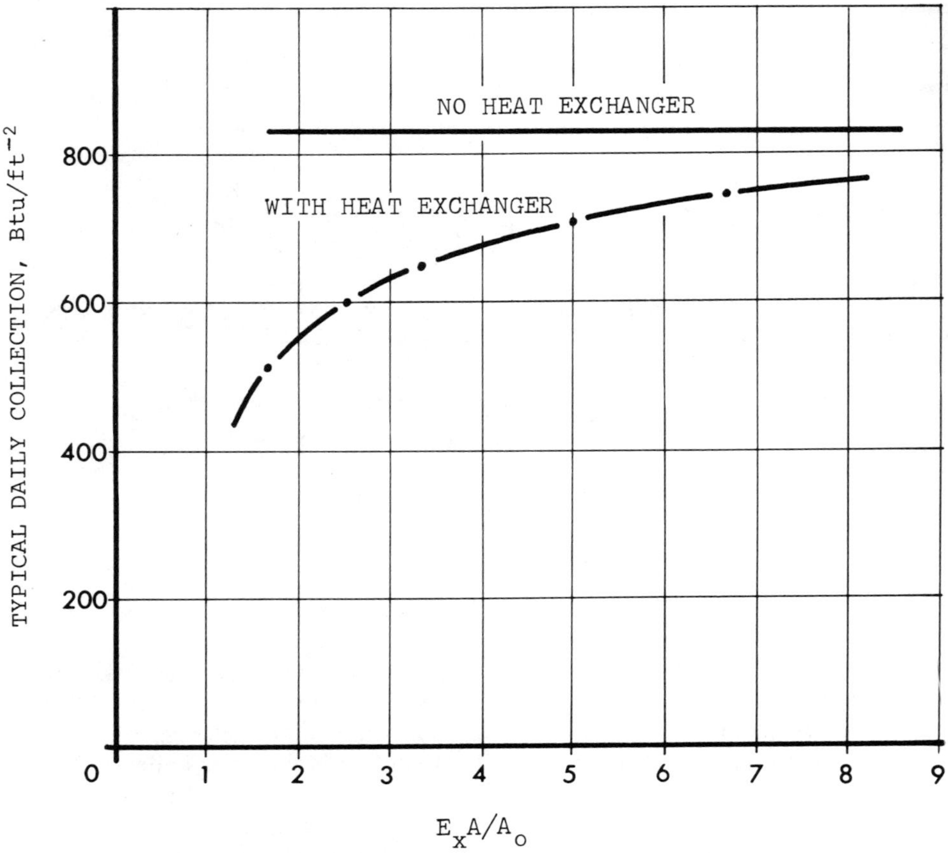

Fig. E.1. Effect of heat-exchanger area (A) on total daily solar heat collection for a typical solar installation. $E_x A/A_0$ is the heat-exchanger parameter; A_0 is the solar-collector area.

F

Example of Effect of a Reflective Surface in Front of Flat-Plate Collectors

Tables F.1 and F.2 give approximate collection values (Btu-hr^{-1}-ft^{-2}) that can be expected in addition to the regular collection if the collectors are aided by reflectors for the following conditions:

ω (reflector tilt): 28° for Table F.1; 0° for Table F.2
ψ (collector tilt): 45° for Table F.1; 60° for Table F.2
γ (geographic latitude): 38° N.
r_0 (reflectance): 0.90[1]
reflector area = collector area

Only sun azimuth values $-60° < \phi < +60°$ were used for obvious geometric reasons.

[1]A glazing transmission times absorber absorptance of $\epsilon = 0.85$ is included in the estimates of Tables F.1 and F.2.

TABLE F.1 Daily Reflector Augmentation for 28° Reflector Tilt Angle and 45° Collector Tilt, Approximate.

Collection Time	Additional Collection per Square Foot Reflector Area at 28° Reflector Tilt (Btu-hr^{-1}-ft^{-2})			
	June 21	May 21 July 21	Apr. 21 Aug. 21	Mar. 21 Sept. 21
Noon	100	110	135	95
11 A.M./1P.M.	125	130	132	90
10 A.M./2 P.M.	75	90	125	85
9 A.M./3 P.M.	0	0	0	27
8 A.M./4 P.M.	0	0	0	0
7 A.M./5 P.M.	0	0	0	0
Total per day	500	550	650	500

TABLE F.2 Daily Reflector Augmentation for 0° Reflector Tilt Angle and 60° Collector Tilt, Approximate

Collection Time	Additional Collection per Square Foot Reflector Area at 0° Reflector Tilt (Btu-hr^{-1}-ft^{-2})				
	Dec. 21	Nov. 21 Jan. 21	Oct. 21 Feb. 21	Sept. 21 Mar. 21	Aug. 21 Apr. 21
Noon	95	90	65	25	0
11 A.M/1 P.M.	90	95	70	35	0
10 A.M./2P.M.	75	85	85	50	0
9 A.M./3 P.M.	55	60	50	0	0
8 A.M./4 P.M.	0	0	0	0	0
Total per day	535	570	475	195	0

G

Computation of Average Collector Temperature Increase Due to Fin Efficiency Losses

The temperature rise due to irradiation of the fin portion, which is dx wide and has the thermal conductivity C and the thickness d_2 (see Fig. G.1) is

$$T(x + dx) - T(x) = d(\delta T) = \frac{qx\,dx}{Cd_2} \tag{G.1}$$

The total temperature increase δT over the entire film width is then

$$\delta T = T(x_0) - T_0 = \int_{x=0}^{x=x_0} \frac{qxdx}{Cd_2} = \frac{1}{2}\frac{qx_0^2}{Cd_2} \tag{G.2}$$

The mean fin temperature increase, $\overline{\delta T} = \overline{T(x)} - T_0$, can be computed as

$$\overline{\delta T} = \frac{\int_{x=0}^{x=x_0} \delta T(x)\,dx}{x_0} = \frac{1}{6}\frac{qx_0^2}{Cd_2} \tag{G.3}$$

If we now set $x_0 = d_1/2$, we obtain

$$\overline{\delta T} = \frac{1}{24}\frac{qd_1^2}{Cd_2} \tag{G.4}$$

Equation G.4 is correct only if all length dimensions are commensurate, i.e., if C is given in Btu-hr^{-1}-°F^{-1}-ft^{-1}. However, if C is expressed in Btu-hr^{-1}-°F^{-1}-ft^{-2}-in., which is normally the case, then equation G.4 must be multiplied by 12, and equation 1.11 results:

$$\overline{\delta T} = 0.5\frac{qd_1^2}{Cd_2} \quad \text{with} \quad \begin{array}{l} d_1,\ d_2 \text{ in ft} \\ C \text{ in Btu-hr}^{-1}\text{-°F}^{-1}\text{-ft}^{-2}\text{-in.} \end{array} \tag{G.5}$$

Note that d_1 is not the entire fin width, but the fin width minus the diameter of the duct for the heat-transport fluid.

[1]A glazing transmission times absorber absorptance of $\epsilon = 0.85$ is included in the estimates of Tables F.1 and F.2.

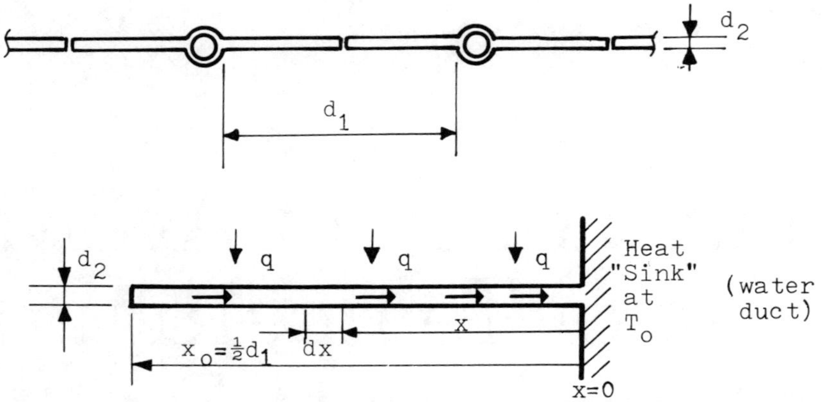

Fig. G.1. Schematic cross section through finned absorber (top) and enlarged view of a portion of it showing nomenclature for calculations (bottom).

H

Tables on Related Subjects

TABLE H.1 Average Monthly and Yearly Degree-F-Days for Selected U.S. Locations, for 65°F base temperature (after [18], rounded).

	Heating-Degree-F-Days, 65°F Base Temperature								
	OCT	NOV	DEC	JAN	FEB	MAR	APR	MAY	YEAR
ALABAMA									
Birmingham	95	365	555	590	460	365	110	10	2550
Huntsville	125	425	665	695	555	435	140	20	3070
Mobile	20	215	355	415	300	210	40	0	1560
Montgomery	70	330	525	545	415	315	90	0	2290
ARIZONA									
Phoenix	20	235	415	475	330	215	75	0	1765
Flagstaff	560	865	1075	1170	990	910	650	435	7150
Prescott	245	580	795	865	710	605	360	160	4360
Tucson	25	230	405	470	345	240	75	5	1800
Winslow	245	710	1010	1055	770	600	290	95	4780
Yuma	0	150	320	365	230	130	30	0	1215
ARKANSAS									
Fort Smith	125	450	705	780	595	455	145	20	3290
Little Rock	125	465	715	755	575	435	125	10	3220
Texarkana	80	345	560	625	470	350	105	0	2535
CALIFORNIA									
Bakersfield	35	280	500	545	365	265	105	20	2120
Bishop	250	575	795	875	665	540	305	145	4225
Blue Canyon	345	580	765	865	780	790	580	395	5505
Burbank	45	175	300	365	275	240	140	80	1645
Eureka	330	415	500	545	470	505	440	370	4645
Fresno	80	340	560	585	405	320	150	55	2490
Long Beach	40	155	290	375	295	265	170	90	1710
Los Angeles	80	180	290	370	300	290	220	160	2060
Mt. Shasta	405	695	900	985	785	740	530	345	5720
Oakland	125	310	480	525	400	355	255	180	2870
Palo Alto	155	355	515	550	420	375	260	165	2985
Red Bluff	55	320	555	605	430	340	170	45	2515

TABLE H.1 Average Monthly and Yearly Degree-F-Days—Continued

	Heating-Degree-F-Days, 65°F Base Temperature								
	OCT	NOV	DEC	JAN	FEB	MAR	APR	MAY	YEAR
Sacramento	80	365	575	615	440	360	215	100	2775
Sandberg	200	480	690	780	660	620	425	265	4210
San Diego	35	125	250	315	250	200	125	85	1440
San Francisco	145	305	460	510	395	365	280	215	3015
San Jose	105	285	450	490	365	315	215	120	2440
Santa Rosa	135	355	550	585	420	405	290	170	3065
Santa Cruz	165	310	445	495	400	375	290	215	3050
COLORADO									
Colorado Spr.	455	825	1030	1130	940	895	580	320	6425
Denver	430	820	1035	1130	940	885	560	290	6285
Grand Junction	315	785	1115	1210	905	730	385	145	5640
CONNECTICUT									
Bridgeport	305	615	985	1080	965	855	510	210	5615
Hartford	370	710	1120	1210	1060	900	495	175	6170
New Haven	345	650	1010	1095	990	870	545	245	5895
DELAWARE									
Wilmington	270	590	925	980	875	735	385	110	4930
FLORIDA									
Jacksonville	10	145	310	330	245	175	20	0	1240
Lakeland	0	55	165	195	145	100	0	0	660
Miami Beach	0	0	40	55	35	10	0	0	140
Orlando	0	70	200	220	165	105	5	0	765
Tallahassee	30	200	360	375	270	200	35	0	1485
Tampa	0	60	170	200	150	100	0	0	685
GEORGIA									
Athens	115	405	630	640	530	430	140	20	2930
Atlanta	125	415	625	640	530	435	170	25	2985
Columbus	85	335	545	550	435	340	95	0	2385
Rome	160	475	700	710	575	470	175	35	3325
Thomasville	25	200	365	395	305	210	35	0	1530
IDAHO									
Boise	415	790	1015	1115	855	720	440	245	5810
Idaho Falls 46W	625	1055	1370	1540	1250	1085	650	390	8475
Idaho Falls 42NW	650	1105	1430	1600	1290	1105	655	390	8760
Pocatello	495	900	1165	1325	1060	905	555	320	7035
Lewiston	405	755	935	1065	815	695	425	240	5540
ILLINOIS									
Cairo	165	515	790	855	680	540	195	45	3820
Chicago	325	755	1115	1210	1045	890	480	210	6155
Rockford	400	835	1220	1335	1135	960	515	235	6830
Springfield	290	695	1025	1135	935	770	355	135	5430
INDIANA									
Evansville	220	605	895	955	765	620	235	70	4435
Indianapolis	315	725	1050	1115	950	810	430	175	5700
South Bend	370	775	1125	1220	1070	935	525	240	6440
IOWA									
Burlington	320	770	1135	1260	1040	860	425	175	6115
Waterloo	430	910	1295	1460	1220	1025	530	230	7320
KANSAS									
Dodge City	250	665	940	1050	840	720	355	125	4985

TABLE H.1 Average Monthly and Yearly Degree-F-Days—Continued

	Heating-Degree-F-Days, 65°F Base Temperature								
	OCT	NOV	DEC	JAN	FEB	MAR	APR	MAY	YEAR
Goodland	380	810	1075	1165	955	885	505	235	6140
Wichita	230	620	905	1025	805	645	270	85	4620
KENTUCKY									
Covington	290	670	985	1035	895	755	390	150	5265
Lexington	240	610	900	945	820	685	325	105	4685
Louisville	250	610	890	930	820	680	315	105	4660
LOUISIANA									
Alexandria	55	275	430	470	360	260	70	0	1920
Baton Rouge	30	215	370	410	295	210	35	0	1560
Burrwood	0	95	215	300	220	170	25	0	1025
New Orleans	20	190	320	365	260	192	40	0	1385
Shreveport	45	295	475	550	425	304	80	0	2185
MAINE									
Caribou	680	1045	1535	1690	1470	1310	860	470	9765
Portland	510	805	1215	1340	1180	1040	675	370	7510
MARYLAND									
Baltimore	265	585	905	935	820	680	325	90	4655
Frederick	305	625	955	995	875	740	385	125	5085
MASSACHUSETTS									
Boston	315	605	985	1090	970	845	515	210	5635
Pittsfield	525	830	1230	1340	1195	1065	660	325	7580
Worcester	450	775	1170	1270	1125	995	610	305	6970
MICHIGAN									
Alpena	580	910	1270	1405	1300	1220	775	445	8505
Detroit	360	740	1090	1180	1060	935	520	220	6230
Escabana	540	925	1295	1445	1295	1205	775	455	8480
Flint	465	845	1210	1330	1200	1065	640	320	7375
Grand Rapids	435	805	1145	1260	1135	1010	550	280	6895
Lansing	430	815	1165	1260	1140	1010	580	275	6910
MINNESOTA									
Minneapolis	505	1015	1455	1630	1380	1165	620	290	8380
Rochester	475	1005	1440	1595	1365	1150	630	300	8295
Saint Cloud	550	1065	1500	1700	1445	1225	665	325	8880
Internat. Falls	700	1235	1725	1920	1620	1414	830	445	10605
MISSISSIPPI									
Jackson	65	315	500	545	415	310	85	0	2240
Vicksburg	55	280	460	510	385	280	70	0	2040
MISSOURI									
Columbia	250	650	965	1075	875	715	325	120	5045
Kansas	220	610	905	1030	820	680	295	110	4710
St. Joseph	285	710	1040	1170	950	770	350	135	5485
St. Louis	250	625	935	1025	850	705	310	120	4900
Springfield	225	600	875	975	780	660	290	105	4560
MONTANA									
Billings	485	895	1135	1295	1100	970	570	285	7050
Glasgow	610	1105	1465	1710	1440	1185	650	335	8995
Great Falls	545	920	1170	1350	1155	1265	640	385	7750
Havre	595	1065	1365	1585	1365	1180	655	340	8700
Helena	600	1000	1265	1440	1170	1040	650	350	8150
Missoula	650	1035	1285	1420	1120	970	620	390	8125

TABLE H.1 Average Monthly and Yearly Degree-F-Days—Continued

	Heating-Degree-F-Days, 65°F Base Temperature								
	OCT	NOV	DEC	JAN	FEB	MAR	APR	MAY	YEAR
NEBRASKA									
Lincoln	300	725	1065	1235	1015	835	400	170	5865
Norfolk	395	875	1235	1415	1180	985	500	235	6980
Omaha	355	830	1175	1355	1125	940	465	210	6610
Valentine	495	940	1235	1395	1175	1045	575	290	7425
NEVADA									
Elko	560	925	1195	1315	1035	910	620	410	7435
Las Vegas	80	385	615	670	485	335	110	5	2710
Reno	490	800	1025	1075	825	730	510	355	6330
Winnemucca	535	875	1090	1170	915	835	575	365	6760
NEW HAMPSHIRE									
Concord	505	820	1240	1360	1185	1030	635	300	7385
NEW JERSEY									
Atlantic City	250	550	880	935	850	740	420	135	4810
Newark	250	575	920	985	875	730	380	120	4860
Trenton	265	575	925	990	885	755	400	120	4980
NEW MEXICO									
Albuquerque	230	640	870	930	705	595	290	80	4350
Clayton	310	700	900	985	810	745	430	185	5160
Raton	430	825	1050	1115	905	835	545	300	6230
Roswell	200	575	805	840	640	480	200	30	3795
NEW YORK									
Albany	440	775	1195	1310	1155	990	565	240	6875
Binghamton (AP)	470	810	1185	1275	1155	1045	645	315	7285
Binghamton (PO)	405	730	1105	1190	1080	950	545	230	6450
Buffalo	440	775	1155	1255	1145	1040	645	330	7060
Central Park	235	540	900	985	885	760	410	120	4870
JFK Intl. Airp.	250	565	935	1030	935	815	480	165	5220
LaGuardia	225	530	885	975	880	750	415	125	4810
Rochester	415	745	1125	1235	1125	1015	595	280	6750
Schenectady	420	755	1160	1285	1130	970	545	210	6650
Syracuse	415	745	1155	1270	1140	1005	570	250	6755
NORTH CAROLINA									
Asheville	245	555	775	785	685	590	275	85	4040
Charlotte	125	440	690	690	580	480	155	20	3190
Raleigh	165	450	715	725	615	485	180	35	3395
Wilmington	75	290	520	545	460	355	95	0	2345
Winston-Salem	170	485	745	755	650	525	205	35	3595
NORTH DAKOTA									
Bismarck	575	1085	1465	1710	1440	1205	645	330	8850
Devils Lake	640	1190	1635	1870	1580	1345	755	380	9900
Fargo	575	1105	1570	1790	1520	1260	690	330	9225
OHIO									
Cincinnati	250	610	920	970	835	700	335	120	4805
Cleveland	385	740	1090	1150	1045	920	550	260	6350
Columbus	345	715	1040	1090	950	810	425	170	5660
Dayton	310	695	1045	1095	955	810	430	165	5620
Mansfield	395	770	1110	1170	1040	925	545	245	6405
Toledo	405	790	1140	1200	1055	925	545	240	6495

TABLE H.1 Average Monthly and Yearly Degree-F-Days—Continued

	Heating-Degree-F-Days, 65°F Base Temperature								
	OCT	NOV	DEC	JAN	FEB	MAR	APR	MAY	YEAR
OKLAHOMA									
Oklahoma City	165	500	765	870	665	525	190	35	3725
Tulsa	160	520	785	895	685	540	215	45	3860
OREGON									
Astoria	375	560	680	755	620	635	480	365	5185
Burns	515	865	1115	1245	990	855	570	365	6955
Eugene	365	585	720	805	625	590	425	280	4725
Medford	370	680	870	920	695	640	430	240	5010
Portland	335	595	735	825	645	585	395	245	4635
Salem	340	595	730	820	645	610	415	275	4755
PENNSYLVANIA									
Allentown	355	695	1065	1115	1000	850	470	165	5810
Harrisburg	298	650	990	1045	905	765	395	125	5250
Philadelphia	290	620	965	1015	890	745	390	115	5100
Pittsburgh	375	725	1065	1120	1000	875	480	195	5985
Reading	255	595	940	1000	885	735	370	105	4945
Williamsport	375	715	1075	1120	1000	855	470	175	5935
RHODE ISLAND									
Block Island	305	595	900	1020	955	875	610	345	5805
Providence	370	660	1025	1110	990	870	535	235	5955
SOUTH CAROLINA									
Charleston	60	280	470	485	390	290	55	0	2035
Columbia	80	345	575	570	470	355	80	0	2485
Florence	80	315	550	550	460	345	85	0	2385
Greenville	110	385	635	650	535	435	120	0	2885
Spartanburg	130	415	665	665	560	455	145	25	3075
SOUTH DAKOTA									
Huron	510	1015	1430	1630	1355	1125	600	290	8225
Rapid City	480	895	1170	1335	1145	1050	615	325	7345
Sioux Falls	460	970	1360	1545	1285	1080	575	270	7840
TENNESSEE									
Bristol	235	575	830	830	700	600	260	70	4145
Knoxville	170	490	725	730	615	495	200	45	3495
Memphis	130	445	700	730	585	455	145	20	3230
Nashville	160	495	730	780	645	510	190	40	3580
Oak Ridge (CO)	190	530	770	780	670	550	230	55	3815
TEXAS									
Austin	30	225	390	470	325	225	50	0	1710
Brownsville	0	65	150	205	105	75	0	0	600
Dallas	60	320	525	600	440	320	90	5	2365
El Paso	85	405	650	685	445	320	105	0	2700
Fort Worth	65	325	535	615	450	320	100	0	2405
Galveston	0	140	270	350	260	190	30	0	1235
Houston	5	185	305	385	290	190	35	0	1395
Lubbock	175	515	745	800	615	485	200	30	3580
Midland	85	380	590	650	470	320	90	0	2590
San Angelo	70	320	535	565	410	290	65	0	2255
San Antonio	30	205	365	430	285	195	40	0	1550
Waco	45	270	455	535	390	270	65	0	2030
Wichita Falls	100	380	630	700	520	380	120	5	2830

TABLE H.1 Average Monthly and Yearly Degree-F-Days—Continued

	Heating-Degree-F-Days, 65°F Base Temperature								
	OCT	NOV	DEC	JAN	FEB	MAR	APR	MAY	YEAR
UTAH									
Milford	445	865	1140	1250	990	820	520	280	6495
Salt Lake City	420	850	1080	1170	910	765	460	235	6050
Wendover	370	820	1090	1180	900	730	410	175	5780
VERMONT									
Burlington	540	890	1350	1515	1335	1185	715	355	8270
VIRGINIA									
Cape Henry	110	360	645	695	635	535	245	55	3280
Norfolk	135	410	700	740	655	535	215	35	3420
Richmond	215	495	785	815	705	545	220	55	3865
Roanoke	230	550	825	835	720	615	260	65	4150
Washington AP	215	520	835	870	760	625	290	75	4225
WASHINGTON									
Olympia	420	635	755	835	675	645	450	305	5235
Seattle	330	545	655	740	600	575	395	240	4425
S.—Boeing	385	625	765	830	655	610	410	240	4840
S.—Tacoma	390	635	750	830	680	655	475	295	5145
Spokane	495	880	1080	1230	980	835	530	290	6655
WEST VIRGINIA									
Charleston	250	590	865	880	770	650	330	95	4475
Elkins	400	730	990	1010	895	790	445	200	5675
Parkersburg	265	605	905	940	825	690	340	115	4755
Huntington	255	585	855	880	765	635	295	100	4445
WISCONSIN									
Madison	475	930	1330	1475	1275	1115	620	310	7865
Milwaukee	470	875	1250	1375	1195	1055	640	370	7635
WYOMING									
Casper	525	940	1170	1290	1085	1020	655	380	7410
Cheyenne	545	925	1100	1230	1055	1010	670	380	7280
Lander	555	1020	1300	1415	1145	1015	655	380	7870
Sheridan	540	950	1200	1355	1155	1055	640	365	7685

TABLE H.2 Survey of Computer Programs for Solar System Performance Calculations[a]

Program Name	Date of Latest Revision	Space Heating	Space Cooling	DWH	Passive Systems	Many System Parameters	Hour-by-Hour Simulation	Economics	Large Computer	Minicomputer	Hand-held Calculator	Address and Cost
BLAST	1981	x	x	x^c	x	$x^{e\text{-}k}$	x^l	x	x			Control Data Corporation, BLAST Computer Service, 8100 34th Avenue South, Minneapolis, Mn. 55440
DEROB	1981	x	x		x	(f,g,k)	x^n	x	x			Solenco, Inc., P.O. Box 7907 Austin, Tx. 78712 (cost $500[p])
DOE 2.1	1980	x	x	x^c	x	x	x^l	x	x			National Technical Information Service U.S. Dep. of Commerce, 5285 Port Royal Rd. Springfield, Va. 22161 (cost $475[p])
FCHART 4.0	1980	x		x^c		(e,i,j)	(m)	x	x		x	FCHART, P.O. Box 5562
FCHART 3.0	1978	x		x		(e,i,j)	(m)	x	x	x	x	Madison, Wi. 53705 (cost $200-300[p])
HISPER	1979	x^b		x^c		(e,j)	(n)		x			R.F. Elkin, EL-55, G.C. Marshall Space Flight Center, Al. 35812 (cost: N/C)
PASCAL II	1980	x^b			x	(f,g)	(n)	x			x	Total Environmental Action, Inc., Church Hill, Harrisville, N.H. 03450 (cost $75[p])
RSVP/2	1979	x^b		x		(j)	(m)	x	x			NTIC, 5285 Port Royal Road, Springfield, Va. 22161 (cost $210[p])
SEEC I	1980	x^b		x		(f,j)		x		x	x	Solar Environmental Engineering Co.,
SEEC II	1978	x^b		x				x			x	2524 E. Vine Drive, Fort Collins, Co. 80524
SEEC III	1978	x^b		x		(f)		x		x	x	(cost $95-200)
SUNSIM-1[o]		x	x	x		(f,g)	x			x		SOLARTEK, P.O. Box 298, Guilderland, N.Y. 12084 (cost $49[p])

TABLE H.2 Survey of Computer Programs for Solar System Performance Calculations[a]**—Continued**

Program Name	Date of Latest Revision	Space Heating	Space Cooling	DWH	Passive Systems	Many System Parameters	Hour-by-Hour Simulation	Economics	Large Computer	Minicomputer	Hand-held Calculator	Address and Cost
SYRSOL	1980	x		x		$x^{d,f,g,i,j}$	x^{m}		x			Manas Ucar, 734 Westcott Street, Syracuse, N.Y. 13210
TEANET III	1980	x^{b}	x^{b}		x	(g)	x^{n}				x	Total Environmental Action, Church Hill, Harrisville, N.H. 03450 (cost $95[p])
TRNSYS	1981	x	x	x^{c}	x	$x^{f\text{-}j}$	x^{l}	x	x			TRNSYS Coordinator, University of Wisconsin, 1500 Johnson Drive, Madison, Wi. 53706 (cost $400[p])
This report	1980	x	x	x^{c}	x	$x^{e\text{-}h,j}$	x^{n}	x			x	Klaus Heinemann, ELORET, 1178 Maraschino, Sunnyvale, Ca. 94087
KH-80	1981	x	x	x^{c}	x	$x^{d\text{-}k}$	x^{m}	x		x		(cost $390 for KH-80[p])

[a] Updated and revised [28].
[b] only residential
[c] also industrial process water heating
[d] includes reflector augmentation
[e] includes heat exchanger
[f] computes building heat loss
[g] computes solar window heat gain
[h] includes second-order collector thermal loss term
[i] includes incident angle modifier
[j] includes variation of collector tilt and azimuth
[k] includes collector shading
[l] processes weather data tapes
[m] weather data for several stations are contained in program
[n] user must input weather data
[o] from "Microcomputer Methods for Solar Design and Analysis," 1st Edition, 1981, Solar Energy Research Institute, 1617 Cole Blvd., Golden, Co. 80401. This report also lists several other microcomputer programs for active solar systems, most of which are based on the FCHART method.
[p] prices may vary.

In Table H.3, we make the following assumptions:

1. Insolation at normal incidence ($I_0 + I_1$), direct plus diffuse (Btu-hr^{-1}-ft^{-2}):
   275 in Nov., Dec., Jan.
   281 in Feb., Oct.
   286 in Mar., Sept.
   292 in Apr., Aug.
   297 in May, July
   303 in June
2. Diffuse component of insolation: $I_1 = 0.1I_0$
3. Collector efficiency parameters:
   $\epsilon = 0.736$
   $a = 0.82$ Btu-hr^{-1}-ft^{-2}-°F^{-1}
   $b = 0.718$ Btu^2-hr^{-2}-ft^{-4}-°F^{-2}
   } for 30 ft^2 gross collector area[1]
4. Temperature between ambient near collectors and ambient distant from collectors (see Sections 5.2.6 and 1.1.2): c = 5°F
5. Azimuth collector orientation: $\phi_0 = 0$ (due south)
6. Declination: 21st of each month (see Fig. I.1, Appendix I)
7. Geographic latitude: 36°, 38°, 40°, 42° north
8. Collector tilt angle: 18.5°, 30°, 45°, 60°
9. ΔT: 40°F, 60°F, 80°F (no heat exchanger[2])
10. Hourly iterations for computation at constant ΔT

This table may be used to estimate the collection with an average solar system for space or domestic water heating. Since during the course of a sunny day the temperature in the heat-storage tank (T_0) increases roughly at the same rate as the ambient temperature (T_a), the average ΔT operating value remains to a rough approximation constant during the collection day, i.e., $\Delta T \approx T_0$(8:00 A.M.) $-$ T_a(8:00). For more accurate collection computations, the program presented in Appendix I should be used.

[1]The calculation was performed for the reference collector used in Section 1.1.2.

[2]Note: The tabulated values do not include the effect of using a heat exchanger between the collectors and the heat-storage tank, nor do they include any heat losses from pipes and storage facilities. To include a heat exchanger, add an appropriate temperature difference to the ΔT value used.

TABLE H.3 Daily Collection with a Good Single-Glazed Flat-Plate Collector at Various Temperatures, Tilt Angles, and Geographic Latitudes (Btu-ft^{-2}-day^{-1})

Month	ΔT (°F)	Geographic Latitude 36°	38°	40°	42°	Collector Tilt Angle
Jan. 21, Nov. 21 (β = 20.24°) $I_0 + I_1$ = 275 Btu-hr^{-1}-ft^{-2}[a]	40	850	815	780	740	18.5
		1015	990	965	935	30
		1165	1150	1135	1115	45
		1220	1215	1210	1205	60
	60	630	595	555	515	18.5
		810	780	750	720	30
		960	945	925	910	45
		1015	1015	1010	1005	60
	80	415	385	350	320	18.5
		570	540	515	490	30
		730	715	700	680	45
		795	790	785	780	60
	100	220	195	170	145	18.5
		370	345	320	290	30
		495	485	470	455	45
		545	545	540	535	60
Feb. 21, Oct. 21 (β = 11.75°) $I_0 + I_1$ = 282 Btu-hr^{-1}-ft^{-2}[a]	40	1000	975	945	915	18.5
		1125	1105	1090	1065	30
		1210	1205	1195	1185	45
		1200	1205	1210	1215	60
	60	790	760	730	700	18.5
		920	900	880	860	30
		1010	1000	1000	980	45
		1000	1005	1010	1015	60
	80	555	525	500	475	18.5
		690	670	650	630	30
		780	780	770	760	45
		770	780	785	790	60
	100	355	330	305	275	18.5
		470	460	440	420	30
		545	540	535	525	45
		530	535	540	545	60
Mar. 21, Sept. 21 (β = 0°) $I_0 + I_1$ = 286 Btu-hr^{-1}-ft^{-2}[a]	40	1165	1150	1130	1115	18.5
		1220	1215	1210	1200	30
		1210	1220	1225	1230	45
		1110	1130	1145	1160	60
	60	960	945	930	970	18.5
		1020	1015	1010	1000	30
		1010	1020	1020	1025	45
		900	925	940	960	60
	80	735	720	700	680	18.5
		800	790	780	770	30
		785	790	800	800	45
		670	690	710	730	60
	100	505	490	475	460	18.5
		560	555	550	540	30
		550	560	560	565	45
		450	470	490	500	60

TABLE H.3 Daily Collection with a Good Single-Glazed Flat-Plate Collector at Various Temperatures, Tilt Angles, and Geographic Latitudes (Btu-ft^{-2}-day^{-1})—Continued

Month	ΔT (°F)	Geographic Latitude 36°	38°	40°	42°	Collector Tilt Angle
Apr. 21, Aug. 21 ($\beta = -11.75°$) $I_0 + I_1 =$ 292 Btu-hr^{-1}-ft^{-2a}	40	1275	1270	1270	1260	18.5
		1260	1270	1270	1275	30
		1155	1175	1190	1210	45
		960	990	1020	1050	60
	60	1080	1075	1070	1060	18.5
		1060	1070	1070	1080	30
		950	970	990	1010	45
		745	780	810	840	60
	80	855	850	850	840	18.5
		840	845	850	855	30
		720	745	765	780	45
		515	545	570	600	60
	100	610	610	600	595	18.5
		595	600	605	610	30
		500	520	535	550	45
		320	350	375	400	60
May 21, July 21 ($\beta = -20.24°$) $I_0 + I_1 =$ 297 Btu-hr^{-1}-ft^{-2a}	40	1330	1335	1340	1340	18.5
		1260	1275	1290	1300	30
		1085	1110	1140	1165	45
		825	860	900	935	60
	60	1130	1135	1140	1145	18.5
		1060	1075	1090	1145	30
		880	905	935	960	45
		605	640	680	720	60
	80	915	920	925	925	18.5
		840	855	870	885	30
		645	675	705	735	45
		395	430	465	495	60
	100	670	680	685	685	18.5
		595	610	625	640	30
		435	460	485	510	45
		205	235	265	300	60
June 21 ($\beta = -23.45°$) $I_0 + I_1 =$ 303 Btu-hr^{-1}-ft^{-2a}	40	1375	1375	1380	1385	18.5
		1290	1300	1310	1325	30
		1095	1115	1135	1155	45
		815	840	865	900	60
	60	1180	1180	1185	1185	18.5
		1090	1105	1115	1125	30
		890	910	930	955	45
		595	620	650	685	60
	80	960	965	965	970	18.5
		870	885	895	905	30
		660	680	900	925	45
		355	395	430	465	60
	100	725	725	730	735	18.5
		630	640	650	655	30
		420	450	475	500	45
		175	205	235	265	60

TABLE H.3 Daily Collection with a Good Single-Glazed Flat-Plate Collector at Various Temperatures, Tilt Angles, and Geographic Latitudes (Btu-ft^{-2}-day^{-1})—Continued

Month	ΔT (°F)	Geographic Latitude 36°	38°	40°	42°	Collector Tilt Angle
Dec. 21 (β = 23.45°) $I_0 + I_1$ = 275 Btu-hr^{-1}-ft^{-2} [a]	40	795	755	720	680	18.5
		980	950	920	890	30
		1150	1130	1110	1090	45
		1230	1225	1215	1210	60
	60	570	530	490	455	18.5
		770	740	705	670	30
		945	925	905	885	45
		1030	1025	1015	1005	60
	80	365	330	295	260	18.5
		530	505	475	450	30
		720	700	675	655	45
		805	800	790	780	60
	100	180	155	125	100	18.5
		330	305	275	245	30
		480	465	450	430	45
		560	550	545	535	60

[a] For locations with different insolation conditions, correct the table values by six times that difference.

The assumptions for Table H.4 are the following:

1. Collector efficiency parameters:
   $\epsilon = 0.63$
   $a = 0.50$ Btu^{-1}-hr^{-1}-ft^{-2}-°F^{-1}
   $b = 0.50$ Btu^{2}-hr^{-2}-ft^{-4}-°F^{-2}
   } for 30 ft^2 gross collector area
2. ΔT: 40°F, 60°F, 80°F, 120°F, 140°F
3. All other parameters same as in Table H.3

TABLE H.4 Daily Collection with a Double-Glass-Glazed Collector at Various Temperatures, Tilt Angles, and Geographic Latitudes (Btu-ft^{-2}-day^{-1})

Month	ΔT (°F)	Geographic Latitude 36°	38°	40°	42°	Collector Tilt Angle
Jan. 21, Nov. 21 (β = 20.24°) $I_0 + I_1$ = 275 Btu-hr^{-1}-ft^{-2a}	40	800	770	740	710	18.5
		940	920	895	870	30
		1065	1055	1040	1025	45
		1110	1110	1110	1105	60
	60	650	630	595	565	18.5
		810	785	760	735	30
		935	925	910	895	45
		985	985	980	975	60
	80	495	465	435	405	18.5
		655	630	605	580	30
		790	780	765	745	45
		840	840	835	830	60
	100	350	325	295	265	18.5
		485	465	440	420	30
		630	615	600	580	45
		680	680	675	670	60
	120	205	185	165	140	18.5
		345	325	300	280	30
		460	445	435	420	45
		505	505	500	495	60
	140	95	75	55	40	18.5
		200	185	170	150	30
		315	305	290	275	45
		350	350	350	345	60
Feb. 21, Oct. 21 (β = 11.75°) $I_0 + I_1$ = 282 Btu-hr^{-1}-ft^{-2a}	40	930	905	880	855	18.5
		1035	1020	1000	985	30
		1105	1100	1095	1085	45
		1095	1100	1105	1110	60
	60	795	770	745	720	18.5
		905	890	870	855	30
		980	975	965	960	45
		970	975	980	980	60
	80	640	615	590	560	18.5
		755	740	720	700	30
		835	830	820	815	45
		825	830	835	840	60
	100	475	450	430	405	18.5
		590	575	555	535	30
		675	665	660	650	45
		665	670	675	680	60
	120	335	310	290	265	18.5
		435	420	405	390	30
		500	495	490	485	45
		485	495	500	500	60
	140	195	180	160	140	18.5
		290	275	260	240	30
		360	355	350	340	45
		340	350	355	360	60

TABLE H.4 Daily Collection with a Double-Glass-Glazed Collector at Various Temperatures, Tilt Angles, and Geographic Latitudes (Btu-ft^{-2}-day^{-1})—Continued

Month	ΔT (°F)	Geographic Latitude 36°	38°	40°	42°	Collector Tilt Angle
Mar. 21, Sept. 21 (β = 0°) $I_0 + I_1$ = 286 Btu-hr^{-1}-ft^{-2a}	40	1065	1055	1040	1025	18.5
		1115	1110	1105	1095	30
		1110	1115	1120	1120	45
		1020	1035	1050	1065	60
	60	940	925	910	895	18.5
		990	985	975	970	30
		980	985	990	995	45
		890	905	920	935	60
	80	795	780	765	745	18.5
		845	840	830	825	30
		835	840	845	850	45
		740	760	775	790	60
	100	630	615	600	580	18.5
		685	680	670	660	30
		675	680	685	690	45
		575	595	610	625	60
	120	465	455	440	425	18.5
		515	510	500	495	30
		505	510	515	520	45
		420	435	450	465	60
	140	325	310	295	275	18.5
		375	370	360	355	30
		365	370	375	380	45
		270	290	305	320	60
Apr. 21, Aug. 21 (β = −11.75°) $I_0 + I_1$ = 292 Btu-hr^{-1}-ft^{-2a}	40	1160	1160	1155	1150	18.5
		1150	1155	1160	1160	30
		1060	1075	1090	1105	45
		890	920	945	970	60
	60	1035	1035	1030	1025	18.5
		1020	1030	1035	1035	30
		930	945	960	975	45
		755	785	810	835	60
	80	895	890	885	880	18.5
		880	885	890	895	30
		785	800	815	830	45
		600	630	660	685	60
	100	735	735	730	725	18.5
		720	730	735	735	30
		620	640	655	670	45
		440	465	490	515	60
	120	565	560	555	550	18.5
		545	550	560	565	30
		460	475	490	505	45
		300	325	350	375	60
	140	410	410	405	395	18.5
		405	410	410	415	30
		315	335	350	365	45
		170	190	210	230	60

TABLE H.4 Daily Collection with a Double-Glass-Glazed Collector at Various Temperatures, Tilt Angles, and Geographic Latitudes (Btu-ft^{-2}-day^{-1})—Continued

Month	ΔT (°F)	Geographic Latitude 36°	38°	40°	42°	Collector Tilt Angle
May 21, July 21 (β = −20.24°) $I_0 + I_1$ = 297 Btu-hr^{-1}-ft^{-2}[a]	40	1210	1210	1215	1220	18.5
		1150	1160	1175	1185	30
		1000	1025	1045	1070	45
		780	810	840	875	60
	60	1085	1090	1090	1095	18.5
		1025	1035	1050	1060	30
		870	890	915	940	45
		635	670	705	735	60
	80	945	950	950	955	18.5
		880	895	905	920	30
		720	745	770	795	45
		475	505	545	580	60
	100	780	795	795	800	18.5
		720	735	750	760	30
		550	580	605	630	45
		336	365	395	425	60
	120	620	625	630	630	18.5
		550	565	580	590	30
		405	425	450	470	45
		195	220	250	285	60
	140	450	455	455	455	18.5
		400	415	425	435	30
		255	280	305	325	45
		85	110	135	160	60
June 21 (β = −23.45°) $I_0 + I_1$ = 303 Btu-hr^{-1}-ft^{-2}[a]	40	1245	1250	1250	1255	18.5
		1175	1185	1190	1200	30
		1010	1025	1040	1060	45
		770	790	815	840	60
	60	1120	1125	1125	1130	18.5
		1050	1060	1065	1080	30
		880	895	910	930	45
		630	650	675	705	60
	80	985	985	990	990	18.5
		910	920	925	935	30
		730	750	765	785	45
		465	490	515	545	60
	100	830	835	835	840	18.5
		750	760	770	780	30
		565	585	600	620	45
		300	330	365	400	60
	120	665	665	670	670	18.5
		580	590	600	610	30
		390	415	440	460	45
		170	195	220	255	60
	140	480	485	485	490	18.5
		405	425	435	450	30
		240	265	295	320	45
		60	85	110	140	60

TABLE H.4 Daily Collection with a Double-Glass-Glazed Collector at Various Temperatures, Tilt Angles, and Geographic Latitudes (Btu-ft^{-2}-day^{-1})—Continued

Month	ΔT (°F)	Geographic Latitude 36°	38°	40°	42°	Collector Tilt Angle
Dec. 21 (β = 23.45°) $I_0 + I_1$ = 275 Btu-hr^{-1}-ft^{-2}[a]	40	750	720	690	655	18.5
		910	885	860	835	30
		1055	1040	1025	1005	45
		1120	1115	1110	1105	60
	60	610	575	540	505	18.5
		775	750	725	695	30
		925	910	895	875	45
		995	990	985	975	60
	80	445	415	385	355	18.5
		620	595	565	535	30
		780	765	745	725	45
		850	845	840	835	60
	100	310	280	245	215	18.5
		455	430	410	385	30
		615	600	580	560	45
		695	690	680	675	60
	120	170	150	125	100	18.5
		310	290	265	240	30
		445	430	420	400	45
		520	515	505	495	60
	140	60	45	30	15	18.5
		180	160	140	120	30
		300	285	270	255	45
		355	355	350	345	60

[a] For locations with different conditioners, correct the table values by five times that difference.

The assumptions for Table H.5 are the following:

1. Insolation at normal incidence ($I_0 + I_1$) is constant during the day, but adjusted to season as indicated (see also Tables H.3 and H.4)
2. Diffuse component of insolation: $I_1 = 0.1I_0$
3. Effective collector absorptivity: $\rho = \epsilon = 0.9$[3]
4. Azimuth collector orientation: $\phi_0 = 0$ (due south)
5. Declination: 21st of each month (see Fig. I.1 of Appendix I)
6. Geographic latitude: 34°, 38°, 42° north
7. Collector tilt angle: 0°, 18°, 30°, 45°
8. ΔT: 10°F, 20°F, 30°F, 40°F
9. Hourly iterations for computation at constant ΔT and v
10. Radiation to surroundings: 1 Btu-hr^{-1}-ft^{-2}-°F^{-1}, const.[3]
11. Black-body radiation into sky: 40 Btu-hr^{-1}-ft^{2}, const.[3]
12. Influence of wind speed, v, on collector temperature-loss coefficient, a_w: $a_w = 0.2v + 0.5$

The table can be used to estimate the collection with unglazed solar-collector systems, in particular for pool-heating applications. For such estimates, the use of an average daytime ambient temperature, $\overline{T}_a$, is suggested, leading to $\Delta T = T_p - \overline{T}_a$, if T_p is the due (constant) pool temperature. Approximated linear interpolation for unlisted geographic latitudes, wind speeds, and temperatures is permissible. The tables can also be studied to assess the effect of prevailing wind conditions on the collection efficiency in order to select the appropriate type of collector (unglazed or glazed) for a given application.

[3]See section 1.1.3. Sometimes product brochures of unglazed collectors indicate that the collectors were tested by an independent testing laboratory, and the test results for the ordinate intercept and the slope are given. These results usually include an (undisclosed and often unknown) amount for the radiation losses and prevailing wind velocities. Because of this ambiguity we suggest to disregard the test results and to establish the performance with unglazed collectors on the basis of the absorptivity of the absorber coating ρ (reduced by fin efficiency losses if applicable, see Section 1.1.2) and the ambient conditions by using Table H.5 or the computational method presented in Appendix K.

TABLE H.5 Daily Collection with a Good Unglazed Flat-Plate Collector at Various Temperatures, Tilt Angles, Geographic Latitudes, and Wind Speeds (Btu-ft^{-2}-day^{-1}, Full-Sunshine Conditions)

Month	ΔT (°F)	Wind Speed (mph) 2	7	12	17	Tilt Angle	Geographic Latitude
Jan. 21, Nov. 21 ($\beta = 20.24°$) $I_0 + I_1 =$ 275 Btu-hr^{-1}-ft^{-2}	10	520	435	365	295	0	34°
		915	825	735	645	18	
		1115	1025	935	845	30	
		1270	1180	1090	1000	45	
	20	370	230	120	35	0	
		745	565	415	275	18	
		940	760	580	435	30	
		1100	920	740	560	45	
	30	240	75	0	0	0	
		575	350	165	35	18	
		770	510	300	130	30	
		930	660	420	220	45	
	40	130	0	0	0	0	
		430	170	10	0	18	
		600	310	85	0	30	
		760	430	175	10	45	
	10	420	350	280	210	0	38°
		835	745	655	565	18	
		1055	965	875	785	30	
		1240	1150	1060	970	45	
	20	285	160	60	0	0	
		665	490	350	215	18	
		880	705	525	385	30	
		1070	890	710	530	45	
	30	165	30	0	0	0	
		495	290	120	0	18	
		710	465	255	100	30	
		800	630	400	200	45	
	40	70	0	0	0	0	
		365	125	0	0	18	
		540	260	60	0	30	
		725	405	160	0	45	
	10	335	265	195	140	0	42°
		750	660	570	485	18	
		985	900	810	720	30	
		1200	1100	1020	930	45	
	20	200	95	15	0	0	
		580	420	280	165	18	
		815	635	475	335	30	
		1030	850	670	500	45	
	30	100	0	0	0	0	
		430	220	70	0	18	
		645	410	210	65	30	
		860	590	370	180	45	
	40	25	0	0	0	0	
		295	75	0	0	18	
		485	215	35	0	30	
		685	375	135	0	45	

TABLE H.5 Daily Collection with a Good Unglazed Flat-Plate Collector at Various Temperatures, Tilt Angles, Geographic Latitudes, and Wind Speeds (Btu-ft^{-2}-day^{-1}, Full-Sunshine Conditions)—Continued

Month	ΔT (°F)	Wind Speed (mph) 2	7	12	17	Tilt Angle	Geographic Latitude
Feb. 21, Oct. 21 (β = 11.75°) $I_0 + I_1$ = 282 Btu-hr^{-1}-ft^{-2}	10	760	670	580	490	0	34°
		1090	1000	910	820	18	
		1235	1140	1055	965	30	
		1315	1225	1135	1045	45	
	20	585	425	285	165	0	
		915	735	555	415	18	
		1060	880	700	535	30	
		1145	965	785	505	45	
	30	430	220	70	0	0	
		745	495	285	125	18	
		890	620	400	210	30	
		975	705	465	255	45	
	40	295	75	0	0	0	
		575	290	80	0	18	
		720	410	165	0	30	
		805	470	210	35	45	
	10	670	580	490	410	0	38°
		1025	935	845	755	18	
		1190	1100	1010	920	30	
		1305	1215	1125	1035	45	
	20	500	350	210	105	0	
		855	675	505	365	18	
		1020	840	660	500	30	
		1135	955	775	595	45	
	30	355	160	30	0	0	
		685	445	235	85	18	
		850	580	370	185	30	
		965	695	455	250	45	
	40	225	35	0	0	0	
		520	240	50	0	18	
		680	375	140	0	30	
		790	465	200	30	45	
	10	575	485	405	335	0	42°
		955	865	780	690	18	
		1145	1055	1105	1015	30	
		1285	1195	1105	1015	45	
	20	410	270	145	0	0	
		785	605	450	310	18	
		975	795	615	460	30	
		1115	935	755	575	45	
	30	280	100	0	0	0	
		615	385	190	55	18	
		800	540	330	155	30	
		945	675	440	240	45	
	40	155	0	0	0	0	
		465	195	25	0	18	
		630	335	110	0	30	
		775	450	195	25	45	

TABLE H.5 Daily Collection with a Good Unglazed Flat-Plate Collector at Various Temperatures, Tilt Angles, Geographic Latitudes, and Wind Speeds (Btu-ft^{-2}-day^{-1}, Full-Sunshine Conditions)—Continued

Month	ΔT (°F)	Wind Speed (mph) 2	7	12	17	Tilt Angle	Geographic Latitude
Mar. 21, Sept. 21 ($\beta = 0°$) $I_0 + I_1 =$ 286 Btu-hr^{-1}-ft^{-2}	10	1055	965	875	785	0	34°
		1270	1180	1090	1000	18	
		1330	1240	1150	1060	30	
		1305	1215	1125	1035	45	
	20	880	700	520	375	0	
		1100	920	740	565	18	
		1160	980	800	620	30	
		1135	955	775	595	45	
	30	710	455	245	85	0	
		930	660	430	230	18	
		990	720	485	275	30	
		965	695	460	255	45	
	40	540	250	50	0	0	
		760	440	185	20	18	
		820	490	225	45	30	
		990	470	210	35	45	
	10	985	895	805	715	0	38°
		1235	1145	1055	965	18	
		1320	1230	1140	1050	30	
		1325	1235	1145	1055	45	
	20	815	635	460	320	0	
		1065	885	705	535	18	
		1150	970	790	610	30	
		1155	975	795	615	45	
	30	645	395	190	50	0	
		895	625	400	210	18	
		995	705	475	265	30	
		980	710	475	265	45	
	40	475	195	25	0	0	
		725	410	165	0	18	
		805	480	215	40	30	
		810	485	220	45	45	
	10	910	820	730	640	0	42°
		1195	1100	1010	925	18	
		1300	1210	1120	1030	30	
		1335	1245	1155	1065	45	
	20	740	560	395	255	0	
		1020	840	660	495	18	
		1130	950	770	590	30	
		1165	985	805	625	45	
	30	570	330	145	20	0	
		850	580	365	180	18	
		960	690	455	250	30	
		990	720	485	275	45	
	40	410	150	0	0	0	
		680	370	135	0	18	
		785	465	205	35	30	
		820	490	225	45	45	

TABLE H.5 Daily Collection with a Good Unglazed Flat-Plate Collector at Various Temperatures, Tilt Angles, Geographic Latitudes, and Wind Speeds (Btu-ft^{-2}-day^{-1}, Full-Sunshine Conditions)—Continued

Month	ΔT (°F)	Wind Speed (mph) 2	7	12	17	Tilt Angle	Geographic Latitude
Apr. 21, Aug. 21 ($\beta = -11.75°$) $I_0 + I_1 =$ 292 Btu-hr^{-1}-ft^{-2}	10	1300	1210	1120	1030	0	34°
		1395	1300	1215	1125	18	
		1365	1275	1185	1095	30	
		1225	1135	1045	955	45	
	20	1125	945	765	585	0	
		1200	1040	860	680	18	
		1190	1010	830	650	30	
		1050	870	690	530	45	
	30	955	685	435	225	0	
		1050	780	525	315	18	
		1020	750	505	295	30	
		880	610	395	205	45	
	40	785	440	180	0	0	
		880	535	225	60	18	
		850	515	240	55	30	
		710	405	160	0	45	
	10	1255	1165	1075	985	0	38°
		1385	1295	1205	1115	18	
		1380	1290	1200	1110	30	
		1270	1180	1090	1000	45	
	20	1085	905	725	545	0	
		1215	1035	855	575	18	
		1210	1030	850	670	30	
		1100	920	740	570	45	
	30	915	645	395	190	0	
		1045	775	520	310	18	
		1040	770	520	310	30	
		930	760	435	235	45	
	40	740	405	145	0	0	
		875	525	245	55	18	
		870	525	250	60	30	
		760	440	190	25	45	
	10	1205	1115	1025	935	0	42°
		1370	1280	1190	1100	18	
		1390	1300	1210	1120	30	
		1310	1220	1130	1040	45	
	20	1035	855	675	495	0	
		1200	1020	840	660	18	
		1220	1040	860	680	30	
		1140	960	780	600	45	
	30	860	590	350	155	0	
		1030	760	505	295	18	
		1050	780	525	315	30	
		970	700	465	260	45	
	40	690	355	110	0	0	
		860	510	235	50	18	
		880	535	255	65	30	
		800	475	215	40	45	

TABLE H.5 Daily Collection with a Good Unglazed Flat-Plate Collector at Various Temperatures, Tilt Angles, Geographic Latitudes, and Wind Speeds (Btu-ft^{-2}-day^{-1}, Full-Sunshine Conditions)—Continued

Month	ΔT (°F)	Wind Speed (mph) 2	7	12	17	Tilt Angle	Geographic Latitude
May 21, July 21 ($\beta = -20.24°$) $I_0 + I_1 =$ 297 Btu-hr^{-1}-ft^{-2}	10	1450	1360	1270	1180	0	34°
		1455	1365	1275	1185	18	
		1360	1270	1180	1090	30	
		1145	1055	965	875	45	
	20	1280	1100	920	735	0	
		1285	1105	925	745	18	
		1190	1010	830	650	30	
		970	796	615	450	45	
	30	1105	835	565	340	0	
		1110	845	575	355	18	
		1020	750	495	285	30	
		800	530	320	150	45	
	40	935	575	275	60	0	
		940	580	290	80	18	
		850	500	230	50	30	
		630	325	105	0	45	
	10	1425	1335	1245	1155	0	38°
		1465	1375	1285	1195	18	
		1390	1300	1210	1120	30	
		1195	1105	1015	925	45	
	20	1255	1075	895	715	0	
		1290	1110	930	750	18	
		1220	1040	860	680	30	
		1025	845	665	505	45	
	30	1085	815	545	315	0	
		1120	850	580	365	18	
		1050	780	525	315	30	
		855	585	375	190	45	
	40	910	550	250	45	0	
		950	590	300	85	18	
		880	530	255	60	30	
		685	380	145	0	45	
	10	1395	1305	1215	1125	0	42°
		1470	1380	1290	1200	18	
		1425	1335	1245	1155	30	
		1260	1170	1080	990	45	
	20	1220	1040	860	680	0	
		1300	1120	940	760	18	
		1255	1075	895	715	30	
		1090	910	730	555	45	
	30	1050	780	510	285	0	
		1130	860	590	365	18	
		1080	810	550	340	30	
		920	650	425	225	45	
	40	880	520	220	30	0	
		955	595	305	85	18	
		910	555	275	70	30	
		745	430	180	20	45	

TABLE H.5 Daily Collection with a Good Unglazed Flat-Plate Collector at Various Temperatures, Tilt Angles, Geographic Latitudes, and Wind Speeds (Btu-ft^{-2}-day^{-1}, Full-Sunshine Conditions)—Continued

Month	ΔT (°F)	Wind Speed (mph) 2	7	12	17	Tilt Angle	Geographic Latitude
June 21 ($\beta = -23.45°$) $I_0 + I_1 =$ 303 Btu-hr^{-1}-ft^{-2}	10	1520	1430	1340	1250	0	34°
		1505	1415	1325	1235	18	
		1395	1305	1215	1125	30	
		1150	1060	970	880	45	
	20	1350	1170	990	810	0	
		1330	1155	970	790	18	
		1220	1040	860	680	30	
		980	800	620	440	45	
	30	1180	910	640	395	0	
		1160	890	620	380	18	
		1050	780	510	290	30	
		810	540	295	135	45	
	40	1005	645	330	95	0	
		990	630	320	45	18	
		880	520	235	50	30	
		640	305	90	0	0	
	10	1505	1415	1325	1235	0	38°
		1515	1425	1335	1245	18	
		1420	1330	1240	1150	30	
		1200	1110	1020	930	45	
	20	1335	1155	975	795	0	
		1345	1165	985	805	18	
		1250	1070	890	710	30	
		1030	850	670	495	45	
	30	1160	890	620	375	0	
		1175	905	635	395	18	
		1080	810	540	325	30	
		860	590	360	180	45	
	40	990	630	315	80	0	
		1000	640	335	110	18	
		910	550	265	65	30	
		690	365	135	0	45	
	10	1480	1390	1300	1210	0	42°
		1520	1430	1340	1250	18	
		1450	1360	1270	1180	30	
		1250	1160	1070	980	45	
	20	1310	1130	950	770	0	
		1350	1170	990	810	18	
		1275	1095	915	740	30	
		1080	900	720	550	45	
	30	1135	865	595	350	0	
		1180	905	640	405	18	
		1105	835	565	355	30	
		910	640	415	220	45	
	40	965	605	290	60	0	
		1005	650	340	110	18	
		935	575	295	80	30	
		735	425	175	15	45	

TABLE H.5 Daily Collection with a Good Unglazed Flat-Plate Collector at Various Temperatures, Tilt Angles, Geographic Latitudes, and Wind Speeds (Btu-ft^{-2}-day^{-1}, Full-Sunshine Conditions)—Continued

Month	ΔT (°F)	Wind Speed (mph) 2	7	12	17	Tilt Angle	Geographic Latitude
Dec. 21 (β = 23.45°) $I_0 + I_1$ = 275 Btu-hr^{-1}-ft^{-2}	10	435	365	295	225	0	34°
		855	765	675	558	18	
		1075	985	895	805	30	
		1260	1170	1080	990	45	
	20	300	170	70	0	0	
		685	505	365	225	18	
		900	720	540	400	30	
		1090	910	730	550	45	
	30	175	40	0	0	0	
		510	300	130	0	18	
		730	475	265	105	30	
		915	645	410	210	45	
	40	80	0	0	0	0	
		380	135	0	0	18	
		560	275	65	0	30	
		745	415	165	0	45	
	10	345	275	205	150	0	38°
		770	680	590	500	18	
		1005	915	825	735	30	
		1220	1130	1040	950	45	
	20	215	105	0	0	0	
		600	435	295	175	18	
		835	655	485	345	30	
		1050	870	690	515	45	
	30	110	0	0	0	0	
		445	235	80	0	18	
		665	425	220	70	30	
		880	610	380	190	45	
	40	0	0	0	0	0	
		310	85	0	0	18	
		500	225	40	0	30	
		705	385	145	0	45	
	10	255	185	135	85	0	42°
		680	590	500	425	18	
		935	845	755	665	30	
		1175	1085	995	905	45	
	20	140	45	0	0	0	
		505	365	220	120	18	
		760	580	430	290	30	
		1000	820	640	475	45	
	30	45	0	0	0	0	
		370	175	40	0	18	
		590	365	175	45	30	
		830	560	345	165	45	
	40	0	0	0	0	0	
		235	45	0	0	18	
		440	180	0	0	30	
		660	350	115	0	45	

I

Program for Solar System Heat Collection Computation[1]

This appendix describes a program for solar heat collection that is applicable for a Hewlett-Packard 67/97 calculator.[2] The limitation to 224 program steps and 26 direct storage registers necessitates omission of influences of collector shading and the incident-angle modifer (see Section 1.1.2) as well as a division of the calculation for a 24-hr day into two segments: the daytime hours, 8:00 A.M. to 5:00 P.M., and the night hours, 5:00 P.M. to 8:00 A.M. Program A (daytime hours) contains the entire solar collection, and its results are written at the end of each day's computation; i.e., the storage-tank temperature at 5:00 P.M., the accumulated collection per square foot of collector area, and the accumulated daily total system collection are printed (flashed with HP 67). All input parameters are also printed for permanent reference. The computation for each day takes 3 min. Program B computes the heating degree-hours between 5:00 P.M. and 8:00 A.M. on the basis of the temperature profile indicated in Fig. 5.1, where T_a is assumed to decrease steadily after 4:00 P.M. by 2°F per hour until the 6:00 A.M. base temperature is reached. It then uses the storage-tank temperature at 5:00 P.M. (i.e., at the end of the daytime calculation) to compute the heat content of the storage tank as compared with the 8:00 A.M. condition. Also computed is the energy that was spent during the day to meet the demand, as well as the amounts and ratio of the total daily energies required and supplied by the solar system. Multiplying this ratio by the probability of sunshine gives the probability of solar coverage for the particular day or month. Program B requires the reentering of approximately five variables for each new calculation and takes only a few seconds of computing time.

[1]See footnote 5 in Chapter 5 concerning a full-size computer program for complete annual computation [25] using a TRS-80 desktop computer.

[2]The program is written for space and domestic water heating, as well as for absorption-chilling air conditioning, and it follows the block diagram of Fig. 5.3. The computation of solar system collection expectance has been the subject of investigations by various other authors. A survey is given in Appendix H, Table H.2.

Program A

The listing of the program steps is given in Table I.1. Steps 1 through 50 calculate cos ϑ;[3] steps 51–103 compute the hourly solar heat collection;[4] steps 104–112 calculate the resulting increase in the storage-tank temperature; steps 113–138 compute the decrease in the storage-tank temperature due to consumption of heat and heat losses. Execution of the program, which is designed for hourly iterations, starts with "LBLa" (step 154) for winter or "LBLA" for summer. The start is at 8:00 A.M. (steps 165, 166), and the last hourly computation is done for 1600 hours (steps 199–201). Steps 202–208 arrange for the data printout. With "LBLC" a small subroutine is internally initiated that calculates the ambient temperatures between 8:00 A.M. and noon. The subroutine starting with "LBLE" shortens the computing time by skipping several time-consuming trigonometric calculations for those hours during which the sun is below the horizon ($\alpha < 0$, step 23). Step 207 sets a flag back to summer condition. The following data must be entered into the registers:

Register	Data to be Entered
P0	I_0 (direct insolation component) (Btu-ft^{-2}-hr^{-1})[5]
P1	ϵ } (collector efficiency parameters)[6]
P2	a }
P3	b }
P4	c (temperature difference between ambient near collectors and ambient distant from collectors)[7]
P5	A_0 (solar-collector area; use gross area if ϵ, a, and b were determined for gross collector area; otherwise use net collector = absorber area) (ft^2)
P6	C_B (building heat loss = heat-gain coefficient) (Btu-hr^{-1}-°F^{-1})[8]
P7	W_2 (heat losses of piping ducts and heat-storage tank[9] plus estimated average hourly DWH consumption[10]) (Btu/hr)
P8	W_1 (direct solar heat gain of building through windows) (Btu/hr)[11]

[3]See Appendix A.

[4]See Section 1.1.2 with $k = 0.1$; ΔT is raised by δT_c (equation 5.1) because of the heat exchanger, using $i = 0.03A_0$.

[5]See Section 5.2.1 and Appendix B, particularly Table B.1.

[6]See Section 1.1.2. The parameters a and b must be expressed in British units, i.e., $[a]$ = [Btu hr^{-1}-°F^{-1}-ft^{-2}] and $[b]$ = [Btu^2-°F^{-2}-hr^{-2}-ft^{-4}]. For conversion of values that are given in metric units, use: a (British) = 0.176 × a (W-m^{-2}-°C^{-1}); b (British) = 0.311 × b (W^2-m^{-4}-°C^{-2}).

[7]See Sections 1.1.2 and 5.2.6. Typical values are 0–10 [°F], depending on ambient wind velocities.

[8]See Appendix J. Divide by chiller efficiency for absorption cooling (summer).

[9]See Sections 1.3 and 1.4.2. Make reasonable estimate of average constant ΔT between pipes (tank) and ambient temperatures.

[10]Daytime consumption only. In most cases the main consumption occurs after 5:00 P.M. and is considered in program B.

[11]For winter heating: Calculate total daily amount, subtract appropriate fraction for shading and poor building thermal distribution capacity, divide by 9 (computation hours), and enter that value in Register P8. For summer cooling: Calculate total daily amount, subtract appropriate fraction for shading, divide by 9, divide by chiller efficiency, and enter result in register P8. Include appliance and people gains. See Appendix J for details of computation (enter positive number).

Register	Data to be Entered
S2	ϕ(collector azimuth deviation from due south; blf$_0$ < u designates deviation toward east; $\vartheta_0 < 0$ toward west) (deg)
S3	γ (geographic latitude) (deg)
S4	β (declination; See Fig. I.1 of Appendix I) (deg)
S5	ψ (collector tilt angle) (deg)
S6	T_a(6:00) (ambient temperature at 6:00 A.M.) (°F)
S7	T_a(noon)
S8	T_a(2:00 P.M.) (see Section 5.2.4 for details on temperature profile)
A	T_0(8:00 A.M.) (storage-tank temperature at 8:00 A.M., equals lowest setting compatible with the type of heating system that the solar system is connected to; see Sections 2.3 and 3.4)
B	V_0 (volume of heat-storage tank) (gal)[12]
C	AE_x (heat-exchanger efficiency; see Section 5.2.5) (Btu-hr^{-1}-°F^{-1})[13]

[12]Water is assumed as the heat-storage medium.

[13]Typically, the value required depends on the collector area A_0. The numerical value of $AE_x = 8A_0$ is a reasonable design goal (see Appendix E). Use $AE_x = 30A_0$ to simulate a system operating without a heat exchanger.

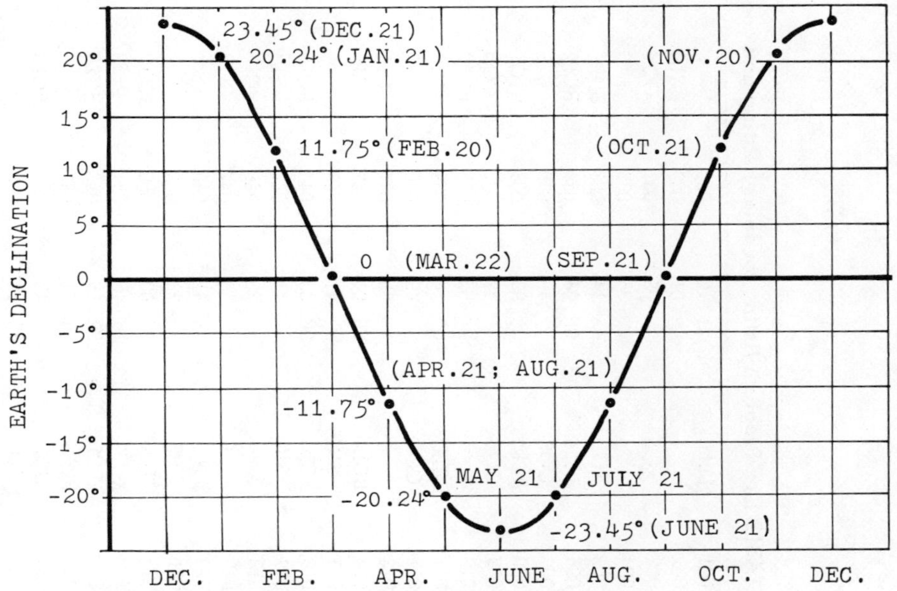

Fig. I.1. The earth's declination during the year.

TABLE I.1 Program listing for solar heating and cooling for computation with an HP 97 calculator.

001	*LBLB	21 12	029	×	−35	057	×	−35	085	×	−35
002	1	01	030	RCL0	36 00	058	+	−55	086	RCL0	36 00
003	5	05	031	COS	42	059	RCL1	36 01	087	.	−62
004	×	−35	032	÷	−24	060	×	−35	088	1	01
005	STOD	35 14	033	SIN^{-1}	16 41	061	RCLA	36 11	089	×	−35
006	RCL4	36 04	034	RCL2	36 02	062	P⇌S	16−51	090	+	−55
007	COS	42	035	+	−55	063	RCL9	36 09	091	RCLI	36 46
008	RCL3	36 03	036	STO1	35 01	064	P⇌S	16−51	092	X^2	53
009	COS	42	037	COS	42	065	−	−45	093	X⇌Y	−41
010	×	−35	038	RCL0	36 00	066	RCL4	36 04	094	÷	−24
011	RCLD	36 14	039	COS	42	067	−	−45	095	RCL3	36 03
012	COS	42	040	×	−35	068	RCLE	36 15	096	×	−35
013	×	−35	041	RCL5	36 05	069	RCLC	36 13	097	−	−45
014	RCL4	36 04	042	SIN	41	070	÷	−24	098	X<0?	16−45
015	SIN	41	043	×	−35	071	+	−55	099	0	00
016	RCL3	36 03	044	RCL0	36 00	072	RCLE	36 15	100	RCL5	36 05
017	SIN	41	045	SIN	41	073	3	03	101	×	−35
018	×	−35	046	RCL5	36 05	074	0	00	102	ST+9	35−55 09
019	+	−55	047	COS	42	075	÷	−24	103	STOE	35 15
020	SIN^{-1}	16 41	048	×	−35	076	RCL5	36 05	104	8	08
021	CHS	−22	049	+	−55	077	÷	−24	105	.	−62
022	STO0	35 00	050	STOD	35 14	078	−	−45	106	3	03
023	X<0?	16−45	051	P⇌S	16−51	079	STOI	35 46	107	÷	−24
024	GTOE	22 15	052	RCL0	36 00	080	RCL2	36 02	108	RCLB	36 12
025	RCL4	36 04	053	×	−35	081	×	−35	109	÷	−24
026	COS	42	054	RCL0	36 00	082	−	−45	110	RCLA	36 11
027	RCLD	36 14	055	.	−62	083	RCL0	36 00	111	+	−55
028	SIN	41	056	1	01	084	RCLD	36 14	112	STOA	35 11

113	RCL6	36 06	140	RTN	24	167	GSBC	23 13	194	GSBB	23 12
114	P⇌S	16−51	141	*LBLE	21 15	168	9	09	195	1	01
115	RCL9	36 09	142	0	00	169	GSBB	23 12	196	ST-9	35−45 09
116	P⇌S	16−51	143	STOD	35 14	170	GSBC	23 13	197	1	01
117	F1?	16 23 01	144	P⇌S	16−51	171	1	01	198	6	06
118	GTOb	22 16 12	145	RTN	24	172	0	00	199	GSBB	23 12
119	6	06	146	*LBLC	21 13	173	GSBB	23 12	200	PREG	16−13
120	8	08	147	RCL7	36 07	174	GSBC	23 13	201	P⇌S	16−51
121	−	−45	148	RCL6	36 06	175	1	01	202	RCL9	36 09
122	×	−35	149	−	−45	176	1	01	203	RCL5	36 05
123	RCL8	36 08	150	4	04	177	GSBB	23 12	204	÷	−24
124	+	−55	151	÷	−24	178	GSBC	23 13	205	STOD	35 14
125	X>0	16−44	152	ST+9	35−55 09	179	1	01	206	PREG	16−13
126	0	00	153	RTN	24	180	2	02	207	SF1	16 21 01
127	RCL7	36 07	154	*LBLa	21 16 11	181	GSBB	23 12	208	RTN	24
128	−	−45	155	CF1	16 22 01	182	RCL8	36 08	209	*LBLb	21 16 12
129	*LBLd	21 16 14	156	*LBLA	21 11	183	STO9	35 09	210	7	07
130	RCLB	36 12	157	0	00	184	1	01	211	8	08
131	÷	−24	158	STO9	35 09	185	3	03	212	−	−45
132	8	08	159	STOE	35 15	186	GSBB	23 12	213	×	−35
133	.	−62	160	P⇌S	16−51	187	1	01	214	X<0?	16−45
134	3	03	161	RCLA	36 11	188	4	04	215	0	00
135	÷	−24	162	PRTX	−14	189	GSBB	23 12	216	RCL8	36 08
136	RCLA	36 11	163	RCL6	36 06	190	1	01	217	+	−55
137	+	−55	164	STO9	35 09	191	St-9	35−45 09	218	RCL7	36 07
138	STOA	35 11	165	8	08	192	1	01	219	+	−55
139	P⇌S	16−51	166	GSBB	23 12	193	5	05	220	CHS	−22
									221	GTOd	22 16 14

A typical result printout for a calculation performed with program A is given in Table I.2.

Note the following additional assumptions made for program A:

1. $I_1 = 0.1I_0$ is used in steps 54–56 and 86–88. This value may be changed to any other reasonable value.
2. A collector-loop flow rate (needed for the heat-exchanger influence, steps 68–78) of approximately $0.03A_0$ (numerical value for gpm flow rate) was assumed and is implicitly contained in the number 30 (steps 73, 74).

TABLE I.2 Printout for Typical Solar Space Heating for Single-Family Residence in January Performed with the HP 97 Program (Fig. I.1)

No.	Designation	Result Printout			Comments
1	T_0(8:00 A.M.)	120.00		**	
	—	9.30	0		Not relevant, intermediate results
	—	−34.12	1		
2	ϕ_0 (deviation)	10.00	2	*	Positive: west of south
3	γ (latitude)	37.50	3	*	
4	β (declination)	20.24	4	*	(Jan. or Nov.)
5	ψ (collector tilt)	60.00	5	*	
6	T_a(6:00 A.M.)	40.00	6	*	Relatively mild climate conditions
7	T_a(12:00 noon)	50.00	7	*	
8	T_a(2:00 P.M.)	51.00	8	*	
9	T_a(4:00 P.M.)	49.00	9		
10	T_0(5:00 P.M.)	140.52	A		
11	V_0 (gal)	1000.00	B	*	Typical size, $\approx 3A_0$
12	AE_x	1500.00	C	*	Typical size, $\approx 5A_0$
	—	0.68	D		Non-relevant intermediate results
	—	9359.49	E		
	—	96.33	I		
13	I_0(Btu-hr^{-1}-ft^{-2})	250.00	0	*	
14	ϵ	0.74	1	*	Alten 100 G collector, DSET Laboratories test data
15	a	0.82	2	*	
16	b	0.72	3	*	
17	c	5.00	4	*	Sheltered location
18	A_0 (ft^2)	300.00	5	*	Typical single-family residence
19	C_B	800.00	6	*	
20	W_2	2000.00	7	*	
21	W_1	20000.00	8	*	
22	Btu/day collection	191148.91	9		
	—	148.52	A		Same results as in Nos. 10, 11, 12
	—	1000.00	B		
	—	1500.00	C		
23	Btu-ft^{-2}-day^{-1}	637.16	D		#22#18
	—	9359.49	E		Not relevant, intermediate results
	—	96.33	I		

*Data to be entered as parameters in program.
**To be entered in register A before each calculation.

3. The inside temperature of the building is set at 68°F for winter operation and 78°F for summer operation. Change steps 119 and 120, and 210 and 211, respectively, if these values are not acceptable.
4. There is no provision in the program to allow for errors due to an undersized system with respect to daytime conditions. If the system is so badly undersized that the solar supply on a sunny day does not cover the heating or cooling demand until 5:00 P.M., the tank temperature will drop below the desirable minimum temperature. This may be of consequence in absorption-chilling applications, where the chiller efficiency becomes very poor if the minimum supply water temperature is not reached.
5. There is no capacity within the registers to allow for chiller efficiency in air-conditioning applications. This efficiency ($0 < E_{ch} < 1$) must therefore be already included in the respective data entered for building conductive heat gain (register P6) and for direct gain (register P8). Divide the numbers to be entered by E_{ch} and enter that result.
6. Run the program by pressing "fa" for winter heating or "A" for summer air conditioning.
7. Make sure that the P register is in front before running the program.
8. Reload register A after each daily calculation.

Program B

The listing of the program steps is given in Table I.3. The program for heating begins with step 18 ("LBLa"), continues through step 26, at which time the hourly loop (steps 1–17) begins, and ends with the "LBLC" routine, which computes the solar coverage. As with program A, the air-conditioning part is initiated with "LBLA" (step 114). The hourly subroutine comprises steps 151–168, and the program ends again with the "LBLC" subroutine. The following data must be entered into the registers:

Register	Data to be Entered
P1	T_a(4:00 P.M.)
P2	T_a(8:00 A.M.)
P3	T_1 between 8:00 A.M. and 10:00 P.M. (inside due temperature)
P4 (for heating)	T_i between 10:00 P.M. and 8:00 A.M. (inside due temperature)
P4 (for cooling)	Chiller efficiency (%/100)
P6	C_B (building heat loss = heat-gain coefficient)[14] (Btu-hr^{-1}-°F^{-1})
P9	Total solar heat collected on a sunny day, per system (Btu)[15]
A	T_0(8:00 A.M.)
B	V_0(gal)
C	T_0(5:00 P.M.)[16]
D	Total domestic water heat load between 5:00 P.M. and 8:00 A.M. (Btu)
E	Sunshine probability[17] (%/100)
I	Total tank and piping heat losses, plus internal heat gain (in summer) (subtract internal heat gain in winter), during 5:00 P.M. to 8:00 A.M. (Btu)

[14]See Appendix J; from line 19 of Table I.2 (register P6 of program A).
[15]From line 22 of Table I.2 (register P9 after computation with program A).
[16]From line 10 of Table I.2 (register A after computation with program A).
[17]See Section 5.2.8 and Appendix C.

TABLE I.3 Program Listing for Part B of Solar System Collection Program for HP 97 Calculator

001	*LBLB	21 12	023	0	00	045	RCL3	36 03	067	−	−45
002	RCLO	36 00	024	STO7	35 07	046	SPC	16−11	068	CHS	−22
003	2	02	025	STO8	35 00	047	PRTX	−14	069	STO5	35 05
004	−	−45	026	GSBB	23 12	048	RCL4	36 04	070	PRTX	−14
005	RCL2	36 02	027	GSBB	23 12	049	PRTX	−14	071	RCLD	36 14
006	X>Y?	16−34	028	GSBB	23 12	050	RCL8	36 08	072	PRTX	−14
007	RCL2	36 02	029	GSBB	23 12	051	PRTX	−14	073	ST+5	35−55 05
008	R↓	−31	030	GSBB	23 12	052	RCL7	36 07	074	ST+7	35−55 07
009	STO0	35 00	031	RCL4	36 04	053	PRTX	−14	075	RCL5	36 05
010	RCL5	36 05	032	STO5	35 05	054	*LBLC	21 13	076	PRTX	−14
011	−	−45	033	GSBB	23 12	055	RCLC	36 13	077	RCL9	36 09
012	CHS	−22	034	GSBB	23 12	056	RCLA	36 11	078	8	08
013	ST+8	35−55 08	035	GSBB	23 12	057	−	−45	079	.	−62
014	RCL6	36 06	036	GSBB	23 12	058	8	08	080	3	03
015	×	−35	037	GSBB	23 12	059	.	−62	081	÷	−24
016	ST÷7	35−55 07	038	GSBB	23 12	060	3	03	082	RCLB	36 12
017	RTN	24	039	GSBB	23 12	061	×	−35	083	÷	−24
018	*LBLa	21 16 11	040	GSBB	23 12	062	RCLB	36 12	084	RCLA	36 11
019	RCL1	36 01	041	GSBB	23 12	063	×	−35	085	+	−55
020	STO0	35 00	042	GSBB	23 12	064	STO0	35 00	086	RCLC	36 13
021	RCL3	36 03	043	RCLI	36 46	065	PRTX	−14	087	−	−45
022	STO5	35 05	044	ST+7	35−55 07	066	RCL7	36 07	088	8	08

089	.	−62
090	3	03
091	×	−35
092	RCLB	36 12
093	×	−35
094	ST+7	35−55 07
095	RCL9	36 09
096	RCL7	36 07
097	÷	−24
098	PRTX	−14
099	RCLE	36 15
100	PRTX	−14
101	×	−35
102	PRTX	−14
103	RCL9	36 09
104	PRTX	−14
105	RCLE	36 15
106	×	−35
107	RCL7	36 07
108	PRTX	−14
109	−	−45
110	CHS	−22
111	PRTX	−14
112	RTN	24
113	RTN	24
114	*LBLA	21 11
115	RCL1	36 01
116	STO0	35 00
117	RCL3	36 03
118	STO5	35 05
119	0	00
120	STO7	35 07
121	STO6	35 08
122	GSBb	23 16 12
123	RCLI	36 46
124	ST+7	35−55 07
125	GSBb	23 16 12
126	GSBb	23 16 12
127	GSBb	23 16 12
128	GSBb	23 16 12
129	GSBb	23 16 12
130	GSBb	23 16 12
131	GSBb	23 16 12
132	GSBb	23 16 12
133	GSBb	23 16 12
134	GSBb	23 16 12
135	GSBb	23 16 12
136	GSBb	23 16 12
137	GSBb	23 16 12
138	GSBb	23 16 12
139	RCL4	36 04
140	ST÷7	35−24 07
141	SPC	16−11
142	RCL3	36 03
143	PRTX	−14
144	PRTX	−14
145	RCL8	36 08
146	PRTX	−14
147	RCL7	36 07
148	PRTX	−14
149	GTOC	22 13
150	RTN	24
151	*LBLb	21 16 12
152	RCL0	36 00
153	2	02
154	−	−45
155	RCL2	36 02
156	X>Y?	16−34
157	RCL2	36 02
158	R↓	−31
159	STO0	35 00
160	RCL5	36 05
161	−	−45
162	X>0?	16−45
163	0	00
164	ST+8	35−55 08
165	RCL6	36 06
166	×	−35
167	ST+7	35−55 07
168	RTN	24

A typical result printout for a calculation performed with program B is given in Table I.4 (for the same system presented in Table I.2). Note the following specific requirements:

1. For air conditioning, enter all heat-loss values (registers P6, D, I) directly. Division by the chiller efficiency (as required for program A) is not needed here (it is performed internally by the program).
2. For air conditioning, when $T_a < T_i$, the program does not compute an effective cooling result for the building, but only assumes that no more air conditioning is required, except to remove the heat that is internally generated.
3. The program requires the exact entries of the results from program A in registers P6, P9, A, B, and C.
4. The entry in register P3 should be 68 for heating and 78 for air conditioning, unless the respective values in program A (steps 119/120 or 210/211) were changed.
5. Run the program by depressing ''fa'' for winter heating and ''A'' for summer air conditioning.

TABLE I.4. Printout for Program B with Explanations (Numerical Results Correspond to Table I.2)[a]

No.	Result	Description
1	68.00	T_i(8:00 A.M.–10:00 P.M.)
2	65.00	T_i(10:00 P.M.–8:00 A.M.)
3	374.00	Heating/cooling degree-F hours, 5:00 P.M.–8:00 P.M.
4	302200.00	Heating/cooling demand (Btu), 5:00 P.M.–8:00 A.M.
5	170316.00	Heat content of storage tank at 5:00 P.M. (Btu)
6	131884.00	Auxiliary heat required for space heating/cooling only
7	50000.00	DWH requirement from 5:00 P.M. to 8:00 A.M. (Btu)
8	181884.00	Total auxiliary heat requirement, including DWH (Btu)
9	0.51	Solar coverage, full sunshine (%/100)
10	0.65	Sunshine probability (%/100)
11	0.33	Average solar coverage, including cloudiness (%/100)
12	191148.91	Total Btu collected on a sunny day
13	373032.91	Total Btu needed during that 24-hr period
14	248786.12	Average daily auxiliary heat requirement (Btu)

[a] 3000 (Btu) was entered in register I.

J

Heat-Loss/Gain Computations for Solar Applications

The purpose of these calculations is to size properly the solar installation. Because every solar heating or air-conditioning system should operate in conjunction with a conventional heating or air-conditioning system, properly sized solar portions for building service systems cover (a desired percentage of) the average demand, not the peak demand. Consequently, the heat-load and cooling-load calculations for solar systems must be performed for average conditions, not for extreme conditions.

The building thermal-loss constant (Btu-hr^{-1}-°F^{-1}) and the direct solar heat gain through the windows are the two main heat-loss/heat-gain parameters that must be known for solar systems sizing. The thermal-loss constant is used in the calculation of the solar heat collection per day, in that it determines the heat taken out of the heat-storage tank for space heating or (being subject to the chiller efficiency) for air conditioning. The window gain offsets some of the space-heating requirements, and it increases the air-conditioning load. For commercial buildings, the additional introduction of the heat gain due to electric lighting and electric utilities, as well as the heat gain due to people, becomes very important and must be considered in the heat-loss/gain computations for solar applications.[1]

The accuracy of the heat-loss/gain computations must be in proportion to their purpose. This means that in practice it is quite sufficient to perform somewhat approximate calculations rather than extremely accurate heat-loss calculations, but on the other hand, care must be used in the assessment of the results. The calculated window gain used to offset the space-heating load may be unrealistic, for example, because of shading effects or because the collected heat cannot be distributed well enough in the building and will therefore in part be vented to the outside, rather than actually utilized to offset the space-heating demand.

[1]Heat gain due to people and electric consumption may easily amount to one-third each of the total air-conditioning demand of a commercial building.

It is adequate to reduce the calculation of the building heat-loss coefficient to a simple addition of areas divided by effective R values,[2] plus a percentage allowance for infiltration. The calculation of the direct solar gain through the windows is usually more important and also more time-consuming. The total daily amount[3] should be calculated and should then be divided by the number of daily iterations for the solar system collection computation; i.e., it is sufficient to use averaged values per computation interval as long as the daily sum of the averaged values equals the total daily window gain.[3] The same should be done with the people and lighting gain.[4]

Program for Heat-Loss and Heat-Gain Calculations

A description of a building heat-loss calculation and detailed direct solar heat-gain calculation is given, applicable for an HP 67/97 calculator. The program first computes the direct solar heat gain successively from south-, east-, and west-facing windows. It then computes the thermal-loss coefficient of the building.

For the direct solar heat-gain calculation, the program can accept the following variables: east-, south-, and west-facing window areas; window transmission coefficient; azimuth orientation of building (off-south orientation of south-facing windows); geographic latitude; date.

For the calculation of the thermal-loss coefficient, the program can accept the following variables: total area of single-glazed windows and/or doors, R factor thereof, total area of double-glazed windows and/of doors, R factor thereof, ceiling area, R factor of ceiling, floor area (perimeter), R factor or perimeter[5] of floor, wall area (net), R factor of walls, and percentage of infiltration. Among the results that are calculated is the hourly average direct window gain that must be used for the solar system collection computation (Appendix I). The following data must be entered into the registers:

Register	Data to Be Entered
P0	I_0 (Btu-hr^{-1}-ft^{-2}), e.g., from Table B.1
P1	R value for single-glazed windows
P2	Ceiling area (ft^2)
P3	R value for ceiling
P6	Wall area, net (ft^2)
P7	R value for walls
P8	Floor area (ft^2), or floor perimeter (ft)
P9	R value of floor, or perimeter loss factor[5]

[2]For example, an R-11 insulated wall may be approximated with R-14, an R-19 insulated ceiling with R-21, single-glazed windows with R-0.9, double-glazed windows with R-1.8, and wooden doors with R-1.25 per inch thickness.

[3]For a sunny day; cloudiness is figured at a different stage of the computations (see program B of Appendix I).

[4]The KH-80 computer program referenced earlier [25] contains the calculation of the building heat-loss coefficient and the direct solar window gain as a subroutine.

[5]R-5 for 1-in. × 24-in. foam insulation around slab, R-2 for 1-in. × 12-in. foam insulation, R-1.2 for uninsulated perimeter. (R is defined here as Btu^{-1}-hr-ft).

Register	Data to Be Entered	
S3	γ (geographic latitude)	
S4	β (declination)	
S5	ϕ_0 (deviation of south-facing windows from true south; $\phi_0 > 0$ designates orientation toward west, $\phi_0 < 0$ toward east)	
S8	Transmission of window for solar radiation (<1)	
S9	R factor for double-glazed windows	
A	Area of south-facing windows[6]	Sum must equal the entire window area (ft^2)
B	Area of east-facing windows[6]	
C	Area of west-facing windows[6]	
D	Area of north-facing windows	
E	Area of double-glazed windows and/or doors (ft^2)	

The program listing is presented in Table J.1. The result printout for a typical calculation is given in Table J.2.

The following additional assumptions were made for the generation of this program:

1. The direct insolation (I_0, in register P0) is constant during the entire day.
2. An additional diffuse radiation of $I_1 = 0.1I_0$ is used.
3. A sunny day is assumed; no solar collection is assumed through north-facing windows.
4. All windows are vertical (skylights must be treated separately).
5. Doors must be treated as single-glazed or double-glazed windows[7], whichever is most appropriate.
6. The infiltration must be keyed in (as a percent figure) after the calculator has come to a halt. Continue the calculation by depressing "R/S."
7. Run the program by depressing "A."

[6] Only those areas should be counted that are unshaded. Count the shaded portions of the east-, south-, or west-facing windows as north-facing windows, and add this to register D.

[7] Solid doors are to be treated as north-facing windows.

TABLE J.1 Program Listing for Heat-Loss/Gain Calculations with an HP 97 Calculator

001	*LBLD	21 14	027	×	−35	053	P⇌S	16−51	079	*LBLd	21 16 14
002	1	01	028	RCL0	36 00	054	RCL0	36 00	080	CF1	16 22 01
003	5	05	029	COS	42	055	×	−35	081	0	00
004	×	−35	030	÷	−24	056	P⇌S	16−51	082	STO2	35 02
005	STO1	35 01	031	SIN⁻¹	16 41	057	ST+7	35−55 07	083	STO7	35 07
006	RCL4	36 04	032	RCL2	36 02	058	P⇌S	16−51	084	5	05
007	COS	42	033	X<Y?	16−35	059	GTO?	22 02	085	STO6	35 06
008	RCL3	36 03	034	GTOb	22 16 12	060	*LBL1	21 01	086	GTOD	22 14
009	COS	42	035	R↓	−1	061	RCL0	36 00	087	*LBLα	21 16 11
010	×	−35	036	STO2	35 02	062	X<0?	16−45	088	1	01
011	RCL1	36 01	037	*LBLe	21 16 15	063	GTO3	22 03	089	ST+6	35−55 06
012	COS	42	038	RCL5	36 05	064	P⇌S	16−51	090	RCL6	36 06
013	×	−35	039	+	−55	065	GTO2	22 02	091	1	01
014	RCL4	36 04	040	COS	42	066	*LBL2	21 02	092	2	02
015	SIN	41	041	RCL0	36 00	067	RCL0	36 00	093	X=Y?	16−33
016	RCL3	36 03	042	COS	42	068	.	−62	094	SF1	16 21 01
017	SIN	41	043	×	−35	069	1	01	095	RCL6	36 06
018	×	−35	044	X<0?	16−45	070	×	−35	096	2	02
019	+	−55	045	0	00	071	P⇌S	16−51	097	0	00
020	SIN⁻¹	16 41	046	STO1	35 01	072	ST+7	35−55 07	098	X≤Y?	16−35
021	CHS	−22	047	X=0?	16−43	073	GTOα	22 16 11	099	GTOB	22 12
022	STO0	35 00	048	GTOI	22 01	074	*LBL3	21 03	100	R↓	−31
023	RCL4	36 04	049	RCL0	36 00	075	0	00	101	GTOD	22 14
024	COS	42	050	X<0?	16−45	076	ST+7	35−55 07	102	*LBLB	21 12
025	RCL1	36 01	051	GTO3	22 03	077	GTOα	22 16 11	103	RCL7	36 07
026	SIN	41	052	R↓	−31	078	RTN	24	104	RCLA	36 11

```
105  ×         -35
106  RCL8      36 08
107  ×         -35
108  PRTX      -14
109  9         09
110  ÷         -24
111  P⇌S       16-51
112  ST+5      35-55 05
113  P⇌S       16-51
114  RTN       24
115  *LBLA     21 11
116  PREG      16-31
117  0         00
118  STO5      35 05
119  P⇌S       16-51
120  PREG      16-13
121  GSBd      23 16 14
122  RCLA      36 11
123  RCLC      36 13
124  +         -55
125  RCLB      36 12
126  +         -55
127  RCLD      36 14
128  +         -55
129  STOI      35 46
130  RCLC      36 13
131  STOA      35 11
132  9         09
133  0         00
134  ST-5      35-45 05
135  GSBd      23 16 14
136  RCLB      36 12
137  STOA      35 11
138  1         01
139  8         08
140  0         00
141  ST+5      35-55 05
142  GSBd      23 16 14
143  P⇌S       16-51
144  RCL5      36 05
145  PRTX      -14
146  RCLI      36 46
147  RCLD      36 14
148  -         -45
149  RCLB      36 12
150  -         -45
151  RCLC      36 13
152  -         -45
153  STOA      35 11
154  P⇌S       16-51
155  RCLE      36 15
156  RCL9      36 09
157  ÷         -24
158  PRTX      -14
159  P⇌S       16-51
160  STO4      35 04
161  RCLI      36 46
162  RCLE      36 15
163  -         -45
164  RCL1      36 01
165  ÷         -24
166  PRTX      -14
167  STO4      35 04
168  RCL2      36 02
169  RCL3      36 03
170  ÷         -24
171  PRTX      -14
172  ST+4      35-55 04
173  RCL6      36 06
174  RCL7      36 07
175  ÷         -24
176  PRTX      -14
177  ST+4      35-55 04
178  RCL8      36 08
179  RCL9      36 09
180  ÷         -24
181  PRTX      -14
182  ST+4      35-55 04
183  RCL4      36 04
184  PRTX      -14
185  R/S       51
186  PRTX      -14
187  %         55
188  ST+4      35-55 04
189  RCL4      36 04
190  PRTX      -14
191  P⇌S       16-51
192  9         09
193  0         00
194  ST-5      35-45 05
195  P⇌S       16-51
196  RCL5      36 05
197  9         09
198  ×         -35
199  PRTX      -14
200  RTN       24
201  *LBLb     21 16 12
202  R↓        -31
203  STO2      35 02
204  1         01
205  8         08
206  0         00
207  F1?       16 23 01
208  GTOc      22 16 13
209  -         -45
210  CHS       -22
211  GTOe      22 16 15
212  RTN       24
213  *LBLc     21 16 13
214  +         -55
215  CHS       -22
216  GTOe      22 16 15
217  RTN       24
218  R/S       51
```

TABLE J.2 Printout with Explanations for Two Typical Heat-Loss/Gain Calculations (June 21, left; Dec. 21, right) Performed with the HP 97 Program (Table J.1)

No.	Designation	Result Printout: First Calculation		Result Printout: Second Calculation		Comments
1	I_0 (Btu-hr^{-1}-ft^{-2})	250.00	0	250.00	0	
2	R (single window)	0.90	1	0.90	1	
3	Ceiling area	2000.00	2	2000.00	2	(ft^2)
4	R (ceiling)	21.00	3	21.00	3	
5	Total heat loss[a]	0.00	4	532.54	4	From previous calculation
6	Average gain/hour[a]	0.00	5	21971.77	5	From previous calculation
7	Net wall area	1500.00	6	1500.00	6	(ft^2)
8	R (walls)	14.00	7	14.00	7	
9	Floor area	2000.00	8	2000.00	8	(ft^2, or perimeter)
10	R (floor)	14.00	9	14.00	9	
11	South window area	100.00	A	100.00	A	(ft^2)
12	West window area	50.00	B	50.00	B	(ft^2)
13	East window area	60.00	C	60.00	C	(ft^2)
14	North window area	40.00	D	40.00	D	(ft^2)
15	Double-glazed window area	125.00	E	125.00	E	(ft^2)
16	Total window area[a]	0.00	I	250.00	I	(ft^2)
		0.00	0	3.09	0	Unimportant intermediate results
		0.00	1	0.99	1	Unimportant intermediate results
		0.00	2	−62.55	2	Unimportant intermediate results
17	γ (geographic latitude)	37.50	3	37.50	3	(deg)

18	β (declination)	−23.45	4	23.45	4	(deg)
19	ϕ_0 (off south)	20.00	5	20.00	5	(20° west of south)
		0.00	6	20.00	6	Unimportant
		0.00	7	1547.06	7	intermediate results
20	Average glazing transmission	0.80	8	0.80	8	(Average for all windows)
21	R (double window)	1.80	9	1.80	9	
		100.00	A	100.00	A	
		50.00	B	50.00	B	
		60.00	C	60.00	C	Same as
		40.00	D	40.00	D	results 11–16
		125.00	E	125.00	E	
		0.00	I	250.00	I	
22	Gain, south	64517.64	***	145693.46	***	From program step
23	Gain, east	71345.89	***	51493.25	***	108, for windows,
24	Gain, west	61882.43	***	19672.73	***	per day (Btu)
25	Average gain/hour	21971.77	***	24095.50	***	From program step 145
26	Loss, double window	69.44	***	69.44	***	From program step 158
27	Loss, single window	138.89	***	138.89	***	From program step 166
28	Loss, ceiling	95.24	***	95.24	***	From program step 171
29	Loss from walls	107.14	***	107.14	***	From program step 176
30	Loss from floor	142.86	***	142.86	***	From program step 181
31	Loss w/o infiltration	484.13	***	484.13	***	Total ($Btu\text{-}hr^{-1}\text{-}°F^{-1}$)
32	% infiltration	10.00	***	10.00	***	From program step 186
33	Total loss	532.54	***	532.54	***	$Btu\text{-}hr^{-1}\text{-}°F^{-1}$
34	Window gain	197745.97	***	216859.47	***	Total (Btu/day)

[a] From previous calculation.
*** Actual results (Nos. 1–21 are mainly input data).

K

Programs for Solar Heat Collection Computations for Pool-heating Systems

Programs are presented with which the thermal analysis of a solar-heated pool during a representative 24-hr period can be performed. In each case the heat loss from the pool and the heat collection with the collectors are calculated.

Programs for the following four cases are given: Program A: unglazed collectors, pool covered with a thermal blanket between 1700 and 800 hours. Program B: glazed collectors, pool covered with a thermal blanket between 1700 and 800 hours. Program C: unglazed collectors, uncovered pool. Program D: glazed collectors, uncovered pool. The program loading instructions are compiled in Table K.1. The program listings are given in Table K.2 (program A), Table K.3 (B), Table K.4 (C), and Table K.5 (D). In all cases the programs are based on an hourly iteration during the daytime period, following the principle discussed in Section 5.3 (see in particular Fig. 5.3). The heat loss at night is estimated as a one-time calculation based on the number of degree-hours that the pool is above ambient temperature. The main assumptions made for the different programs are summarized in Table K.6 and the typical printing format of the results is presented and discussed in Table K.8.

TABLE K.1 Loading Instructions for Programs A–D[a]

	Program A		Program B		Program C		Program D	
	P	S	P	S	P	S	P	S
Insolation (Btu-hr^{-1}-ft^{-2}), direct component	0		0		0		0	
Collector absorptivity × transmissivity (%/100)	1		1		1		1	
Collector area (ft^2)	2		2		2		2	
Pool surface area (ft^2)	3		3		3		3	
Wind factor (slope, m (usually $m = 0.2$)	4		4		4		4	
Wind factor, w_0 (intercept) (usually $w_0 = 0.5$)	5		5		5		5	
Wind velocity, v (mph), average during daytime	6		6[b]		6		6[b]	
Collector linear temperature-loss coefficient (Btu-hr^{-1}-ft^{-2}-°F^{-1})[c]			7				7	
R value of pool cover	8		8					
Wind velocity, v (mph), average during night[d]					8		8	
Collector orientation, azimuth (deg) (west > 0)		2		2		2		2
Geographic latitude (deg)		3		3		3		3
Declination (deg)		4		4		4		4
Collector tilt (deg)		5		5		5		5
T_a(6:00 A.M.) (°F)		6		6		6		6
T_a(noon) (°F)		7		7		7		7
T_a(16:00) (°F)		9		9		9		9
T_p at 8:00 A.M (°F)	A		A		A		A	
Volume of pool (U.S. gal)	B		B		B		B	
$200(P_w - P_a)$[e] (see equation 4.1)	E		E		E		E	

[a] Note that register P must be first and that register A must be reloaded before each calculation.
[b] Note that $c = 0$ is assumed (see equation 1.1). Make appropriate adjustments in the ambient temperatures entered in registers S6, S7, and S9 to account for stagnant air conditions near the collectors.
[c] The program is designed to take only the linear loss coefficient. In case the collector performance is known only for a second-order fit, one must first convert the coefficients $\epsilon^{(2)}$, $a^{(2)}$, and b into new coefficients $\epsilon^{(1)}$, $a^{(1)}$, as described in Section 1.1.2.
[d] Register P6 must be reloaded after each calculation for programs C and D. The original data entered in register P6 are neither preserved nor printed.
[e] See Table K.7 for suggested values to be entered.

TABLE K.2 Program Listing for Computation of Solar Heat Collection for Pool Systems Using Unglazed Flat-Plate Collectors and a Pool Blanket During the Night Hours (HP 67/97, Program A)

001	*LBLB	21 12	027	+	−55	053	COS	42	079	P⇌S	16−51
002	RCL4	36 04	028	RTN	24	054	×	−35	080	RTN	24
003	RCL6	36 06	029	*LBLa	21 16 11	055	RCLD	36 14	081	*LBLb	21 16 12
004	×	−35	030	COS	42	056	COS	42	082	1	01
005	RCL5	36 05	031	RCL0	36 00	057	×	−35	083	ST+8	35−55 08
006	+	−55	032	COS	42	058	RCL4	36 04	084	RCL7	36 07
007	1	01	033	×	−35	059	SIN	41	085	RCL6	36 06
008	+	−55	034	RCLC	36 13	060	RCL3	36 03	086	−	−45
009	RCLA	36 11	035	SIN	41	061	SIN	41	087	4	04
010	P⇌S	16−51	036	×	−35	062	×	−35	088	÷	−24
011	RCL1	36 01	037	RCL0	36 00	063	+	−55	089	ST+1	35−55 01
012	P⇌S	16−51	038	SIN	41	064	SIN^{-1}	16 41	090	RTN	24
013	−	−45	039	RCLC	36 13	065	CHS	−22	091	*LBLE	21 15
014	×	−35	040	COS	42	066	STO0	35 00	092	RCL5	36 05
015	X<0?	16−45	041	×	−35	067	RCL4	36 04	093	STOC	35 13
016	RCL4	36 04	042	+	−55	068	COS	42	094	RCL8	36 08
017	2	02	043	STOD	35 14	069	RCLD	36 14	095	GSBD	23 14
018	3	03	044	RTN	24	070	SIN	41	096	.	−62
019	+	−55	045	*LBLD	21 14	071	×	−35	097	1	01
020	STO7	35 07	046	1	01	072	RCL0	36 00	098	0	00
021	RTN	24	047	5	05	073	COS	42	099	+	−55
022	*LBLC	21 13	048	×	−35	074	÷	−24	100	RCL0	36 00
023	RCL4	36 04	049	STOD	35 14	075	SIN^{-1}	16 41	101	×	−35
024	RCL6	36 06	050	RCL4	36 04	076	RCL2	36 02	102	RCL1	36 01
025	×	−35	051	COS	42	077	+	−55	103	×	−35
026	RCL5	36 05	052	RCL3	36 03	078	GSBa	23 16 11	104	STOI	35 46

105	GSBB	23 12	133	RCLE	36 15	161	*LBLA	21 11	189	RCL1	36 01
106	CHS	−22	134	×	−35	162	0	00	190	RCL6	36 06
107	RCLI	36 46	135	−	−45	163	STO9	35 09	191	−	−45
108	+	−55	136	RCL3	36 03	164	8	08	192	3	03
109	X<0?	16−45	137	×	−35	165	P⇌S	16−51	193	÷	−24
110	0	00	138	RCLI	36 46	166	STO8	35 08	194	RCL6	36 06
111	ST+9	35−55 09	139	+	−55	167	PREG	16−13	195	+	−55
112	RCL2	36 02	140	RCLB	36 12	168	RCL6	36 06	196	−	−45
113	×	−35	141	÷	−24	169	STO1	35 01	197	P⇌S	16−51
114	STOI	35 46	142	8	08	170	GSBE	23 15	198	RCL3	36 03
115	P⇌S	16−51	143	.	−62	171	GSBb	23 16 12	199	×	−35
116	0	00	144	3	03	172	GSBE	23 15	200	RCL8	36 08
117	STOC	35 13	145	÷	−24	173	GSBb	23 16 12	201	÷	−24
118	GSBα	23 16 11	146	RCLA	36 11	174	GSBE	23 15	202	1	01
119	P⇌S	16−51	147	+	−55	175	GSBb	23 16 12	203	.	−62
120	.	−62	148	STOA	35 11	176	GSBE	23 15	204	8	08
121	1	01	149	P⇌S	16−51	177	GSBb	23 16 12	205	×	−35
122	0	00	150	RTN	24	178	GSBE	23 15	206	CHS	−22
123	+	−55	151	*LBLc	21 16 13	179	GSBc	23 16 13	207	RCLB	36 12
124	RCL0	36 00	152	1	01	180	GSBE	23 15	208	÷	−24
125	×	−35	153	ST+8	35−55 08	181	GSBc	23 16 13	209	RCLA	36 11
126	.	−62	154	RCL9	36 09	182	GSBE	23 15	210	+	−55
127	8	08	155	RCL7	36 07	183	GSBc	23 16 13	211	STOA	35 11
128	0	00	156	−	−45	184	GSBE	23 15	212	PREG	16−13
129	×	−35	157	4	04	185	GSBc	23 16 13	213	RTN	24
130	RCL7	36 07	158	÷	−24	186	GSBE	23 15	214	R/S	51
131	−	−45	159	ST+1	35−55 01	187	RCLA	36 11			
132	GSBC	23 13	160	RTN	24	188	PRTX	−14			

TABLE K.3 Program Listing for Computation of Solar Heat Collection for Pool Systems Using Glazed Flat-Plate Collectors and a Blanket During the Night Hours (HP 67/97, Program B)

001	*LBLB	21 12	028	*LBLα	21 16 11	055	COS	42	082	ST+8	35−55 08
002	RCL4	36 04	029	COS	42	056	×	−35	083	RCL7	36 07
003	RCL6	36 06	030	RCL0	36 00	057	RCL4	36 04	084	RCL6	36 06
004	×	−35	031	COS	42	058	SIN	41	085	−	−45
005	RCL5	36 05	032	×	−35	059	RCL3	36 03	086	4	04
006	+	−55	033	RCLC	36 13	060	SIN	41	087	÷	−24
007	1	01	034	SIN	41	061	×	−35	088	ST+1	35−55 01
008	+	−55	035	×	−35	062	+	−55	089	RTN	24
009	RCLA	36 11	036	RCL0	36 00	063	SIN^{-1}	16 41	090	*LBLE	21 15
010	P⇌S	16−51	037	SIN	41	064	CHS	−22	091	RCL5	36 05
011	RCL1	36 01	038	RCLC	36 13	065	STO0	35 00	092	STOC	35 13
012	P⇌S	16−51	039	COS	42	066	RCL4	36 04	093	RCL8	36 08
013	−	−45	040	×	−35	067	COS	42	094	GSBD	23 14
014	×	−35	041	+	−55	068	RCLD	36 14	095	.	−62
015	X<0?	16−45	042	STOD	35 14	069	SIN	41	096	1	01
016	RCL4	36 04	043	RTN	24	070	×	−35	097	0	00
017	2	02	044	*LBLD	21 14	071	RCL0	36 00	098	+	−55
018	3	03	045	1	01	072	COS	42	099	RCL0	36 00
019	+	−55	046	5	05	073	÷	−24	100	×	−35
020	RTN	24	047	×	−35	074	SIN^{-1}	16 41	101	RCL1	36 01
021	*LBLC	21 13	048	STOD	35 14	075	RCL2	36 02	102	×	−35
022	RCL4	36 04	049	RCL4	36 04	076	+	−55	103	RCLA	36 11
023	RCL6	36 06	050	COS	42	077	GSBα	23 16 11	104	P⇌S	16−51
024	×	−35	051	RCL3	36 03	078	P⇌S	16−51	105	RCL1	36 01
025	RCL5	36 05	052	COS	42	079	RTN	24	106	P⇌S	16−51
026	+	−55	053	×	−35	080	*LBLb	21 16 12	107	−	−45
027	RTN	24	054	RCLD	36 14	081	1	01	108	RCL7	36 07

109	×	−35	137	RCLE	36 15	165	*LBLA	21 11	193	RCL1	36 01
110	X<0?	16−45	138	×	−35	166	0	00	194	RCL6	36 06
111	CLX	−51	139	−	−45	167	STO9	35 09	195	−	−45
112	−	−45	140	RCL3	36 03	168	8	08	196	3	03
113	X<0?	16−45	141	×	−35	169	P⇌S	16−51	197	÷	−24
114	0	00	142	RCLI	36 46	170	STO8	35 08	198	RCL6	36 06
115	ST+9	35−55 09	143	+	−55	171	PREG	16−13	199	+	−55
116	RCL2	36 02	144	RCLB	36 12	172	RCL6	36 06	200	−	−45
117	×	−35	145	÷	−24	173	STO1	35 01	201	P⇌S	16−51
118	STOI	35 46	146	8	08	174	GSBE	23 15	202	RCL3	36 03
119	P⇌S	16−51	147	.	−62	175	GSBb	23 16 12	203	×	−35
120	0	00	148	3	03	176	GSBE	23 15	204	RCL8	36 08
121	STOC	35 13	149	÷	−24	177	GSBb	23 16 12	205	÷	−24
122	GSBα	23 16 11	150	RCLA	36 11	178	GSBE	23 15	206	1	01
123	P⇌S	16−51	151	+	−55	179	GSBb	23 16 12	207	.	−62
124	.	−62	152	STOA	35 11	180	GSBE	23 15	208	8	08
125	1	01	153	P⇌S	16−51	181	GSBb	23 16 12	209	×	−35
126	0	00	154	RTN	24	182	GSBE	23 15	210	CHS	−22
127	+	−55	155	*LBLc	21 16 13	183	GSBc	23 16 13	211	RCLB	36 12
128	RCL0	36 00	156	1	01	184	GSBE	23 15	212	÷	−24
129	×	−35	157	ST+8	35−55 08	185	GSBc	23 16 13	213	RCLA	36 11
130	.	−62	158	RCL9	36 09	186	GSBE	23 15	214	+	−55
131	8	08	159	RCL7	36 07	187	GSBc	23 16 13	215	STOA	35 11
132	0	00	160	−	−45	188	GSBE	23 15	216	PREG	16 −13
133	×	−35	161	4	04	189	GSBc	23 16 13	217	RTN	24
134	GSBB	23 12	162	÷	−24	190	GSBE	23 15	218	R/S	51
135	−	−45	163	ST+1	35−55 01	191	RCLA	36 11			
136	GSBC	23 13	164	RTN	24	192	PRTX	−14			

TABLE K.4 Program Listing for Computation of Solar Heat Collection for Pool Systems Using Unglazed Flat-Plate Collectors for an Uncovered Pool (HP 67/97, Program C)

001	*LBLB	21 12	028	RTN	24	055	RCLD	36 14	082	1	01
002	RCL4	36 04	029	*LBLa	21 16 11	056	COS	42	083	ST+8	35−55 08
003	RCL6	36 06	030	COS	42	057	×	−35	084	RCL7	36 07
004	×	−35	031	RCL0	36 00	058	RCL4	36 04	085	RCL6	36 06
005	RCL5	36 05	032	COS	42	059	SIN	41	086	−	−45
006	+	−55	033	×	−35	060	RCL3	36 03	087	4	04
007	1	01	034	RCLC	36 13	061	SIN	41	088	÷	−24
008	+	−55	035	SIN	41	062	×	−35	089	ST+1	35−55 01
009	RCLA	36 11	036	×	−35	063	+	−55	090	RTN	24
010	P⇄S	16−51	037	RCL0	36 00	064	SIN^{-1}	16 41	091	*LBLE	21 15
011	RCL1	36 01	038	SIN	41	065	CHS	−22	092	RCL5	36 05
012	P⇄S	16−51	039	RCLC	36 13	066	STO0	35 00	093	STOC	35 13
013	−	−45	040	COS	42	067	RCL4	36 04	094	RCL8	36 08
014	×	−35	041	×	−35	068	COS	42	095	GSBD	23 14
015	X<0?	16−45	042	+	−55	069	RCLD	36 14	096	.	−62
016	RCL4	36 04	043	STOD	35 14	070	SIN	41	097	1	01
017	2	02	044	RTN	24	071	×	−35	098	0	00
018	3	03	045	*LBLD	21 14	072	RCL0	36 00	099	+	−55
019	+	−55	046	1	01	073	COS	42	100	RCL0	36 00
020	STO7	35 07	047	5	05	074	÷	−24	101	×	−35
021	RTN	24	048	×	−35	075	SIN^{-1}	16 41	102	RCL1	36 01
022	*LBLC	21 13	049	STOD	35 14	076	RCL2	36 02	103	×	−35
023	RCL4	36 04	050	RCL4	36 04	077	+	−55	104	STOI	35 46
024	RCL6	36 06	051	COS	42	078	GSBa	23 16 11	105	GSBB	23 12
025	×	−35	052	RCL3	36 03	079	P⇄S	16−51	106	CHS	−22
026	RCL5	36 05	053	COS	42	080	RTN	24	107	RCLI	36 46
027	+	−55	054	×	−35	081	*LBLb	21 16 12	108	+	−55

109	X<0?	16−45	138	RCLI	36 46	167	PREG	16−13	196	−	−45
110	0	00	139	+	−55	168	RCL6	36 06	197	RCLE	36 15
111	ST+9	35−55 09	140	RCLB	36 12	169	STO1	35 01	198	+	−55
112	RCL2	36 02	141	÷	−24	170	GSBE	23 15	199	P⇌S	16−51
113	×	−35	142	8	08	171	GSBb	23 16 12	200	RCL8	36 08
114	STOI	35 46	143	.	−62	172	GSBE	23 15	201	STO6	35 06
115	P⇌S	16−51	144	3	03	173	GSBb	23 16 12	202	R↓	−31
116	0	00	145	÷	−24	174	GSBE	23 15	203	GSBC	23 13
117	STOC	35 13	146	RCLA	36 11	175	GSBb	23 16 12	204	×	−35
118	GSBα	23 16 11	147	+	−55	176	GSBE	23 15	205	2	02
119	P⇌S	16−51	148	STOA	35 11	177	GSBb	23 16 12	206	3	03
120	.	−62	149	P⇌S	16−51	178	GSBE	23 15	207	+	−55
121	1	01	150	RTN	24	179	GSBc	23 16 13	208	RCL3	36 03
122	0	00	151	*LBLc	21 16 13	180	GSBE	23 15	209	×	−35
123	+	−55	152	1	01	181	GSBc	23 16 13	210	RCLB	36 12
124	RCL0	36 00	153	ST+8	35−55 08	182	GSBE	23 15	211	÷	−24
125	×	−35	154	RCL9	36 09	183	GSBc	23 16 13	212	1	01
126	.	−62	155	RCL7	36 07	184	GSBE	23 15	213	.	−62
127	8	08	156	−	−45	185	GSBc	23 16 13	214	8	08
128	0	00	157	4	04	186	GSBE	23 15	215	×	−35
129	×	−35	158	÷	−24	187	RCLA	36 11	216	CHS	−22
130	RCL7	36 07	159	ST+1	35−55 01	188	PRTX	−14	217	RCLA	36 11
131	−	−45	160	RTN	24	189	RCL1	36 01	218	÷	−55
132	GSBC	23 13	161	*LBLA	21 11	190	RCL6	36 06	219	STOA	35 11
133	RCLE	36 15	162	0	00	191	−	−45	220	PREG	16−13
134	×	−35	163	STO9	35 09	192	3	03	221	RTN	24
135	−	−45	164	8	08	193	÷	−24	222	R/S	51
136	RCL3	36 03	165	P⇌S	16−51	194	RCL6	36 06			
137	×	−35	166	STO8	35 08	195	+	−55			

TABLE K.5 Program Listing for Computation of Solar Heat Collection for Pool Systems Using Glazed Flat-Plate Collectors for an Uncovered Pool (HP 67/97, Program D)

001	*LBLB	21 12	028	*LBLα	21 16 11	055	COS	42	082	ST+8	35−55 08
002	RCL4	36 04	029	COS	42	056	×	−35	083	RCL7	36 07
003	RCL6	36 06	030	RCL0	36 00	057	RCL4	36 04	084	RCL6	36 06
004	×	−35	031	COS	42	058	SIN	41	085	−	−45
005	RCL5	36 05	032	×	−35	059	RCL3	36 03	086	4	04
006	+	−55	033	RCLC	36 13	060	SIN	41	087	÷	−24
007	1	01	034	SIN	41	061	×	−35	088	ST+1	35−55 01
008	+	−55	035	×	−35	062	+	−55	089	RTN	24
009	RCLA	36 11	036	RCL0	36 00	063	SIN^{-1}	16 41	090	*LBLE	21 15
010	P⇄S	16−51	037	SIN	41	064	CHS	−22	091	RCL5	36 05
011	RCL1	36 01	038	RCLC	36 13	065	STO0	35 00	092	STOC	35 13
012	P⇄S	16−51	039	COS	42	066	RCL4	36 04	093	RCL8	36 08
013	−	−45	040	×	−35	067	COS	42	094	GSBD	23 14
014	×	−35	041	+	−55	068	RCLD	36 14	095	.	−62
015	X<0?	16−45	042	STOD	35 14	069	SIN	41	096	1	01
016	RCL4	36 04	043	RTN	24	070	×	−35	097	0	00
017	2	02	044	*LBLD	21 14	071	RCL0	36 00	098	+	−55
018	3	03	045	1	01	072	COS	42	099	RCL0	36 00
019	+	−55	046	5	05	073	÷	−24	100	×	−35
020	RTN	24	047	×	−35	074	SIN^{-1}	16 41	101	RCL1	36 01
021	*LBLC	21 13	048	STOD	35 14	075	RCL2	36 02	102	×	−35
022	RCL4	36 04	049	RCL4	36 04	076	+	−55	103	RCLA	36 11
023	RCL6	36 06	050	COS	42	077	GSBα	23 16 11	104	P⇄S	16−51
024	×	−35	051	RCL3	36 03	078	P⇄S	16−51	105	RCL1	36 01
025	RCL5	36 05	052	COS	42	079	RTN	24	106	P⇄S	16−51
026	+	−55	053	×	−35	080	*LBLb	21 16 12	107	−	−45
027	RTN	24	054	RCLD	36 14	081	1	01	108	RCL7	36 07

109	×	−35	138	−	−45	167	8	08	196	÷	−24
110	X<0?	16−45	139	RCL3	36 03	168	P⇌S	16−51	197	RCL6	36 06
111	CLX	−51	140	×	−35	169	STO8	35 08	198	+	−55
112	−	−45	141	RCLI	36 46	170	PREG	16−13	199	−	−45
113	X<0?	16−45	142	+	−55	171	RCL6	36 06	200	RCLE	36 15
114	0	00	143	RCLB	36 12	172	STO1	35 01	201	+	−55
115	ST+9	35−55 09	144	÷	−24	173	GSBE	23 15	202	P⇌S	16−51
116	RCL2	36 02	145	8	08	174	GSBb	23 16 12	203	RCL8	36 08
117	×	−35	146	.	−62	175	GSBE	23 15	204	STO6	35 06
118	STOI	35 46	147	3	03	176	GSBb	23 16 12	205	R↓	−31
119	P⇌S	16−51	148	÷	−24	177	GSBE	23 15	206	GSBC	23 13
120	0	00	149	RCLA	36 11	178	GSBb	23 16 12	207	×	−35
121	STOC	35 13	150	+	−55	179	GSBE	23 15	208	2	02
122	GSBa	23 16 11	151	STOA	35 11	180	GSBb	23 16 12	209	3	03
123	P⇌S	16−51	152	P⇌S	16−51	181	GSBE	23 15	210	+	−55
124	.	−62	153	RTN	24	182	GSBc	23 16 13	211	RCL3	36 03
125	1	01	154	*LBLc	21 16 13	183	GSBE	23 15	212	×	−35
126	0	00	155	1	01	184	GSBc	23 16 13	213	RCLB	36 12
127	+	−55	156	ST+8	35−55 00	185	GSBE	23 15	214	÷	−24
128	RCL0	36 00	157	RCL9	36 09	186	GSBc	23 16 13	215	1	01
129	×	−35	158	RCL7	36 07	187	GSBE	23 15	216	.	−62
130	.	−62	159	−	−45	188	GSBc	23 16 13	217	8	08
131	8	08	160	4	04	189	GSBE	23 15	218	×	−35
132	×	−35	161	÷	−24	190	RCLA	36 11	219	CHS	−22
133	GSBB	23 12	162	ST+1	35−55 01	191	PRTX	−14	220	RCLA	36 11
134	−	−45	163	RTN	24	192	RCL1	36 01	221	+	−55
135	GSBC	23 13	164	*LBLA	21 11	193	RCL6	36 06	222	STOA	35 11
136	RCLE	36 15	165	0	00	194	−	−45	223	PREG	16−13
137	×	−35	166	STO9	35 09	195	3	03	224	R/S	51

TABLE K.6 Main Assumptions and Explanations for Programs A–D

	Program A (Unglazed, with Blanket)	Program C (Unglazed, No Blanket)	Program B (Glazed, with Blanket)	Program D (Glazed, No Blanket)
Calculation of direct solar heat collection of pool	The pool is assumed to be exposed to full-sunshine conditions throughout the day. The ratio of diffuse component to direct component of solar radiation is assumed to be $I_1/I_0 = 0.1$. The absorptivity of the pool for solar radiation is assumed to be 0.8. The hourly direct solar heat received by the pool (area A_p) is therefore $I_p = 0.80A_p{}^{I}{}_0[\cos \vartheta_0(t) + 0.10]$.[a]			
Calculation of heat loss from pool during daytime	A constant black-body radiation of 23 $Btu\text{-}hr^{-1}\text{-}ft^{-2}$ from the pool surface is assumed. The radiation to the surroundings is figured ΔT ($Btu\text{-}hr^{-1}\text{-}ft^{-2}$) (see Section 4.3.1). The temperature-dependent heat loss due to evaporation is assumed to be $200(P_w - P_a)$ (see Section 4.3.1).[b] The temperature-dependent heat loss by convection is $\alpha_w \Delta T$ (see Section 1.1.3 for definition of α_w). The hourly heat loss from the pool is then (compare with equation 4.2) $L_p = A_p[23 + (\alpha_w + 1)\Delta T + 200(P_w - P_a)]$.			
Heat loss from pool during night	Subtract $1.8A_p\delta T/VR$ from pool temperature at 1700 hours.[c] R = R value of pool cover.	Subtract $\frac{1.8A_p}{V}[\alpha_w \delta T + 200(P_w - P_a) + 23]$ from pool temperature at 1700 hours.[c]	Subtract $1.8A_p\delta T/VR$ from pool temperature at 1700 hours.[c]	Subtract $\frac{1.8A_p}{V}[\alpha_w \delta T + 200(P_w - P_a) + 23]$ from pool temperature at 1700 hours.[c]
Collector deviation from due south	To be entered in register S2 (positive number designates deviation toward west of south, negative deviation means east of south).			

Insolation	The diffuse component is assumed to be 10% of the direct component; the direct component must be entered in register P0. The ratio of I_1 and I_2 may easily be changed in the program by replacing the steps 97/98 (program A), 96/97 (program B), 121/122 (program C), and 125/126 (program D) with the desired new ratio.	
Collector efficiency equation	The hourly heat collection I_c is assumed to be $I_c = A_0\{\epsilon I_0[\cos \vartheta(t) + 0.1] - (\alpha_w + 1)\Delta T - 23\}$. See equation 1.2 with $k_1 = 0.1$ and equation 1.16 with a constant black-body radiation of 23 Btu-hr^{-1}-ft^{-2}. This value may easily be adjusted in the program by changing steps 17/18 and 205/206 (program C only).	The hourly heat collection I_c is assumed to be $I_c = A_0\{\epsilon I_0[\cos \vartheta(t) + 0.1] - \alpha\Delta T\}$. See equation 1.2 with $k = 0.1$ and $b = 0$, which necessitates a linear-fit approximation for ϵ and α as described in Section 1.1.2 (equations 1.6 to 1.8).
Consideration of wind on collector efficiency	Included in $\alpha_w = mv + w_0$ (see Fig. 1.16), where m and w_0 must be entered in registers P4 and P5. Normally, $m = 0.2$ and $w_0 = 0.5$. The wind velocity v is (as daytime average figure) to be entered in register P6.	Not considered because of lack of storage registers, i.e., $c = 0$ (see equation 1.1). If the ambient temperature near the collectors differs significantly from the table values for $T_a(t)$, adjust registers S6, S7, and S9 accordingly.
Fin efficiency	Included in factor ϵ, which is the instantaneous efficiency at $\Delta T = 0$ and must be entered in register P1.	
Ambient temperature data	Three data must be known and entered for each month: T_a(6:00) in register S6, T_a(noon) in register S7, and T_a(16:00) in register S9. The program uses $T_a(6{:}00) = T_a(8{:}00)$ and interpolates linearly between 8:00 and noon and noon and 16:00.	
Pool temperature	The program starts with a given pool temperature at 8:00 (to be entered in register A before each calculation). It then computes hourly the pool temperature as it develops during the day, ending with T_p(17:00), which is printed as first main result. The program then subtracts the heat loss incurred between 17:00 and 8:00 from the heat content of the pool and prints the pool temperature that would result at 8:00 A.M. of the next day. This result and the pool temperature originally entered in register A give the amount of heat to be made up by conventional auxiliary heat.	

[a] Change the factor 0.80 in the program (steps 126–128 or 130–132) if the pool is not exposed to full-sunshine conditions.
[b] For numerical values of $200(P_w - P_a)$, see Table K.7.
[c] $\delta T = T_p(17{:}00) - \overline{T_a} \approx T_p(17{:}00) - \{T_a(6{:}00) + \frac{1}{3}[T_a(17{:}00) - T_a(6{:}00)]\}$, with $1.8 = 15$ hr/8.34 Btu-gal^{-1}-°F^{-1} as factor. Radiation losses to surroundings are omitted (programs A and C).

TABLE K.7 Values of $200(P_w - P_a)$[a]

Relative Humidity	Pool Temperature 74 (°F)	76 (°F)	78 (°F)	80 (°F)	82 (°F)	84 (°F)
10%	76	81	87	95	102	109
20%	68	73	79	85	91	98
30%	60	64	69	75	80	86
40%	52	55	60	65	69	74
50%	44	46	50	55	59	62
60%	35	37	40	45	48	50
70%	26	28	30	33	36	38
80%	18	19	21	23	25	26
90%	9	10	11	12	13	14

[a] Approximate values for $200(P_w - P_a)$ (see equation 4.1) to be used in register E of programs A, B, C, and D; generated from Fig. 4.7.

TABLE K.8 Typical Printout Format for Results of Computations with Programs A, B, C, and D

	Program A (Unglazed Collector, with Blanket)[d]	Program B (Glazed Collector, with Blanket)[d]	Program C (Unglazed Collector, without Blanket)[d]	Program D (Glazed Collector, without Blanket)[d]
·/·[c]				
1 Deviation of collector azimuth from due south	0.00	0.00	0.00	0.00
2 Geographic latitude	37.50	37.50	37.50	37.50
3 Declination	−20.24	−20.24	−20.24	−20.24
4 Collector tilt angle	18.50	18.50	18.50	18.50
5 Ambient temperature at 6:00 A.M.	60.00	60.00	60.00	60.00
6 Ambient temperature at noon	70.00	70.00	70.00	70.00
·/·	8.00	8.00	8.00	8.00
7 Ambient temperature at 4:00 P.M.	72.00	72.00	72.00	72.00
8 Pool (set) temperature at 8:00 A.M.	80.00	80.00	80.00	80.00
9 Volume of pool (gal)	15000.00	15000.00	15000.00	15000.00
·/·	0.00	0.00	0.00	0.00
·/·	0.00	0.00	0.00	0.00
10 Value used for $200(P_w - P_a)$ (see Table K.7)	60.00	60.00	60.00	60.00
·/·	0.00	0.00	0.00	0.00
11 Pool temperature at 5:00 P.M.	82.73	83.31	82.73	83.31
12 Insolation, direct computation ($Btu\text{-}hr^{-1}\text{-}ft^{-2}$)	275.00	275.00	275.00	275.00
13 ϵ	0.90	0.80	0.90	0.80
14 Collector area (ft^2)	300.00	300.00	300.00	300.00
15 Pool surface area (ft^2)	400.00	400.00	400.00	400.00
16 m (wind factor, usually $m = 0.2$)	0.20	0.20	0.20	0.20
17 w_0 (wind factor, usually $w_0 = 0.5$)	0.50	0.50	0.50	0.50
18 Wind velocity during daytime (mph)[a]	10.00	10.00	2.00	2.00
19 Collector linear temperature-loss coefficient α[b]	60.37	1.20	60.37	1.20
20 R value of cover (A, B), or wind speed (C, D)	1.00	1.00	2.00	2.00
21 Btu/ft^2 per day collected	1402.10	1655.62	1402.10	1655.62
22 Pool temperature at 8:00 A.M., next day	81.83	82.38	78.23	78.78
·/·[c]				

[a] For programs C and D, wind velocity at night is printed.
[b] Only for programs B and D.
[c] Discard first two and last five numbers printed out.
[d] Sample calculations performed for 10-mph daytime wind velocity.

Index